Farm Animal Behaviour

Characteristics for Assessment of Health and Welfare

Farm Animal Behaviour

Characteristics for Assessment of Health and Welfare

Ingvar Ekesbo

Department of Animal Environment and Health
Faculty of Veterinary Medicine and Animal Science
Swedish University of Agricultural Sciences
SE 532 23 SKARA
Sweden

www.cabi.org

CABI is a trading name of CAB International

CABI Head Office	CABI North American Office
Nosworthy Way	875 Massachusetts Avenue
Wallingford	7th Floor
Oxfordshire OX10 8DE	Cambridge, MA 02139
UK	USA
Tel: +44 (0)1491 832111	Tel: +1 617 395 4056
Fax: +44 (0)1491 833508	Fax: +1 617 354 6875
E-mail: cabi@cabi.org	E-mail: cabi-nao@cabi.org
Website: www.cabi.org	

A catalogue record for this book is available from the British Library, London, UK.

Library of Congress Cataloging-in-Publication Data

Ekesbo, Ingvar.
    Farm animal behaviour : characteristics for assessment of health and welfare / Ingvar Ekesbo.
        p. cm.
    Includes bibliographical references and index.
    ISBN 978-1-84593-770-6 (alk. paper)
1. Domestic animals--Behavior. 2. Livestock--Behavior. 3. Animal health. 4. Animal welfare. I. Title.

SF756.7.E38 2011
636--dc22

                                                            2010027357

ISBN-13: 978 1 84593 770 6

Commissioning editor: Sarah Hulbert
Production editor: Kate Hill

Typeset by SPi, Pondicherry, India.
Printed and bound in the UK by Cambridge University Press, Cambridge.

Contents

Preface

Our forefathers lived in close association with animals for millennia and evidence from the earliest available drawings and texts suggests that our main methods of caring for farm animals seem to have changed only slightly from the time when the first records were available until the early 20th century.

The great changes in farm animal husbandry since the 1950s have probably made the past 5–6 decades without parallel in the joint history of humans and farm animals. New technology and management have changed the environments for farm animals radically. Some academics have stopped using the term 'animal husbandry', meaning both an ethic of care and the techniques used to provide this care, and have advocated instead the term 'animal production'. Breeding for high production combined with intense feeding has changed the phenotype of several species. Traditional farms with several species kept in small groups have been replaced by specialized farms with one species or age category and larger groups. Manual labour has been replaced with automation, reducing contact between farmers and their animals. This reduced contact has also diminished our ability to learn about the animal's behaviour. The experience-based knowledge of our forefathers that came from spending their lives working closely with animals has been lost, even though today's farmers theoretically are better educated.

Simultaneous with these changes in farm animal practice was the emergence of a range of new health and welfare problems for farm animals. Breeding for high production led to increases in a range of so-called 'production diseases'. Injuries and diseases caused by factors in housing or management, as well as abnormal behaviour and stereotypies that were almost never seen previously, started to became more common. Research increasingly suggests that many of these health problems are caused by housing or management systems not designed to meet the essential biological needs of the animal and which prevent the performance of highly motivated behaviour. All this makes new and increased demands on the veterinarian compared with some decades ago.

Examination of the patient's behaviour and analysis of any discrepancies are a cornerstone in clinical diagnosis. Each deviation from the signs of health, including abnormal behaviour, must be regarded as a symptom of a disorder or disease but, without knowledge of the normal behaviour of the species, it is not possible to identify what is aberrant. Knowledge of natural behaviour helps in understanding the behavioural needs of farm animals and thus informs changes to housing and management to meet these needs better. Intimate knowledge of the altered behaviour of sick animals is also important to diagnosis.

After diagnosis can come treatment, but problems in animal health should be addressed by examining the causes, not the symptoms. Interventions that target aberrant behaviour rather than the underlying causes are liable to delay, distort or even prevent the return to health.

The classical veterinary curriculum provided knowledge of diagnosis and treatment of physical injuries and diseases, but today's veterinarians require knowledge of behaviour that they can use to improve their diagnostic skills and for more general assessments of the health and welfare of individuals and groups of animals. As described above, the veterinarian can no longer rely on the producer to provide a profound understanding of the animal's behaviour and its needs. The modern farmer may not always recognize early signals because he may not have learned to do so, may not have the time to watch for these or may not understand the value of using these behaviours to assess diseases and disease risk.

In order to meet these new demands of training, veterinary schools increasingly are offering coursework in farm animal ethology. Unfortunately, until now, there has been no textbook on farm animal behaviour focused for veterinary students and allied programmes (e.g. for veterinary nurses and technicians). Existing ethology texts are typically too comprehensive for the short and intense courses in the veterinary curriculum and lack a focus on animal health. Teachers and students of animal science, animal welfare and ethology will also find this book a useful alternative or complement to other textbooks. The growing interest in ethology and its importance to the understanding of animal health will also make the book valuable to general practitioners and veterinarians in administrative or advisory services, as well as to farmers and agricultural advisers, animal welfare inspectors and other professionals in the farm animal sector.

This book is organized such that, for each species, identical headings are used to summarize the best available scientific information. These headings are selected to capture a wide range of behaviours. Behavioural information is supplemented with physiological data of interest for clinical examinations and characterizing the healthy animal. The most common abnormal behaviours and stereotypies are also described, as well as examples of injuries and diseases caused by housing or management systems that do not meet health and welfare needs. Where relevant, I also provide examples of disorders caused by breeding for high production. Finally, a short description is provided of useful restraint and handling techniques for each species and the importance of the stockman's behaviour. To help make this text more accessible to readers of a variety of backgrounds, I have also provided a short glossary with explanations of terms frequently used in medical, ethological and agricultural literature.

Ingvar Ekesbo

Acknowledgements

Above all, I express my gratitude to Dr Roger Ewbank, OBE, former head of UFAW, for his tireless dedication and efficiency in scrutinizing and correcting the English in this book. I am also very grateful for the many insights he provided as a scientist, veterinarian and ethologist.

I thank my colleagues at SLU, especially Professor Bo Algers and Drs Lotta Berg, Stefan Gunnarsson, Jan Hultgren and Lena Lidfors, for carefully reading different chapters of this book and providing their frank and valuable points of view. I also thank Dr Karl-Erik Hammarberg for comments on the goat and sheep chapters, and Professor Lennart Bäckström for comments on the swine chapter.

My research for this book was helped greatly by the professional and untiring assistance of the librarians at the Hernqvist Library, who made it possible to study even the most difficult-to-access publications.

The book is made more interesting, and the examples better illustrated, by all those who have allowed me to use their photographs.

Irma, my dear wife, has read and commented on the entire book and, as always, has provided immense support throughout this project.

PART I
Domesticated Mammals

1 Horses (*Equus caballus*)

1.1 Domestication, Changes in the Animals, their Environment and Management

At the end of the Ice Age, wild horses existed in Europe and north Asia, from Spain in the west to the Bering Strait in the east. The southern boundary was the huge Asiatic mountain chains. Wild horses did not exist in America, Africa, south Asia and Australia. They originally were inhabitants of temperate, well-watered grasslands.

The horse was the last of the five most common livestock animals to be domesticated, which may have happened about 2500 BC (Mason, 1984). However, Anthony (1986) and Budiansky (1997) claimed that horses might have been kept in the Ukraine in 4000 BC, first for meat and later for riding. Findings in Kazakhstan demonstrate domestication in the Eneolithic Botai culture, dating to about 3500 BC. Pathological characteristics indicate that some Botai horses were bridled and perhaps ridden, and archaeological remains of processed milk indicate that mares were milked (Outram *et al.*, 2009). At archaeological sites, chopped horse bones have been found, indicating that horses originally were hunted and maybe husbanded as a food animal. However, it cannot have been long before the early farmers found that horses could be loaded, ridden and even pull carts (Clutton-Brock, 1999). Horse remains from around 2000 BC have been found throughout Europe, from Britain to Greece, indicating that horses were in general use (Clutton-Brock, 1999). Their ancestors were wild horses, *Equus ferus*, and among them the tarpan, *E. ferus gmelini* (Zeuner, 1963; Clutton-Brock, 1999). Although some authors (Kowalski, 1967) claim that the Przewalski horse, *E. ferus przewalski*, is linked directly to the ancestors of the domestic European horses, this seems unlikely (Clutton-Brock, 1999). There are thus only slight differences between scientists regarding the origin of horses. Domestic horses arose from wild stock distributed over a moderately extensive geographical region, large enough to have contained considerable pre-existing haplotype diversity (Lister *et al.*, 1998; Lister, 2001). Several distinct horse populations were involved in the domestication of the horse (Vilà *et al.*, 2001; Jansen *et al.*, 2002). Apparently, domestic horses derive from several wild horse gene pools over a large geographic area, and possibly over a long period of time.

The first domesticated horses might have been the size of large ponies, with upstanding manes. According to archaeological findings, the first horses in the north were smaller and sturdier than those in the south, which were larger and more slender-limbed. Most domestic horses in the ancient world were less than 145 cm in withers height, in Eastern Europe about 135 cm and in Western Europe 125–130 cm. The great majority were less than 125 cm. At the onset of domestication, early domestic horses showed less change in size and general morphology from their wild counterparts than did other species of livestock, e.g. cattle (Clutton-Brock, 1999).

During the mutual history of horse and humans, the horse has had a very high-ranking status, being an aristocrat of livestock, irrespective of culture, age and people. In the first place, humans exploited the horse's speed and strength for riding and draught. In addition, some nomadic people in Central Asia used mare's milk for consumption. The consumption of horse meat was common in the early domestication period, but diminished during the Roman age and was discontinued during the transition to the Christian age. However, it still occurs, for example in Iceland and Hungary. Horse meat is eaten as 'salt meat' in many countries.

The horse was, over thousands of years and until the 1950s, important to humans as a draught animal in agriculture and forestry. Several breeds could accomplish an impressive traction force (Fig. 1.1). A very strong bond often developed between a single horse and his owner. Since World War II, horses have played an increasing role in sports and recreation for a great number of people. Riding and looking after a horse has taught many children and young people to respect animal behaviour and to take responsibility for an animal (Fig. 1.2).

The housing and management of the horse has not undergone a dramatic change, as is the case with cattle, swine and hens. The way horses are kept in different countries is based mainly on traditional practice. Horses are kept socially isolated from conspecifics, maybe more than any other domestic animal, and humans therefore are an important part of the horse's social environment. The behaviour, manner and treatment of a horse by humans influence its social life to a very

Fig. 1.1. The horse, for thousands of years up until the 1950s, has been important to humans as a draught animal in agriculture and forestry (image courtesy of photo archive HMH Department).

Fig. 1.2. Since World War II, horses have played an increasing role in sports and recreation (image courtesy of Gitte Häggander-Ekesbo).

great extent. Unsuitable housing or keeping a horse isolated indoors day after day is thus detrimental and will often result in abnormal behaviour.

The skeleton of a horse's extremities indicates its specialization in running fast. In the horse, three of the original four toes (Budiansky, 1997) are regressive. The hoof represents the third toe. The forelegs and hind legs are equipped with tendon arrangements that make it possible for the horse to rest, and even sleep, standing. In the large intestine, and especially in the caecum, cellulose fermentation takes place.

In the upper and lower jaw, there are, on both sides, three front and six back teeth: 36 teeth in total. The permanent set of teeth is complete at 5 years of age (Sisson and Grossman, 1948). Between the front and back teeth there is a gap, the diastema, where the bit of the bridle is applied when the horse is bridled. It is possible to judge the age of a horse by means of the teeth's appearance and the bite.

The horse is one of the most differentiated of all domesticated animals. The horses of the 2000s vary in size from 71 to 76 cm in withers height of the smallest breed, Falabella, to about 190 cm in height of the largest breed, Shire. There are more than 680 breeds in the world and the number of horses was estimated at 62 million in the early 2000s (Hall, 2005). There is no evidence that any single behaviour necessary for survival has disappeared during the domestication process. Thus, when domesticated horses are kept in confined areas, e.g. paddocks, they try to use the grounds in the same way as wild horses – to use certain surfaces for defecation and urinating, others for grazing and yet others for rolling on the ground or other similar body care (Fraser, 1992).

At 3–5 years of age, the horse is full-grown and it can usually live to be 20–30 years old.

The male horse is described as a 'stallion', the female as a 'mare' and the offspring as a 'foal'. The expression 'colt' is used generally for a foal that is more than about 6 months old. 'Colt' is sometimes used for young males and 'filly' for young females.

1.2 Innate or Learned Behaviour

In horses, as in other animals, certain behaviours are innate, others learned. Innate behaviours are, for example: the suckling of a foal, a foal following its mother, getting up and lying down behaviour, body positions at resting, standing or lying, at urination and defecation, or during intense rain and wind. Some innate behaviour, body positions, for example, will usually be performed irrespective of external conditions.

There are innate behavioural differences between breeds, for example for anxiousness and excitability (Lloyd *et al.*, 2008).

Studies indicate that horses can learn to eat or avoid different sorts of feed (Pfister *et al.*, 2002). Examples of learned behaviour are, for example, manipulating devices in order to get feed (Winskill *et al.*, 1996). However, the most conspicuous example of learning in horses is the different behaviours taught by humans, from lifting the hoof to obeying a rider's, coachman's or driver's many different signals.

By early human contact, the young foal learns that there is no danger in being contained by human force. If a horse has been frightened, hurt or beaten in a particular situation or place and/or by a specific person, it usually remembers this and may show a behavioural reaction several years later if it gets into the same situation again. This can be seen if a horse has been badly treated the first time it entered a trailer.

Horses have, however, a very limited short-term memory (McLean, 2004). This must be considered in all horsemanship. Therefore, when a horse has performed well, for example in a riding manoeuvre, it must be rewarded immediately.

Stereotypic-prone horses are less successful in learning than non-stereotypic horses (Hausberger *et al.*, 2007).

1.3 Different Types of Social Behaviour

Social behaviour

Two types of social organization have been observed in wild Equidae: partly the family group of a stallion with as many mares as he can muster and their foals, and partly separate groups of stallions without mares. The stallion tries, often by kicks and bites, to keep competitors away from his group. The groups roam about, overlapping each other's grazing areas, and exchange of individuals occurs between the groups (Klingel, 1974).

Horses show a form of social order when they live in groups, or bands, and a social hierarchy becomes established within these groups. The older and larger animals are usually found to be high in the dominance order. Stallions do not necessarily dominate geldings or mares, but have a significant role in the defence of the group. While being a typical herd species, horses also show a marked preference for certain individuals of their own species (Broom and Fraser, 2007).

Communication

Horses communicate with each other mainly via visual and acoustic signals. The visual signals are many and are extremely sophisticated. Some are obviously for all horses, for example ears turned backwards, indicating aggressiveness. Others are subtler, as with movements in the musculature, chiefly the muscles round the nostrils, the mouth and the posterior end of the mandible (Fraser, 1992).

Behavioural synchronization

Horse herd behaviour is not as synchronized as in sheep or cattle. They do not, like cattle, graze at the same time or in the same direction. However, they often rest simultaneously.

Dominance features, agonistic behaviour

Within groups, stable dominance hierarchies are reported among mares (Houpt *et al.*, 1978). As with other animals, dominance relationships change when new horses are introduced into the group (Houpt, 1982). Aggression is also easily triggered when unfamiliar horses are fed next to each other. As in some other species, the subdominant animal lowers its head, turns it aside and thus shows submission when a dominant individual is approaching.

Like other animals, the horse at aggression shows an increasingly pronounced behaviour, indicating an increased warning of approaching attack. A horse with its ears laid back is a horse prepared to fight. Another social aggressive signal is sideways beats with the tail, which can indicate irritation and, if ignored, can be followed by increased threat like lifting a hind leg and possibly a kick. A highly irritated horse lifts head and tail and moves with high steps, giving the impression of being larger than it is. A head swing and an open mouth with bared teeth are a prelude to a bite; a lifted hind leg is a prelude to a kick (Goldschmidt-Rothschild and Tschanz, 1978; Budiansky, 1997). A bite is usually preceded by wrinkling of the muzzle, while a hanging lower lip is seen in a relaxed horse at rest (Waring *et al.*, 1975). Serious fights often develop from skirmishing by colts. When one horse eventually succumbs, it takes flight. Serious adult fighting can occur in mare groups or between stallions.

Horses attack with bites, kicks with their hind legs or strikes with their foreleg. When attacking by the hind feet, the horse backs up close to the opponent. The kick with the hind leg is usually precisely directed.

1.4 Behaviour in Case of Danger or Peril

If a horse is attacked, for example by a dog, it can try to protect itself with a kick – mainly with the hind legs – chase the dog, or simply flee.

Horses, like other animals, are frightened by sudden and unknown sounds and visual impressions. Horses used to traffic do not usually react to meeting or overtaking cars, but can react to a whirling piece of paper. A horse's reaction can then vary from stopping, shrinking back, stepping or jumping aside, backing, or falling into uncontrolled bolting.

Horses paired with a calm companion horse show less fear-related behaviour and lower heart-rate responses compared to horses with companions not exhibiting calm behaviour (Christensen *et al.*, 2008).

1.5 Some Normal Physiological Frequency Values of Interest at Clinical Examination

The normal rectal temperature of adult horses is 37.5°C, of lighter breeds is up to 38.5°C and of foals is 38.5°C. The critical point in adult animals is > 39°C.

Pulse frequency per minute is 26–40 for adult animals at rest, 40–50 for foals and up to 80 for very young foals of some breeds (Reece, 2004).

For adult animals, the frequency of respiration is 10–14 breaths per min (Reece, 2004).

1.6 Active and Resting Behaviour Patterns

Activity pattern, circadian rhythm

Horses are active for 80% of the day and 60% of the night, but drowse for 20% of the night in several separate periods. Rest as sleep or as total muscular relaxation is a basic need for horses, which must be met. Stabled horses are recumbent for 2 h per day in four or five periods. During the daytime, horses have four to five rest and sleep periods distributed over the day and approximately one-third of the day and night is devoted to rest or sleep (Fraser, 1992).

Exploratory behaviour

The horse, like other animals, shows a strong motivation to examine new environments, or new items in a well-known environment. However, exploratory behaviour occurs only as long as the emotions of fear or apprehension are not present. Horses will explore any new field they are put in and will pay most initial attention to the field's boundaries. They are likely to follow the boundary before exploring the interior of their enclosure (Fraser, 1992).

When it comes to new objects, the horse shows a behaviour that could be characterized as curiosity. It approaches the object, or the animal, with quivering nostrils and sniffs with a rate of locomotion that varies with the degree of uncertainty it seems to feel. The horse sniffs the object and may lick it, or even chew it. If it becomes frightened, it stops and often utters a sounding snort and then either advances carefully or shies away suddenly, returning after a while.

Horses hesitate when confronted with unfamiliar situations and environments, e.g. when entering a horse trailer or a new stable for the first time (Kiley-Worthington, 1987). They hesitate to move from light to darker rooms. They fear unexpected sounds and movements. A piece of paper caught by the wind or a pattering flag can startle a horse.

Two horses encountering each other for the first time show much mutual exploratory behaviour – this involves an investigation of the other's head, body and hindquarters using the olfactory sense (Broom and Fraser, 2007).

Diet, food searching, eating and eating postures (feeding behaviour)

According to dietary profiles compiled from the world literature, the average horse dietary botanical composition for all seasons consists of 69% grass, 15% forb (herbs) and 16% shrub (Alcock, 1992).

Horses eat in total for 12 h or more (Broom and Fraser, 2007), both during daytime and at night.

Therefore, when kept indoors, they must have some forage for the night. Grazing and browsing bouts are interrupted by other behaviours. In housed horses given feed *ad libitum*, 17 feeding bouts were noted during 24 h (Laut *et al.*, 1985).

Biting insects can disturb grazing during daytime. When there are many insects and no wind, horses may gather together closely and, with tail switching and by agitation of the head, try to defend themselves against the insects (Wells and Goldschmidt-Rothschild, 1979).

Horses graze by collecting grass with their prehensile upper lip and biting it off near the ground with the front teeth. When grazing, they move ahead slowly with one leg at a time. They usually take only about two mouthfuls before they step further forward. They avoid grazing on sites with horse droppings (Houpt, 2005; Broom and Fraser, 2007). On the other hand, they graze the grass round cow dung, which is rejected by cattle. They like to take leaves and minor twigs, sometimes bark, from trees. Sometimes, horses in pastures with trees start biting the trunks and chew bark and wood, often thereby totally debarking the lower parts of the trees. This seemingly aberrant form of eating behaviour is not understood, but it is possible that some wood eating is normal (Fraser, 1992).

The young foal does not graze very efficiently until it is several weeks old. By about the end of the first week of life, however, the foal has begun to nibble the herbage in association with its dam. Young foals learn the essential feeding habits of their mares (Fraser, 1992).

A medium-sized horse may need to drink up to about 40 l water per day. Unlike other animals, horses do not drink often, many horses not more than twice a day and night, except when temperatures are high (SCAHAW, 2002). Horses drink large quantities with each gulp. Free-ranging horses in arid environments may drink only once every day or two (Houpt, 2005).

Selection of lying areas, lying down and getting up behaviour, resting postures, sleep

Horses, when in good health and given space to do so, avoid lying down on ground soiled by their dung. When a horse lies down, it puts its legs close together under the belly, the carpus and hock are bent and the breastbone and chest and forequarters make contact with the ground before the hindquarters. The horse gets up by stretching both forelegs ahead, whereupon, with a forward-moving action of its body by a thrust from the hind legs, the animal finally gets to its hind feet. Once a horse has got up, it often shakes its body.

Young horses spend more time lying down than do adults (Duncan, 1980). Some adult horses lie down to a slight extent only, whereas others always lie down during some time of the day. Mares with young foals tend to lie longer than usual when the foal is nearby and is sleeping in full lateral recumbency. Mature horses never lie in full lateral recumbency for long periods of time.

Horses often rest in a standing position, without sleeping. They have a unique ability to rest, and even to sleep, in a standing position by means of a unique interplay between the knee joint and the hip joint. When either the stifle or the hock flexes or extends, the other joint reciprocates with a similar action (Fraser, 1992).

Horses sleep in short periods for 6–7 h a day and night. They are in a waking state for 80% of the daytime and for 60% of the night-time. The horse shows both forms of deep sleep: slow-wave sleep (SWS), and paradoxical or rapid eye movement sleep (REM sleep). They are able to drowse and engage in SWS while standing, but REM sleep occurs almost always when lying down (Fraser, 1992: SCAHAW, 2002; Broom and Fraser, 2007). All forms of rest represent a major need in equine self-maintenance and well-being (Dallaire, 1986). When a person approaches a horse that is lying down, it almost always gets up.

Locomotion (walking, running)

The natural gaits of a horse are: slow and fast walk, trot and gallop. Other seemingly inherent gaits occur in certain breeds, e.g. rack in Icelandic horses.

Few horses are enthusiastic jumpers until they are taught. Untaught horses may avoid obstacles only 60 cm high rather than clear them by jumping voluntarily. Normally, horses avoid jumping over ditches and in general show reluctance to jump over horizontal obstacles (Fraser and Broom, 1990).

Swimming

Horses are natural swimmers. However, they do not actively seek to swim when close to water.

1.7 Behaviour at Defecation and Urination (Eliminative Behaviour)

Grazing horses as a rule perform eliminating behaviour on an area chosen specifically for this and will return regularly to defecate at this site (Fig. 1.3). When moving, for example when being ridden, they might defecate when walking, but they prefer to stop during the defecation act. They always stand still during urination. They often urinate at the edge of restricted grazing areas. On urination, horses stand with the hind legs spread apart. Stallions usually take this position in a more pronounced manner than mares.

While urinating, the stallion (and the gelding) adopt a characteristic position: the hind legs are abducted and extended and the back becomes hollowed. The mare, when urinating, does not show the same marked straddling posture as the stallion.

Horses seem to avoid urinating on hard surfaces. Instead, they choose surfaces like soft soil, grass or indoor bedded areas. It happens that horses, after having been out on frozen land for a long time, urinate in the bedding of the box immediately after having been taken indoors. Urination occurs 3–6 times per day and defecation about 12 times per day (Fraser, 1992).

1.8 Body Care, Cleanliness

Horses groom their bodies by rolling on the ground, by rubbing against their own legs or against trees, food or water troughs etc., or by nibbling or snapping on their hips, flanks or limbs. They groom eye, face, nose and nostrils by rubbing their face up and down the side of one of their forelegs, which then may be held out in front of the other. After strenuous exercise, the horse may rub its head on a person.

Fig. 1.3. Defecation heaps on an area of the pasture chosen specifically by the horses for excretory behaviour (image I. Ekesbo).

Horses do not use their tongues to clean out their nostrils, as do cattle. They snort to do so.

Horses on pasture often perform grooming by rolling on an area chosen specifically for this (Fraser, 1992). When rolling, the horse lies down in a normal fashion and then proceeds to kick itself over on its back and rub its back against the ground, while keeping its feet up in the air. After several rubs, it rolls back on to its side again. The horse then rises briskly, adopts a stance with all limbs slightly splayed out and begins to shake its body vigorously.

Pairs of horses perform mutual grooming. This allogrooming occurs in all age groups. Mostly, it is performed by individuals usually matched for age and size. However, mares and their foals often groom each other. In the normal grooming position, two horses face each other; one extends its head past the side of the other's neck and nibbles vigorously over the associate's withers and back. This is also part of their social behaviour. In the wild, mares will have one or two preferred social partners with which mutual grooming is performed, thus reinforcing affiliative relationships (Wells and Goldschmidt-Rothschild, 1979; Feh and de Mazières, 1993; Feh, 2005).

1.9 Temperature Regulation and Climate Requirements

Temperature regulation

The horse regulates its temperature mainly through sweating but, after having been given lengthy exercise, it supplements this by means of the breathing apparatus (McConaghy, 1994). The whole skin has sweat glands and a horse sweats very much during physical exertion. Panting stands for about 17% and sweating for about 83% of the relative proportion of total evaporative heat loss in horses (Robertshaw, 2004).

Climate requirements

The lower critical temperature (LCT) values for horses are affected by their age, breed, acclimatization and the plane of nutritional condition (Cymbaluk and Christison, 1989). An LCT of 5°C is noted for Standard bred horses acclimatized to an indoor temperature of 15–20°C (Morgan *et al.*, 1997). The thermoneutral zone of mature Quarter horses, accustomed to mild winter temperatures, ranges from approximately –15°C to 10°C

(McBride *et al.*, 1985). In newborn pony foals, the average LCT is about 20°C, with a variation between 13°C and 26°C, depending on body insulation (Ousey *et al.*, 1992). The LCT for Quarter horse weanlings acclimatized to a cold environment is estimated at –11°C (Cymbaluk and Christison, 1989). Young horses need 1.3% more maintenance energy per Celsius degree decrease in temperature below 0°C. To sustain a constant moderate gain, daily DE intake needs to be increased by 0.7% per Celsius degree decrease in ambient temperature below 0°C (Cymbaluk, 1990).

Adult horses can tolerate different kinds of weather conditions, even though racial differences exist. Horses cease grazing in rain and wind and direct their hindquarters into the wind with the tail closely pressed against the body. They are most comfortable in mild temperatures in the range of 10–20°C. Horses adapted to cold tolerate temperatures in windless cold down to –20°C, even down to –40°C if given shelter (Fraser, 1992). Weather with prolonged rain and wind around 0°C is a greater stressor to horses than windless cold. They stand rather than lie down when it rains and standing increases by 20 min per day for every degree Celsius drop in temperature during the winter, when the mean temperature is –2°C (Duncan, 1985).

Racial differences exist. A thoroughbred foal is more sensitive to rain and wind at temperatures near 0°C than a cold-blooded foal of the Gotland pony, Iceland horse or Norwegian pony breeds. The temperature in the stable for tied horses must therefore vary with the breed, so for an Arab blood horse it should be about 12°C, whereas for Norwegian ponies 8°C may be sufficient. When the horses are kept loose in boxes with plenty of bedding, the temperature in the stable can be held lower than for tied horses in stalls.

Studies of 40 Iceland horses kept outdoors in Norway during winter showed that they used an open shed in severe cold and in rains and wind. Snowfalls did not increase the use of the shed (Mejdell and Bøe, 2005). A small Swedish pilot study showed similar results (Michanek and Ventorp, 1996). As horses are more pronounced flight animals than cattle, it is less natural for them to use the forest, and thereby also sheds, for protection.

To avoid direct solar radiation in hot weather, they will stand in the shade during the warmest parts of the day (Crowell-Davies, 1994). In the pasture, shade from direct sunlight is normally sought in temperatures over 25°C if there is no air movement. The relative humidity in horse stables should be 70% ± 10%.

1.10 Vision, Behaviour in Light and Darkness

As a prey animal, the horse has good vision. With its large eyes on the sides of its head, it has a broad visual angle, i.e. 330°, and when keeping its head high, nearly 357°. The eyes of the horse are arranged such that it is able to see almost entirely to its rear and completely on both sides simultaneously. Some breeds, notably the Arabian, have more prominent eyes, allowing them to have better visual scope than the majority of other horses. The position of the horse's eyes can thereby help it to detect predators easily and keep visual contact with other horses in the group. The horse has the ability, at the same time, to register divided views with each eye, although it may be able to see an object with one eye only for a while. It has been proposed that horses use binocular vision when the ears are erect and facing forward (Fraser, 1992; Roberts, 1992).

The horse has a blind spot behind itself when the head is held alertly forward, for example where the rider is sitting, and another just in front of its muzzle (Fraser, 1992). The latter means that a horse does not see exactly what it is eating.

The horse has a very limited ability to focus between distant and close objects, which the lens of the horse cannot accommodate in order to focus the picture on the retina. By raising or lowering the head, the horse adjusts to the correct distance between the lens and the retina to obtain the best focusing. This means that it can see close objects by lowering its head and objects at a distance by raising its head. However, when an object is in focus, the horse seems to detect even very slight movements. When it concentrates on focusing its eyes to maximum ability, the horse appears momentarily to lose the ability to observe consciously to the rear and sides (Klemm, 1984).

Much of the horse's communication system consists of its ability to consider tiny changes in body movements, for example insignificant changes in the ear position of another horse.

The horse has good scotopic vision, which is important both for detecting predators in the night and for keeping the group together at night.

Horses have some colour vision (Fraser, 1992; Pick *et al.*, 1994), possibly depending on the necessity to detect predators having protective coloration. It is not known whether the ability to see colour plays any role in horses identifying members of their own group (Budiansky, 1997).

1.11 Acoustic Communication

Hearing

Horses are considered to have good hearing and probably can also hear sounds within the inaudible, to humans, ultrasound sphere. The hearing range is from about 60 Hz to about 33,500 Hz, with maximum sensibility at 2000 Hz (Heffner and Heffner, 1983, 1990, 1992a,b).

The ears of the horse can move independently of each other and can thereby be slanted around a lateral arc of 180° in order to catch where the sound comes from. The horse must be able to locate the sound source in time to detect predators, but also in order to find out where the rest of the group is located. The horse can use body positioning as part of acoustic detection in an outdoor environment (Fraser, 1992).

Some predators and humans are able to locate a sound accurately within an angle of 1°, whereas horses can only locate a sound within an angle of about 27° (Heffner and Heffner, 1992b).

Vocalization and acoustic communication

The acoustic signals used by horses are neighing, blow or loud snort, low snort, squeal and moan or groan. Budiansky (1997) has described most of these and has also made attempts to interpret their purposes and senses. The low neigh (nicker) is a chattering tonal sound (about 100 Hz) formed with the mouth shut and is used to maintain contact between mare and foals, when two horses meet, or when something pleasant occurs, e.g. feeds (Kiley-Worthington, 1987). A high neigh, a whinny, is loud and tonal, starts high, up to 2000 Hz, and then drops to half the starting frequency. It is used most in order to establish contact over longer distances. Mares and foals that lose sight of each other will whinny. The blow – a loud, non-tonal snort – is made by blowing a rapid pulse of air through the nose and can be heard at a distance of 200 m. It is emitted in excitement, when the horse is frightened, for

example if something unknown or frightening appears on or at the side of the horse's road. It is a short, percussive, non-tonal sound, containing a lot of different pitches blended together. A horse that is startled will turn towards the source, often freeze and stare at it for about 20 s, then blow and possibly make a cautious approach to investigate. It is also an alarm signal at danger and, in feral horses, is emitted by the stallion. The squeal is much louder and contains a tonal component at around 1000 Hz, but it still carries a lot of non-tonal harshness. Horses squeal in confrontation with other horses and a squeal is emitted in aggressive situations, for example, between two stallions or when a non-receptive mare is approached by a stallion. Low snorting occurs when the horse comes across certain smells, when it cleans the respiratory tract, but also in conflict situations (Budiansky, 1997). Moans or groans are emitted in pain, when a mare foals for example.

1.12 Senses of Smell and Taste, Olfaction

The horse's long nasal meatus facilitates registering of faint scents. The horse frequently makes use of its olfactory sense in order to examine its environment and to identify feeds. This is seen when horses are given new feeds, such as hay of a different composition. The vomeronasal organ (Sisson and Grossman, 1948), located in the dorsal part of the oral cavity, registers various substances, e.g. pheromones. The organ can register pheromones, from other horses as well as from other species, including humans (Whitten, 1985; Marinier *et al.*, 1988). The mare can thus recognize its own foal. The 'flehmen' behaviour, a raised head and a simultaneous characteristic curl of the upper lip shown by the stallion towards mares in heat, is considered to facilitate registration of information from pheromones (Crowell-Davis and Houpt, 1985; Fraser, 1992).

Horses unknown to each other, or horses separated for a time, usually sniff each other carefully when they are brought together.

Stallions give off scent markings. On pasture, horses usually defecate in a relatively limited area. In the wild, stallions defecate over faecal piles from mares in their own group, as well as over other dung piles from other stallions (McDonnel, 2002).

1.13 Tactile Sense, Sense of Feeling

Tactile touch

In horses, the tactile sense is well developed. Tactile communication seems to be important between mare and foal, as well as between other closely bonded horses. Grooming between pairs of horses is an example of this.

Horses show positive reactions when stroked by someone to whom they are habituated. They seem to prefer to be stroked before being patted.

Sense of pain

In response to pain, reflex escape or withdrawal efforts are made, in fear or rage flight or fight. The memory of pain is very durable in horses (Fraser, 1992).

1.14 Perception of Electric and Magnetic Fields

This does not seem to be described in horses.

1.15 Heat and Mating Behaviour, Pregnancy

Sexual maturity sets in at about 2 years of age. Mating may occur for mares and stallions at 3 years of age. The mare comes into heat in early spring as the hormone balance is influenced by increased day length and light. The heat lasts about 5 days, with highly active heat lasting about 1–2 days, sometimes even shorter. If not pregnant, the mare will come into heat again after about 3 weeks. The mare comes into heat about 7–9 days after parturition. At the beginning of the heat, the mare shows aggression against the stallion and can even kick against him. During the intense phase of the heat, this behaviour will change and the mare accepts the courting of the stallion and may even actively apply for mating.

The gestation period is about 340 days, with a variation between 325 and 350 days.

Traditionally, mares gave birth almost without exception in the spring. This natural time of parturition is arranged so as to be centred on a period in the early summer when the nutritional potential of the environment is usually greatest. Nowadays, mating/insemination is made so that the parturition can occur at the most favourable financial time for the horse owner.

1.16 Before and After Parturition

Behaviour before and during parturition

When parturition is near, a mare, if in a group, will separate herself from the group some hours before the birth and select a site where the birth will occur. Most mares, about 80%, foal at night, which might be a hereditary strategy for avoiding predators during the foal's most vulnerable first hours. The practical experience that even stabled mares deliberately try to avoid the hours of supervision when the birth is going to occur is supported by scientific studies (Fraser, 1992).

A few days before parturition, the udder becomes swollen and there is a wax-like discharge from the teats – although this may sometimes occur weeks before actual foaling. About 4 h before parturition, sweating is evident at the elbows and on the flanks. Sometimes, there is unusually high elevation of the head at this time.

Increasing restlessness and other evidence of a build-up of pain constitute predominant indications of the late prepartum phase and its transition to the first stage of labour. In the first stage of labour, the behaviour of the mare includes intermittent restlessness, circling movements, looking round at the flanks, raising and whisking of the tail, repeatedly getting up and lying down, and often urinating. Feeding ceases abruptly with the onset of this stage. Increasingly, the mare rises and lies down, rolls on the ground and slaps the tail against the perineum, shows restlessness and aimless walking, tail swishing, kicking and pawing at bedding; later, crouching and straddling, kneeling and, finally, the mare lies down. The amniotic sac may then appear. This first stage may take from one to several hours.

The transition from the first to the second stage of labour is dependent on the opening of the cervix and the escape of uterine content into the vaginal passage. As a result of this development, the outer fetal membrane, the chorion, still adherent to the uterine wall, extrudes into the canal and tears under the pressure. When this occurs, the chorionic fluid is released through the birth canal and helps to lubricate it. Some mares investigate the allantoic fluid discharge and may then exhibit flehmen (Fraser, 1992). In some cases, the mother is recumbent throughout birth; in some, there is alternate lying and standing; and, in others, lying, standing and crouching. The duration of the second stage of labour is usually much shorter than the first stage.

The inner sac and the fetus now slip further into the vaginal passage. This allows the amniotic sac, containing the fetus, to bulge through the vagina and enforce more dilatation so that the fetus enters the birth canal. These latter events induce the second stage of labour and accelerate the expulsive efforts of the mare. The fetus advances by a combination of voluntary and involuntary muscular contractions of the mare's abdomen and uterus. Repeated straining is therefore a major feature of the second stage. The manner and degree of straining vary from one mare to another, but in all cases the expulsive efforts become intensified and soon the water bag, or amniotic sac, becomes extruded, unburst. Resting intervals, each of a few minutes' duration, separate straining sessions. Bouts of strong uterine contractions force the forelegs of the foal into the vagina with its muzzle inserted between or adjacent to the legs. The bouts of straining become progressively more vigorous until the foal's feet and nose have appeared outside the vulva. The mare strains powerfully and regularly to expel the fetus, and often rotates from one side to the other.

Many mares stand and change positions as the fetal forelegs and muzzle protrude from the vulva. Pain is most evident in the second stage of labour. Such pain secures the undivided attention of the animal and its total participation in the birth process and associated straining. Further extrusion of the fetus is not necessarily achieved with every straining bout, the course of extrusion being subject to halts and even the partial withdrawal of the foal back into the dam.

An obstacle in the course of the birth is the passage of the foal's forehead through the taut rim of the mare's vulval opening. Once the head is born, the rate of passage of the foal is accelerated. The shoulders of the foal follow the head after a few minutes.

The mare's vigorous straining usually ceases when the foal's trunk has been born. Soon after this, the remainder of the foal except its hind feet slip out of the vagina. When the fetal hindquarters have been expelled, there occurs a long resting period. At this point in the birth process, both hind feet of the foal are still in the recumbent mother's pelvis. The second stage is then complete, after an average of 10–20 min. The amniotic sac is usually still unruptured, unlike in cattle, and the umbilical cord is still intact.

At this time, the third stage of labour, the post-partum phase, begins. The actions of the almost wholly born foal rupture the amniotic sac, the foal breathes and its further movements withdraw its hind feet from the dam. Mares usually lie on their side, in apparent exhaustion, for 10–20 min after the birth.

This third stage can last from half an hour to more than an hour, but usually it takes a shorter time than the previous stage. Gradually, the mare turns her head towards the foal and the foal frees itself, whereupon the navel cord ruptures about 3 cm from the foal's abdomen. The foal lifts its head and its ears, which have lain along the head, are raised.

Expulsion of the afterbirth occurs on average 60 min after delivery (Rossdale and Ricketts, 1980) and concludes the third stage of labour. Mares usually show oestrus by the 9th day ('foal heat').

Number of foals born

The birth of twins is unusual. If not interrupted by abortion, it often results in the death of one or both foals.

After the birth

Mare and foal after the birth

After having lain in extension following its expulsion at birth, the foal raises its head and neck, then rotates on to its sternum and gathers the hind legs towards the body. The head is shaken and the ears become mobile. When the mare eventually rises, she begins a period of licking her foal, usually beginning on the head. Continuous licking may last 30 min. During this grooming, the mare develops her strong attachment to her foal and this period is essential for building up the mare–foal bond. From this time on, the mare will care for the young animal and defend it with much intensity. Mares do not know their own foals immediately after birth, but bonding is established firmly within 2 h (Fraser, 1992).

The foal flexes its forelegs and it attempts to rise by standing on its knees. At this point, defecation of meconium may occur before the foal makes a vigorous effort to stand. On rising, it may succeed in taking one or two steps before falling down. Gradually, the foal makes further attempts before it succeeds in getting up on its legs. The mare grooms the foal and during this period she frequently vocalizes.

The mare has, compared to other farm animals, a uniquely close relationship with the newborn, refusing to leave it from the moment of birth. However, this maternal attention is not constant. When not attending the foal, a mare may nibble at hay or straw, smell and lick objects smeared with birth fluids. Unlike some other species, it is rare for horse mares to eat the placenta (McDonnel, 2002).

Nursing and suckling behaviour

While the newborn is exploring its immediate environment in the course of seeking the mare's teats, the mare receives its approaches passively. She also shows positive behaviour in accommodating the foal. The foal locates and investigates the mare's limbs and ventral regions, the mare's inguinal region and ultimately her mammary gland. The foal's teat-seeking behaviour is often extended over a substantial period of time. The newborn foal is a good model for recognizing the basic biology of empirical or trial-and-error activities, showing how they add to experience and success (Fraser, 1992). The mare takes a position adjacent to the newborn and will hold this position, permitting the progressive exploratory approaches of the foal. Experienced mares often position themselves in order to facilitate nursing by standing with their hind legs apart or by bending a hind leg to enable the foal to find the udder. In the teat-seeking activity of the foal, the udder is apparently identified by its tactile characteristics and the protruding teat will be found quickly and sucking begun. After its first successful suck, the foal sucks over an average interval of about 15–20 min.

The horse fetus receives only a slight amount of antibodies via the placenta. Instead, these are, to a greater extent than in other farm animals, delivered via the colostrum (Fraser, 1992). During the first days, nursing occurs about 7 times per hour, while, at 6 months, the frequency is usually once an hour (Kiley-Worthington, 1987).

The foal will have had its first sleep about 3 h after birth. Sleep usually occurs after feeding and a small amount of investigative behaviour or sudden bursts of activity. The latter often take the form of little dances, bucking, short runs or jumps. Sleep episodes are usually less than 30 min in duration. During the first week, foals spend most of the day resting and sleeping. In the second week, they rest and sleep half of the time.

The first urination follows several sleeps and sucks, and urination occurs about four to ten times daily (Fraser, 1992).

The foal and the mare are always close to each other during these early days and foals follow their mares; they are 'followers'. Already during the first week, the mare begins to increase the distance from the foal, but not at the end of a nursing, when the foal usually follows her (Fraser, 1992). Mare and foal seem to recognize each other by smell, vision and sound. However, early foal recognition is based largely on smell.

The foal begins nibbling grass after about 1 week, but does not graze, in a proper sense, before it is several weeks old.

Play behaviour

Foals perform solitary play very early in their life. Even the day-old foal may show sudden bursts of playful activity in the form of leaping.

Later, during their first weeks, foals frequently play with their mothers. Often, it starts with nibbling the legs of the mother. They learn early to avoid other mares with foals, as well as stallions. At about 2 months of age, this behaviour, as a rule, ceases. Instead, male foals try to get in touch with other foals, chiefly with other male foals, while female foals are more inclined to play by themselves, for example, by performing fast solo rushes (Waring et al., 1975). There is, as a rule, an obvious difference between male and female foals in that male foals bite and manipulate things a lot more intensively.

The young foal often shows sudden bursts of playful activity in the form of leaping. In group-kept young horses, pushing, biting and even rearing occur. For both sexes, play is an important part of their behaviour, and nipping and biting may be a part of this. It also involves running about alone or in groups, chasing with much head tossing, sudden stops and starts, and kicking of the hind legs in the air.

Weaning

For free-living or feral horses, weaning takes place naturally at around 8–9 months (Waran et al., 2008). However, if foal and mare are kept together and the mare is not pregnant, suckling may continue sporadically until the foal is about 1 year old. If pregnant, the mares nurse the foals for 35–40 weeks and multiparous mares wean them at about 15 weeks and primiparous mares at 10 weeks before the next foaling (Duncan *et al.*, 1984). If mare and foal are not kept separated, the foal will stick to the mare until she gives birth the next time.

Under domestic conditions, weaning tends to take place earlier, typically between 4 and 6 months of age (Waran *et al.*, 2008).

1.17   The Foal and the Young Horse

There are sex differences in the behaviour of foals. Although nipping and biting occur in both sexes, it is more common in colts than in fillies.

1.18   Assessment of Horse Health and Welfare

In order to assess whether or not a horse is healthy, special attention should be paid to general behaviour, bodily condition, condition of coat, eyes and mucous membranes, posture and movements of head, ears, tail, legs and feet, notice should be given to eating and drinking behaviour, sounds, activity, movements and postures in rest and motion, appropriate to its age, sex, breed or physiological condition.

The healthy horse

A healthy horse is watchful towards its environment, stands with straight legs and straight spine and observes its environment with clean and bright eyes. For its age, sex, breed or physiological condition, it should have normal posture, movements and sounds, clean and bright coat, normal flesh, sound and strong legs and hooves; normal behaviour at eating, drinking, resting, suckling or sucking; it can walk, trot or gallop, lie down, get up, defecate and urinate. The appearance, smell and amount of manure alternate according to the type of feed. Oestrus occurs with normal intervals and of normal intensity.

The sick horse

Symptoms

Free-ranging horses growing up in a natural environment have never been observed with the stereotypic behaviour, self-aggression or other problems that might occur in stabled horses (Feh, 2005).

Each deviation from the signs of health mentioned above, including abnormal behaviour patterns, must be regarded as a symptom of disorder and may be a symptom of disease. Unlike cattle and pigs, horses often eat even when affected by disorders. On the other hand, there may appear to be a fast change to the coat, which becomes lustreless, and the horse often shows drowsiness or apathy. A sick horse does not groom. Sweating is a sign that can indicate disease in a horse. The body movements can be changed, sometimes the horse can be hypersensitive to sound or touch. A lowered head and tail and low body attitude may indicate pain, exhaustion and weakness. Together with these general symptoms, local symptoms may occur, such as discharge from nose or vulva, swellings and perhaps tenderness in the udder or at joints in other parts of the body. The temperature can be increased, but also lower than normal.

Examples of abnormal behaviour and stereotypies

The failure of humans to meet the horse's behavioural needs has caused disease symptoms in the state of abnormal behaviour. Among abnormal behaviours in horses, the most well known are continual scratching or scraping, sometimes hitting one foreleg against the ground, head shaking, weaving and crib biting (Fraser and Broom, 1990; Broom and Kennedy, 1993; Cooper and Mason, 1998; Minero *et al.*, 1999; Bachmann and Stauffacher, 2002; Mills and Taylor, 2003). These abnormal behaviours are regarded as stereotypies and have often been linked with possible understimulation, e.g. lack of staying outdoors or motion. However, the basis of stereotypies may already have been laid in that weaning and social and nutritional factors may play an underlying role (Nicol, 1999; Nicol *et al.*, 2002; Clegg *et al.*, 2008). In crib biting, a fixed object is grasped with the incisor teeth, the lower neck muscles contract to retract the larynx caudally and air is drawn into the cranial oesophagus, producing a characteristic grunt. Air is swallowed. There is a view that crib biting is harmful, and restrictive devices and/or surgery are widely employed to stop the behaviour, with limited success and often against the best interests of the individual horse (McGreevy and Nicol, 1998). One theory is that crib-biting horses have a slower intestine feed passage than normal horses, possibly depending on poorer saliva production (McGreevy *et al.*, 2001). This might indicate that the behaviour is the horse's attempts to increase saliva production (Nicol, 1999) and thereby counteract an inadequate digestion function.

In weaving, the horse sways from side to side repetitively, shifting weight and moving its head and neck back and forth, sometimes moving the forelegs sideways. It might be caused by social isolation and a barren environment (Cooper *et al.*, 2000). The behaviour seems to decrease if the horse sees its own mirror picture (Mills and Davenport, 2002).

It has been asserted among many horsemen and horsewomen that these abnormal stereotypic behaviours are a consequence of imitation from one horse to another. Although several studies show that the environmental factors mentioned above may be the main triggering factor, there are studies indicating that exposure to a stereotypic neighbour is a significant risk factor for performing stereotypy (Nagy *et al.*, 2008).

It sometimes happens that horses try to scrape their tail root by pressing the buttock against furnishings, outdoors against walls, trees, etc. The reasons usually are due to attacks by parasites, either in the anus or at the tail root. The behaviour ceases if the cause is eliminated.

Horses kept alone in small paddocks sometimes show stereotypical behaviour by walking up and down monotonously along the fence – behaviour similar to that shown by bears in inappropriately designed zoological parks.

Examples of injuries and diseases caused by factors in housing and management

Horses are affected by injuries or diseases. Some horse injuries or diseases may be a result of accidents or poor husbandry. Several of these are located on the joints, hooves or the skin.

The cause of acute joint disease is injury to the joint, such as a twist or sprain. Additional predisposing factors for joint disease are exercise on hard or uneven ground, poor conformation, poor shoeing, lack of fitness and overexercise. Exercising a fat, unfit horse will also predispose to injury. The pathological effect of trauma on joints is to cause synovitis and capsulitis, which, in turn, create physical and biochemical damage to the articular cartilage.

Hoof injuries are caused primarily by trauma, such as when a foreign body, for example a chip of stone or a nail, gets into the hoof, or improper shoeing and trimming of the hoof may cause injury. Incorrectly applied horseshoe nails are a common cause. The result of such injuries is usually an infection, often followed by the formation of an abscess. Such abscesses may be found inside the hoof capsule, above the sole of the hoof and even on the coronary band above the hoof. The symptoms are significant lameness.

Hoof diseases may be caused by keeping the horse in an unsuitable environment. Thrush is a bacterial infection that occurs on the hoof, specifically in the region of frogs. It may affect horses kept in wet, muddy or unsanitary conditions, such as an unclean box or stall. The most obvious sign of thrush is usually the odour that occurs when picking out the feet.

Skin swelling caused by incorrect saddling, saddle gall, may lead to sores. Similar injuries caused by wrongly adapted harnesses were seen in draught horses in former times.

Symptoms of pain

A horse in pain often gets a characteristic facial expression with eyelids slightly wrinkled; usually, there is a fixed stare with the eyes and the eyes are not as mobile within the orbit as in the healthy horse. The ears of horses in pain are usually held back slightly for longer periods. In addition, a horse in pain may have dilated nostrils. These facial signs collectively give a horse a facial appearance of concern.

In colic or abdominal pain, the horse shows various abnormalities of posture, sometimes combined with their looking towards the flanks. Horses with persistent pain may adopt an unusual stance or show unusual recumbent behaviour.

1.19 Capturing, Fixation, Handling

As with all animals, horses must be dealt with in a calm manner and capturing must take place carefully. To catch a horse requires patience. One must approach the horse on the flank, not stare the horse in the eyes and never make sudden movements.

A nervous horse can often, through the application of a twitch, be calmed so that it will accept an invasive procedure, e.g. a hoof examination or an injection. It seems as if the attention of the horse will then be concentrated to this painful point so that discomfort or short pain in other parts of the body are not registered and thereby any aggressive or defensive countermeasures from the horse will be avoided. Canali *et al.* (1996), however, explains the effect with the release of β-endorphins.

1.20 The Importance of the Human–Horse Relationship

Human–animal interactions can involve visual, tactile, olfactory and auditory perception, and human contact on farms can be subdivided into five main types: (i) (stationary) visual presence; (ii) moving between the animals without tactile contact (but maybe using vocal interactions); (iii) physical contact; (iv) feeding (rewarding); and (v) invasive, obviously aversive handling (Waiblinger et al., 2006).

Foals must have human contact early; it should begin straight away with the day-old foal. A foal should be touched all over each day to learn to accept human contact in its future life. Although there are individual differences between horses, such regular human contact usually makes the foal calm in its future contact with humans. Handling throughout the first 6 weeks of life increases a foal's performance in training tasks compared to handling between the 7th and the 12th week of age, implying the existence of a critical handling period during the first period of life (Mal and McCall, 1996). Sessions of touching, bodily restraint, lifting the feet and fixation of the hooves, etc. allow the foal to learn successively to accept and adapt to human control over its behaviour. Foals reared without human contact before weaning may have continuous difficulty in accepting handling by humans in their later life. They appear to be 'insecure' under human control. Acquisition of undesirable behavioural states can be avoided by early handling.

When horses were still used as draught horses, one could sometimes discover the friendly and deep relations that could be established between a horse and its owner if the owner understood that the horse should be treated in a good manner during the often very long working days.

Horses trained under a positive reinforcement schedule are more motivated to participate in training sessions and exhibit more exploratory behaviours in novel environments than horses trained under negative reinforcement (Innes and McBride, 2008).

While learning the ability to interact with any horse is crucial in some professions (farriers, veterinarians), learning to develop a relationship is especially crucial for breeders, caretakers, horse owners and trainers. No recipe-based method can offer the required capacity to adapt to horses and/or to situations. Only well-trained observational skills allied with an advanced knowledge of horse behaviour can bring about the safe handling of horses (Hausberger et al., 2008).

2 Swine (*Sus scrofa*)

2.1 Domestication, Changes in the Animals, their Environment and Management

The wild boar

In the literature, the terms 'wild boar(s)' and 'feral pig(s)' are sometimes used alternatively for the origin of domestic pigs. In this text, the expression 'wild boars' will be used consistently.

The European wild boar (*Sus scrofa*), the origin of domestic pigs, can be 185 cm in length with a height of 100 cm from withers to front hoof, but usually are not that high. The wild boar (the feral pig) is capable of adjusting itself to different natural conditions. Its coat is shaggy, especially in the winter, and, as a rule, greyish-black to black-brown. The piglets are pale beige with longitudinal streaks. The head is large and the neck is coarse. Wild boars prefer living by alternating between wooded and open grounds. The size of the wild boars' living area is determined by access to food, but varies widely from fewer than 100 to more than 2500 ha (Jensen, 2002).

The domesticated pig

Pig domestication is thought to have started between 8000 and 7000 BC, independently and in different places, from wild boar subspecies in Europe and Asia (Clutton-Brock, 1999; Giuffra *et al.*, 2000). According to examination of mitochondrial DNA sequences, pig domestication took place at least seven times in areas across Eurasia, including in previously unknown centres in India, Burma–Thailand, Central Italy and Wallacea–New Guinea (Larson *et al.*, 2005). Domestic swine were brought to North America in the 1500s by Fernando de Sota, and wild boars were brought to the USA in the early 1900s (Graves, 1984).

Domestication has resulted in a smaller head in proportion to the body, less growth of hair, a mostly curled tail (Fig. 2.1), a thicker fat layer, aseasonal instead of mostly – but not always – seasonal breeding, increased litter size and more than one litter per year, although wild boars under special circumstances may have two per year. The skull of the pig, like that of the dog, shows more variation than is seen in other species of domestic animals (Sisson, 1975). According to Jensen (2002), wild boars do not have curled tails. The greatest changes in pigs have occurred during the past 200 years because of selection by humans. From the 1700s, pigs were selected for their large size and fat production. From the 1950s, selection has been aimed at faster growth, leanness, i.e. more flesh and less fat, increased size and larger litter size. Despite these phenotype changes, the domesticated pig has kept most of the wild boar's behaviour and patterns of reaction – evidence for this has been shown in studies of domesticated pigs kept under semi-natural conditions (e.g. Stolba and Wood-Gush, 1984, 1989; Jensen, 1988a,c). On the contrary, as far as we know, no single behavioural pattern has disappeared from the wild boar repertoire during the domestication process (Gustafsson *et al.*, 1999).

When kept in large areas, domesticated pigs, like the wild boar, prefer alternating between wooded and open grounds.

Until the 1950s, several farm animal species were kept on most farms. Usually, only one or a few fattening pigs and one or a few sows were kept together with a number of cattle, horses, poultry and sometimes other species as well. The sows were usually kept outdoors during summer (Fig. 2.2). They normally had access to a wallowing pit and protection against the sun. When kept indoors, dry sows were kept loose in pens prepared to function also as farrowing pens with bars along the walls (Fig. 2. 3), or the sows were transferred to such a farrowing pen shortly before farrowing. Both types of pens usually had bedding, mainly straw. In some countries, specialized fattening pig herds were

Fig. 2.1. A sow of the old Danish Landrace breed from the mid-19th century. Until well into the 19th century, pigs in many countries had the characteristics of the wild boar, e.g. the long snout and stiff bristles (image courtesy of photo archive HMH Department).

Fig. 2.2. When most farms still had one or a few sows, they were kept outdoors during summer. They usually had access to a wallowing pit and protection against the sun (image I. Ekesbo).

Fig. 2.3. Until the beginning of the 1960s, most sows were kept in farrowing pens like this, with straw bedding and wooden rails along the walls as protection for the piglets if the sow lay down against the wall (image I. Ekesbo).

established from the beginning of the 20th century, sometimes in connection with dairies, where cheap feed like whey or skimmed milk was available. The piglets, usually about 8–10 weeks of age, were purchased from various farms and brought into the fattening pig farms.

After World War II, when animal husbandry often became specialized by keeping one species only, larger fattening pig or sow herds were established. This led to a radical change of housing systems. The dry sows were kept restrained more and more often, tied or in cages or crates (Fig. 2.4), and were kept indoors all year round. This is still the case in several countries. Sows with litters were, and in several countries still are, kept in crates with

a perforated floor without any litter (Fig. 2.5). During the 1970s, early weaning of the piglets was started in several countries. The young piglets were put in cages from 3 weeks – at an even earlier age in some countries (Fig. 2.6). At about 8 weeks, the piglets were put in the fattening pens with a perforated floor without bedding (Fig. 2.7).

At the end of the 1900s, such methods were questioned more and more. International animal welfare legislation in the Council of Europe and in the EU has laid down minimum standards for the protection of pigs, which are binding to all holdings in all EU member states from 2013. The use of individual stalls for pregnant sows and gilts will then not be allowed during a period starting from 4 weeks after service to 1 week before the expected time of farrowing. The use of tethers will not be allowed. Sows and gilts should have permanent access to materials for rooting. The minimum weaning age for piglets should be 4 weeks and rules regarding light requirements and maximum noise levels must be complied with. In countries, Sweden for example, where animal welfare rules did not accept permanent restraint in crates and totally perforated flooring without bedding, specialized pig farms kept the dry sows in groups, with access to bedding and with individual feeding stalls (Fig. 2.8). Sows with piglets are kept loose in pens, with a separate dunging alley and with access to straw bedding (Fig. 2.9). When the piglets are about 3 weeks old, some farmers transfer the sows and litters into group pens with bedding (Fig. 2.10). Some farmers already apply group keeping some days before farrowing. Each sow is offered an

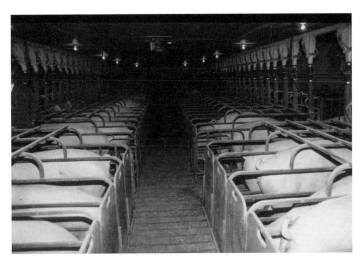

Fig. 2.4. In specialized pig production farms, the dry sows are still often kept restrained indoors year-round, tied or in crates, usually on slatted or perforated floor without bedding, and sometimes in houses without windows, as in this photograph (image I. Ekesbo).

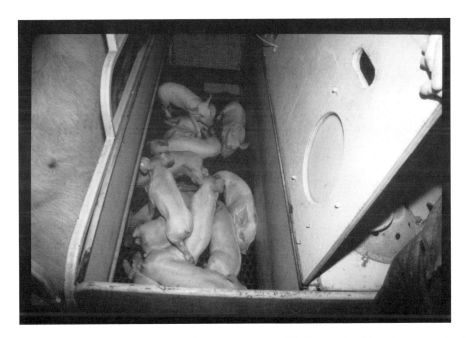

Fig. 2.5. In specialized pig production farms, the farrowing sows are still often confined in crates with perforated floors, without bedding, in order to reduce labour. The piglets are often weaned before 4 weeks or earlier. Note the perforated steel floor under the piglets, the wound on the back of the sow and the docked tails of the piglets. The cover with a heat lamp is intended to keep the pen at a suitable temperature for the piglets. To the right is a similar pen with a cover and heat lamp in place (image I. Ekesbo).

individual farrowing pen constructed so that the piglets usually cannot leave this before they are about 10 days old (Fig. 2.11).

In popular farming journals, modern pigs are often described as homogeneous animals. However, experienced farmers and veterinarians

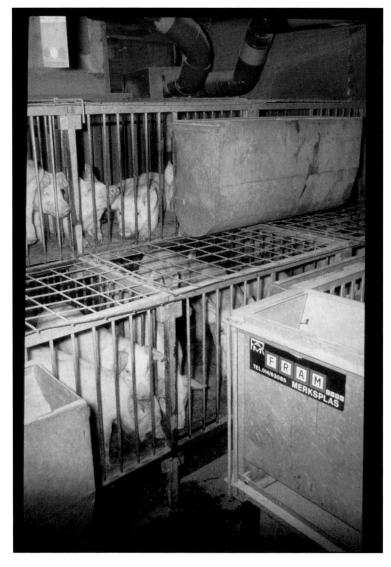

Fig. 2.6. The piglets were taken from the mother at 4 weeks or younger and put in cages with perforated floors, without bedding, in well-insulated rooms without windows and at, for them, a suitable ambient temperature. Note the piglet on the top floor biting the steel bar (image I. Ekesbo).

know well that there are great individual differences, which has also been proved by scientific studies (e.g. Lammers and de Lange, 1986; Mount and Seabrook, 1993; Hemsworth and Coleman, 1998). As in cows and horses, the reactions and behaviour of pigs to humans indicate that individuality is the norm and not the exception (Seabrook and Mount, 1995a). This should always be considered, although it often seems to be unknown or easily forgotten in industrialized pig production.

A sow can reach an age of more than 10 years, although nowadays most sows are culled before having their tenth litter. The main reasons for culling in the USA are reproductive failure (26%), old age (37%) and performance (small litter size, high pre-weaning mortality or low birth weight, 13%) (USDA, 2005, 2007a,b,c).

Fig. 2.7. In many fattening pig farms, the pigs are kept in pens with a perforated floor without bedding (image I. Ekesbo).

Fig. 2.8. Dry sows in a straw-bedded group pen. The individual feeding stalls to the right are closed during feeding in order to protect each sow against competition from others (image I. Ekesbo).

The male swine or pig is described as a 'boar', the female as a 'sow' and the offspring as 'piglets'. The expression 'piglets' is used until weaning; thereafter, they are described as pigs, 'weaners', growing to 'finishing pigs', according to age and size. The young sow pig is called a 'gilt', from its sexual maturity at about 6 months of age until it has weaned its first litter. After this point in time, the term is 'sow'.

2.2 Innate or Learned Behaviour

Examples of innate behaviour are the sow bedding or nesting behaviour before parturition, or the piglets' behaviour to perform urination and defecation away from the nest. Pigs will quickly learn if more information can be obtained by exploration (Studnitz *et al.*, 2007). The pig's ability to learn from experience, memorize and combine new memories with previously available information is outstanding. Pigs can learn to distinguish between a familiar and an unfamiliar person by relying on various cues, including the colour of their clothing (Spinka, 2009).

Pigs quickly learn to manipulate devices, for example, the 'Edinburgh foodball', in order to get feed (Young *et al.*, 1994), or to discriminate

Fig. 2.9. A lactating sow kept loose in a pen with bedding. She has the possibility to turn around and access to the manure alley. Mostly, the piglet creep area to the right is equipped with a lamp as a source of heat (image I. Ekesbo).

between food sites, choosing the one with more food or less difficulty of access (Held *et al.*, 2005).

2.3 Different Types of Social Behaviour

Social behaviour

The wild boar

Pigs are gregarious animals and spend most of their lives in groups. However, these groups do not usually consist of large numbers of animals. Wild boars establish herds which mostly consist of 2–4 sows and their offspring from the previous year (Graves, 1984; Kaminski *et al.*, 2005a). It seems as if the bond between the sow and her offspring remains even after the offspring have reached adult age.

The young boars stay in the sow herd until they are 1–2 years old. Thereafter, they abandon these groups since they have great difficulties in competing with the older boars. Sometimes, they form their own groups. Older boars, more than 3–4 years, live solitarily – boars are loners – and roam larger areas than the sows' family groups. When the sows are in heat, the boars seek the sow herds

Fig. 2.10. Group housing of 8–10 sows and their piglets. New straw bales are present. The feeding trough is to the right. At weaning, the sows are taken away and the piglets remain in the pen (image I. Ekesbo).

Fig. 2.11. Group housing of sows and piglets. The sows are put into the group pen before farrowing. Each sow has access to a separate farrowing pen constructed so that the piglets cannot leave it until they are over 2 weeks old (image courtesy of Bo Algers).

and then there are often fights between different boars (Jensen, 2002).

Pigs interact closely with other group members, nose one another and often lie together. Piglets always lie close to their mother, even when it is warm.

Within groups, pigs form stable social structures, which are maintained with little aggressive behaviour (Jensen, 1988c; Stolba and Wood-Gush, 1989). The dominance hierarchy is a result of the interaction of a number of factors (Ewbank and Meese, 1971; Meese and Ewbank, 1973a,b). The dominant animals direct most of their aggression to the rank immediately below. Within sow groups, older animals are dominant over younger ones, and in young animals, there is a strong relation between size and dominance status (Jensen, 2002). The social organization of groups includes the establishment of various friendly relationships and a social hierarchy (Jensen and Wood-Gush, 1984). The size of a group and the space allocated to it are important for the social hierarchy to function properly.

The domesticated pig

The same herd behaviour as in wild boars is reported from domesticated pigs kept in large enclosures (Jensen, 1988c). Outdoors, agonistic behaviour is seldom a problem. A pig outdoors seldom seeks to feed close to the head of another pig.

On condition that they have enough space and that the environment otherwise fulfils basic biological demands, i.e. is not barren, pigs kept indoors in groups as a rule develop a stable social hierarchy (Jensen, 2002). Within a stable group, agonistic behaviour occurs mainly in connection with feeding. In domestic pigs, however, dominance relationships are less stable than in cattle (Spinka, 2009).

Communication

The wild boar

Pigs' social communication is performed via acoustic, olfactory and visual signals and there is much social communication by vocalization. The olfactory signals are used for recognizing group members. The members of the group must be capable of prompt recognition of each other. Sensory clues such as olfactory stimuli are involved in the maintenance of the social structure. By visual and olfactory cues, pigs in an established group are quickly able to recognize an alien in the group (Jensen, 2002).

Vocal social communication ranges from contact grunts to warning calls and screams when attacked. Important vocal contacts are the grunting repertoire the sow emits at suckling. Other examples are the grunts emitted by group members when seeking for food, as well as warning signals when predators are approaching.

The domesticated pig

Pigs in groups maintain body, vocal and visual contact with other pigs (van Putten and Elshof, 1978).

There exists a very intricate vocal communication between sow and piglets during suckling. This is described under the heading 'Vocalization and acoustic communication', in section 2.11.

Behavioural synchronization

The wild boar

Even if pigs are not as gregarious an animal as sheep and cattle, they show an obvious behavioural synchronization and also a following behaviour when danger threatens. Oestrus is highly synchronized in wild boar females within a group, but not necessarily between groups (Spinka, 2009).

The domesticated pig

When kept in groups, the domesticated pig shows the same behaviour. Oestrus synchronization similar to that in wild boar females is also seen in domesticated sows kept in groups.

Dominance features, agonistic behaviour

The wild boar

As in some other species, the subdominant animal lowers its head, turns it aside and thus shows submission when a dominant individual is approaching.

Aggression between piglets hardly ever occurs in nature (Archer, 1988) or under semi-natural conditions (Jensen, 1988c).

The domesticated pig

If two piglets in a litter have selected and settled on the same teat after parturition, there often arises an intense fight between the two. Such aggression can go on for days, or even weeks, and they often inflict face and ear wounds on each other. The screams of aggression can disturb milk ejection, especially in nervous sows.

If unacquainted pigs are mixed, fighting starts immediately (Meese and Ewbank, 1973a). Fights for dominance between non-adult animals are normally won by the bigger animal (Jensen, 2002). When weaned piglets from two litters are put together in a limited space with few environmental stimuli available, the agonistic behaviour soon begins and can go on for days. However, if they are given more available total area, with straw provided for explorative behaviour, the frequency of agonistic behaviour may not always be reduced, but its intensity and duration will lessen. The frequency and intensity of exploratory behaviour increase with rooting, biting and chewing the straw.

The signs of imminent aggression in pigs are a short moment of distinct eye fixation on the opponent at the same time as the ears are twisted forwards and upwards. The next moment there can be a violent, rapid and, if it is an adult animal, often a very dangerous bite. Adult boars use their tusks, causing flesh wounds. Aggression is practically always, irrespective of the age of the animals, accompanied by loud screams from both the combatants. Aggressive boars champ their jaws, causing the tusks to clack, and great amounts of foamy saliva can be generated.

If a sow with piglets, or a sick or injured sow or boar, is showing signs of aggression against a person, it can quickly develop to a violent biting attack. Thus, it is important for the stockman and the veterinarian to be familiar with these signs.

2.4 Behaviour in Case of Danger or Peril

The wild boar

When pigs are frightened, they emit a short strong 'alarm' sound and stand still for a very short moment, looking in the direction from which they have identified the threat. They then usually take flight at a gallop in a direction away from the threat (Fig. 2.12). However, after a while, the curiosity of the pigs often gets them to stop, turn towards the supposed threat and even approach warily (Fig. 2.13).

Pigs can communicate danger by the use of smell. Thus, a stressed pig releases alarm substances in its

Fig. 2.12. The pigs first emitted a short strong 'alarming' scream, then stood still for a second, looking at the veterinarian, then they took to panic flight out and away from the threat (image I. Ekesbo).

Fig. 2.13. Some minutes after the photograph in Fig. 2.12 was taken, the pigs' strong motivation to carry out exploratory behaviour, their curiosity, got the better of their fear, so they came back carefully into the pen (image I. Ekesbo).

urine that can be detected by other individuals as a signal warning of danger (Spinka, 2009).

Antipredator behaviour involves hiding and running, but sows and boars may fight back with biting. Screams by piglets or adults may elicit assistance from the mother or other group members. If attacked, grown-up boars use their tusks and can thereby cause very serious injuries.

The domesticated pig

The domesticated pig, especially if not familiar with humans – as often is the case in large fattening units – shows exactly the same alarm and flight behaviour as wild boars if frightened. Adrenalin levels are significantly higher during squeals and screams than in other calls (Schrader and Rohn, 1997).

2.5 Some Normal Physiological Frequency Values of Interest at Clinical Examination

Normal rectal temperature in adult animals is 38.5–39°C. The critical point in resting adult animals is > 39.0°C.

In adult animals, the resting pulse frequency is 60–80 per min and c.100 in young pigs.

In adult animals, the frequency of respiration is 8–18 breaths per min; it is 32–58 breaths per min, with a mean of 40, in pigs of about 25 kg body weight (Robertshaw, 2004).

2.6 Active and Resting Behaviour Patterns

Activity pattern, circadian rhythm

The wild boar

Wild boars and feral pigs are not territorial but individuals or groups live in home ranges, usually made up of wooded areas combined with open spaces.

They are largely diurnal but can be more nocturnal if exposed to predators, or during very warm periods. The nocturnal activity, which many scientists report as typical for wild pigs, may be a result of disturbances from humans, e.g. hunting. Even when food availability is good, most pigs in extensive and varied conditions are active for more than half of the day. Exploratory and foraging behaviour occupies a substantial amount of time – up to 75% of a pig's daily activity. In addition to food finding, daily activities include antipredator behaviour and social behaviour. Out of 24 h, adult pigs are active for about 12 h, sleep about 7 h and the rest of the time is spent in a 'half-awake', drowsing state (Robert *et al.*, 1987).

Swine seem to be highly motivated to perform different behaviour in different places.

The domesticated pig

When domestic pigs have been studied in large enclosures, they have been found to spend much time in wooded areas and some time in open areas, as do wild boars when not subject to predation.

Studies of locomotor and rest activities indicated differences between wild and domesticated pigs. Domestic pigs were at rest more often and consequently showed less locomotor activity than wild boars (Robert *et al.*, 1987).

Domestic pigs living in a semi-natural environment spent 52% of the daylight period foraging (rooting and grazing) and another 23% in locomotion and direct investigation of the environmental features (Stolba and Wood-Gush, 1989).

Behavioural studies were performed in three groups of dry sows, one confined in cages (a), one in group pens with laying/feeding stalls with a dunging alley behind (b), and a third in group pens with a bedded lying area, feeding stalls and a dunging area (c). The degree of inactivity during the daytime measured as lying time was 68% in confined sows and 41% in sows kept loose. During the time the sows were active, differences in the behaviour pattern were observed. The confined sows (a) thus showed significantly higher incidences of bar biting, eating/drinking and sitting than those kept loose. Each sow was provided with 0.3 kg straw per day, which the confined sows immediately consumed, whereas the sows in the group with separate lying, dunging and feeding areas (c) consumed only a smaller part thereof. The latter sows performed 12 distinct behaviours with the straw; the confined sows performed mostly only two, sometimes three, of these behaviours and, within 10 min, had consumed all the straw. The conclusion of these findings is that sows loose in straw-bedded pens perform more explorative behaviour than sows confined in crates. The loose sows spent 3% of their time in the dunging area but were never seen lying there – the confined sows, of course, were forced to lie there. The loose sows spent 25% of their time in the feeding stalls. However, it should be noted that in this experiment they were confined in the feeding stalls for 3 h per day in connection with feeding periods (Ekesbo et al., 1979).

Thus, there are reasons to presume that the design of swine housing and management, and especially the design of the traditional pigsty with its partition in eating, lying and dunging areas, is the result of conclusions drawn from observing swine behaviour over several hundreds of years.

Exploratory behaviour

The wild boar

Pigs make considerable efforts to explore all aspects of their environment. Exploratory behaviour by rooting, smelling and chewing occupies a substantial amount of time in the life of a pig, and they seem to be strongly motivated to carry out exploratory behaviour (Wood-Gush and Vestergaard, 1993). Its functions, in addition to finding food, relate both to antipredator behaviour and social behaviour. Even if pigs have quite good auditory ability, much of the time taken in exploration involves movements which increase the chances of coming into contact with important odours.

The domesticated pig

Pigs, like other animals, initially react cautiously to new or strange objects and locations.

Included in the effects of environmental enrichment is a decreased general fearfulness or fear of novelty. Grandin et al. (1987) found that fattening pigs given novel objects to manipulate, such as rubber hoses, were easier to handle. 'Environmental enrichment' improves the behavioural and physiological maturation of laboratory animals, presumably through a stress response. The expression 'infantile stimulation' has been used (Schaefer, 1968; Hinde, 1970).

Rooting is an exploratory behaviour of high priority in pigs. In order for something to be a suitable rooting material, it must stimulate this behaviour in pigs for an extended length of time. Exploratory behaviour in pigs is stimulated best by materials that are complex, changeable, destructible and manipulable and which contain sparsely distributed edible parts (Studnitz et al., 2007).

It has been observed that pigs are startled more easily and are more difficult to handle when reared in semi-darkness (Warriss et al., 1983) or darkness (Grandin, 1991). In the former study the control pigs were reared outdoors, whereas in the latter they were reared indoors with greater access to light. It is possible that these observed effects of general fearfulness and ease of handling may have been a consequence of increased environmental stimulation (Hemsworth, 2000).

Diet, food searching, eating and eating postures (feeding behaviour)

The wild boar

Pigs are omnivorous, eating grass, roots, fruit, berries, seeds, earthworms, frogs, small rodents and other material of plant and animal origin. Most food searching involves rooting in the ground, but grazing and browsing also occur. The snout of the pig is especially adapted for rooting. With the upper part of the snout, adult pigs can move stones and similar heavy objects in order to access roots, seeds and similar feed. The snout is tapered and the nasal disc is rigid enough to withstand considerable force, but is richly supplied with widely spaced, short vibrissae, connected to sensory receptors. The skin of the snout, except for the upper part of the disc, is covered with short hairs. The surface of the skin of the snout is divided by shallow grooves, which are characteristic to each individual and imprints of them can be used for identification purposes (Sisson, 1975). The snout is kept moist by the secretions of serous glands deep in the dermis. These olfactory abilities enable foraging and rooting to be combined with smelling and chewing edible objects.

The domesticated pig

Pigs have a strong preference for rooting, even when satiated. Thus, rooting may constitute a basic need for pigs and it is shown that ringed sows prevented from rooting show evidence of frustration (Horrell *et al.*, 2001), as do sows in crates.

Pigs are susceptible to lack of water. The water requirements of domestic pigs vary with weight and age. Pigs weighing 21–46 kg require, at 20°C, about 0.12 l water per kg body weight, and at 52–66 kg about 0.09 l per kg body weight. The requirements increase up to 70% at 30°C (Mount *et al.*, 1971). Adult sows may require up to 25 l water per day at high ambient temperatures. Thus, pigs need access to water.

The pig's dentition is complete with permanent teeth at 18–20 months of age (Sisson, 1975). In the piglet, the milk teeth set is complete at 3 months of age.

Selection of lying areas, lying down and getting up behaviour, lying postures, sleep

Selection of lying areas

In the wild, pigs rest in specific hiding and resting areas away from their foraging areas. They select dry areas for protection from precipitation and hidden places for safety (Graves, 1984).

Likewise, domestic pigs kept outdoors choose dry lying areas. If bedding material, straw for example, is available, they arrange their lying areas there. They avoid draughty lying areas. If part of a lying area indoors is draughty, it might be transformed into a dunging area.

Resting postures

Piglets prefer to lie close to the sow, even when temperatures in the pen are high. The lying behaviour, parallel with each other, seen in young piglets remains during the whole life, even in grown-up sows. However, lying behaviour is influenced by the ambient temperature. Thus, when cold, adult animals as well as young piglets lie close together, whereas when it is warm, they lie without direct body contact.

Lying down and getting up behaviour, sleep

Calm and experienced sows with piglets show a careful lying down procedure. Before lying down, a pig often turns round, investigating the ground or floor, then goes down on the fore knees (carpal joints). The hindquarters are first lowered vertically and then moved on to one side, after which the sow lowers herself to rest on first the elbow joints then the sternum, and either keeps this position or lies down flat on her side. A sow with a litter often acts initially like a plough in the bed or nest, thus avoiding crushing any piglets, then performs the lying procedure slowly and carefully. If intending to suckle, after having ended flat on one side, she makes a characteristic jerk upwards with the udder, thus also exposing the lower row of teats for the piglets.

Of all farm animals, pigs spend most time resting and sleeping. Pigs confined in stalls may be recumbent in rest or sleep for as much as 19 h per day. However, pigs free to move around sleep much less. Of the total sleep time, SWS occupies 6.11 h per day on average and REM sleep averages 1.75 h in about 33 periods (Broom and Fraser, 2007).

Pigs show extreme muscle relaxation during sleep (Houpt, 1998). This is difficult to discover in adults, but if a sleeping piglet can be picked up without it waking, it is totally relaxed.

Locomotion (walking, running)

A pig's gaits are walk, trot and gallop. They walk when seeking food. A trot is used when pigs are frightened and leave a spot; or when they suddenly discover something to eat, for example when piglets discover the sow preparing to suckle. A gallop is used at a higher degree of fright or possibility of getting food. Galloping often occurs as part of piglet play.

Swimming

Pigs are natural swimmers. Wild boars often swim in order to reach new territories.

2.7 Behaviour at Defecation and Urination (Eliminative Behaviour)

When the pig defecates, it bows the back slightly upwards, lifts the tail and as a rule stands still.

However, it occurs sometimes that pigs defecate when walking forward. At urination, the female pig always stands still, bends its back in a fairly marked upwards bow and lifts the tail at the same time as the hind legs are put forward under the belly just before urination starts. The boar always stands still when urinating.

When kept indoors, healthy pigs always perform eliminative behaviour in the same place, thereby preferring moist places of the pen if a dunging alley is not available (Fig. 2.14). When kept outdoors in vast enclosures, pigs do not seem to use separate places for eliminative behaviour; however, they never eliminate in their lying areas.

The healthy pig, even when newborn, never performs eliminative behaviour in its lying area if not forced to because of being kept in a cage. Studies of domestic pigs in semi-natural enclosures show that, like the sow, the piglets leave the nest for urination and defecation (Fig. 2.15). The characteristic of keeping a sharp distinction between the lying and dunging area seems to be an instinctive behaviour in pigs. The practical experiences that housed pigs, if given the opportunity, perform excretory behaviour in special areas are confirmed by several studies (e.g. Baxter, 1983;

Petherick, 1983; Marx and Buchholz, 1991). Even during their first day, indoor piglets, if given the opportunity, defecate in the dunging alley or in the dunging area. Urination is also performed there, but can also happen on the floor at water cups, etc., since pigs usually choose to urinate on moist surfaces.

It can be concluded that indoor pigs, and outdoor pigs kept in restricted areas, use well-defined areas for excretion, seek isolation for excretory behaviour, do not lie in any space required for excretion, nor do they excrete on any space required for lying, and they usually lie away from a drinker and excrete near it.

Prior to parturition, the sow defecates and urinates normally. If loose-kept sows are restricted in stalls, which may occur just before parturition, they often refrain from defecation and urination over a longish time, apparently because they avoid performing eliminative behaviour in their lying area or because they are prevented from moving to a separate dunging area. This often leads at first to unspecified symptoms of disorder and later, after the parturition, to increased risk of disease, e.g. agalactia toxaemia. If they are let loose some time after the confinement, they may make great amounts of urine.

Fig. 2.14. Housed pigs use well-defined areas for urination and defecation and they seek isolation for excretory behaviour. Even very young piglets leave the lying area to urinate and defecate (image I. Ekesbo).

Fig. 2.15. A piglet, 2 days old, urinates outside the nest built by the sow in an outdoor semi-natural environment study (image courtesy of Bo Algers).

However, constipation is often a more persistent complication.

2.8 Body Care, Cleanliness

In order to maintain the condition of the body surface, pigs scratch themselves but do not otherwise spend much time in grooming. They do not lick themselves or each other – mutual grooming does not occur in wild or domesticated pigs.

The term 'dirt' in the following account means that the swine is soiled by urine or manure. Sows kept outdoors which wallow in mud or water for controlling their body temperature, and therefore are covered by wet mud, are not defined as dirty.

Dirtiness in sows or pigs can be a symptom of disease, fever for example, but mostly it is a result of inadequate environment or management. When sow confinement was introduced, it was apparent that the frequency of dirty sows increased in herds. However, as in other species, there are invariably single animals that have a tendency always to become dirty. Even under ideal conditions, the share of dirty sows in a herd seems to be about 5%. However, in herds where sows are kept confined, often 80–85% of the sows are dirty on the hind part. Apparently, the possibility for the pig to choose a separate dunging area is a factor which promotes cleanliness.

Sows kept loose in pens with an adequate bedded lying area and with a separate dunging area are, in more than nine cases out of ten, absolutely clean as long as the indoor climate does not force them to use the dunging area for their thermoregulation (Ekesbo, 1981).

2.9 Temperature Regulation and Climate Requirements

Temperature regulation

The wild boar

Pigs have limited temperature regulatory abilities. For pigs, high temperatures can be a problem, low not. Pigs cannot cool their skin by transpiration because their sweat glands are very few and their distribution is confined almost entirely to the snout. An accumulation of body heat occurs in warm surroundings. The relatively poor body cover of hair makes the animal particularly vulnerable to the effects of direct solar radiation.

Since evaporation of fluid from the skin is very effective for cooling, pigs seek places to wet themselves when they are hot. They rely on wallowing for cooling in hot weather and, when conditions are cold, often huddle with others in a sheltered, well-insulated place, such as a nest, which they have built. In warm weather, pigs kept outdoors try to get cool in water or mud baths – they regulate their body temperature through skin evaporation. Wetting the skin is effective for about 1 h, depending on the relative humidity. However, wallowing in a mixture of mud and water is more effective. Pigs smeared with mud and water showed evaporative cooling over a longer period of time, i.e. about 2 h, than pigs made thoroughly wet with water alone (Ingram, 1965, 1972; Ingram and Legge, 1972). Pigs prefer wallowing, because it combines cooling and grooming. After the mud has dried on the skin, they rub it off against bushes or rocks. This method of grooming is important, because pigs cannot lick themselves like other farm animals.

Panting accounts for about 20%, sweating for about 20% and wallowing for about 60% of the relative proportion of total evaporative heat loss in swine (Robertshaw, 2004).

The newborn piglet is very susceptible to cold. Coldness impairs the development of thermostability and induces hypothermia. Glycogen and fat reserves are used as major energy substrates for heat production within the first 24 h of life (Close, 1992). During the first weeks of life, the piglets easily lose body heat to the environment. They compensate for this by lying close together and close to their dam. During cold conditions, pigs huddle together, using about 60% of the space they would use under warm conditions (Baxter, 1989). In nature, nest construction is important to their survival. The main factors for avoiding piglet hypothermia are a warm lying area and the colostrum, which is their main energy source.

The domesticated pig

An important cooling system for pigs is evaporation of water from the snout and larynx. The domesticated pig has a shorter snout and fewer possibilities to use the nose cavity for temperature regulation than the wild boar. Ingram and Legge (1970, 1972) found a difference in blood temperature of 3°C between the carotid artery and the jugular vein in pigs in a hot environment. Domestication has played a damaging trick on pigs. Most domesticated animals are selected for a short face in comparison with the rest of the skull – this has a deleterious effect in the pig. The snout and larynx of a wild boar are much longer than those of a domesticated pig. Due to reduced surface area for evaporation, one can conclude that the nasal and laryngeal cooling mechanisms of a domesticated pig have only one-quarter the capacity of those of a wild boar. The cooling problems in domesticated pigs are compounded further by much heavier adult body weights compared to wild boars. Domestic pigs have less efficient evaporative cooling than wild boars, because they have less skin area per kilogram of live weight (Clutton-Brock, 1999).

In warm weather, pigs kept outdoors try to cool down in water or mud baths. Pigs covered with clay or mud outdoors in the summertime is not a sign of dirtiness, it reveals that they have regulated their body temperature. The old expression 'dirty pig' comes from people's misinterpretation of this behaviour. Pigs kept indoors often prefer lying in the wet dunging area if the indoor temperature is too high.

Another method of cooling is radiation of heat via the skin. Small blood vessels in the skin become dilated to cool the blood. In comparison to the wild boar, domesticated pigs have an advantage due to the absence of a hair coat. However, their disadvantage of a relatively small skin area compared to wild boars plays a large role in decreasing the effectiveness of cooling by radiation.

A third method of losing heat is direct contact with a heat-conducting surface, such as wet earth/mud. In hot weather, pigs attempt to maximize body contact with the wet dirt. They will stretch out on their sides, with their heads and feet on the ground. If a cool surface is not available on a hot day, domesticated pigs will suffer from greater heat stress than their wild conspecifics.

Climate requirements

The domesticated pig

It is a well-known clinical experience that indoor sows kept permanently confined are at risk of heart failure because of their inability to cope with high temperatures (Drolet et al., 1992; D'Allaire et al., 1996). As the swine heart weight to body weight ratio, < 0.23–0.28% in adults (Sisson, 1975), is much smaller than that of most other mammals (Stünzi et al., 1959; Lee et al., 1975), high temperatures, like other stressors (Johansson and Jönsson, 1977; Jönsson and Johansson, 1979), may lead to overload of the circulation and to acute heart failure. Sows confined in cages and exposed to direct sunshine for hours without the possibility of avoiding the heat run the risk of heart failure and acute death.

Pigs affected by draught in the lying area try to find better-protected lying quarters. Too high temperatures in the lying area often result in them trying to regulate their body temperature by lying on moist places, near water dispensers or bowls, or even in the dunging area.

If they are not undernourished, pigs are able to stand low temperatures. It is possible to keep adult sows and fattening pigs outdoors at low temperatures provided that they have access to a dry lying area in straw protected against precipitation and draught. Even piglets older than 1–2 weeks can stand low temperatures. In cold ambient temperatures, adult pigs during their resting periods – and piglets always, apart from when sucking – nestle down in the bedding material. If such bedding does not exist, they creep close together; young piglets even try to lie on the belly of the sow. In their studies of the thermal microclimate in winter farrowing nests of free-ranging domestic sows and their litters, Algers and Jensen (1990) found that, when the outdoor temperature varied between –17°C and 7°C, the nest temperature, measured 5 cm from the piglets, varied between 11°C and 26°C, with an average of 20.3°C.

Studies of litters, with and without access to bedding, show that pigs not only prefer bedding but also that piglets are significantly quicker (13 versus 20 min) to reach the udder after birth in premises with bedding than in those without (Stamatopoulis et al., 1993).

The type of floor influences the thermal resistance between the pig body temperature and the floor temperature. Verstege and Vanderhe (1974) report that the effective critical temperature of

40 kg pigs is 11.5–13°C on straw bedding, 14–15°C on asphalt and 19–20°C on concrete slats. With a temperature under the lower critical temperature (LCT), the pig must use a larger part of its body energy turnover to increase its total heat production. When pigs are embedded in straw, the LCT can be quite low. Thus, a single 34 kg pig fed 3.3 times its maintenance requirement has an LCT of –5 when it is 70% embedded in straw (Sällvik and Wejfeldt, 1993). A 90 kg pig on a concrete floor has a thermoneutral zone between 17 and 26°C, without a wallowing opportunity, between 17 and 23°C (Spinka, 2009). Sows kept in cages without bedding in uninsulated buildings and without body contact, and thereby some warmth from other sows, are reported to be subject to abortions during the cold season (L. Bäckström, Wisconsin, 2010, personal communication).

Loose-kept pigs usually lie close together, but when the ambient temperature is high, they lie separately. However, piglets prefer to lie close to the sow even when the temperatures in the pen are very high.

2.10 Vision, Behaviour in Light and Darkness

Pigs have a wide angle of vision, 310°, which enables them to react fast to movements in their surroundings. They have a binocular field of vision covering 30–50°. Their distant vision is limited (Prince, 1977) and their vision is not as good as that of humans (Spinka, 2009).

Studies of pigs kept under three different light conditions, darkness (< 0.1 lux), twilight (1 lux) and light (25 lux), indicate that pigs kept in darkness show pathological and behavioural aberrations (van Putten and Elshof, 1983).

Swine have colour vision (Myers and Coulter, 2004).

2.11 Acoustic Communication

Hearing

Swine have a good hearing capacity but, as in other farm animals, their hearing range still has not been investigated completely. Pigs are unable to localize high-frequency tones. They have a hearing range between about 40 Hz and about 35,000 Hz, with maximum sensibility at about 8000 Hz (Heffner and Heffner, 1990, 1992b).

Some predators and humans are able to locate a sound accurately within an angle of 1°, whereas pigs can locate a sound only within an angle of about 4°. However, pigs are more accurate sound localizers than cattle and goats, where corresponding figures are 30° and 18°, respectively (Heffner and Heffner, 1992b).

Vocalization and acoustic communication

There is much social communication by vocalization between pigs, ranging from contact grunts to warning calls and screams when attacked. Short sounds, grunts, are often heard when the group is seeking food. It is the animals' way of keeping contact with each other. Other typical calls are the guttural one emitted by the boar when courting a sow in heat and the complicated grunts emitted by the sow in connection with suckling (Jensen and Algers, 1982; Algers and Jensen, 1985; Houpt, 1998). The sow grunts with distinct grunt frequency peaks at 4 h after delivery, and, after 8 h, the peaks occur at regular intervals (Algers, 1989). When pigs are frightened, they emit a short, strong 'alarming' sound.

If a piglet is lost from the group, the sow will be excited and when she finds it she will touch it with her snout, at the same time emitting a complicated series of vocalizations (Clutton-Brock, 1999). Pigs can discriminate between individual pigs on the basis of their calls (McLeman et al., 2008).

Heffner and Heffner (1990) found that, when they determined the behavioural audiograms of domestic pigs, the hearing ranged from 42 Hz to 40,500 Hz, with a region of best sensitivity from 250 Hz to 16,000 Hz.

Studies of acoustic communication in domestic pigs show that an important part of the nursing–suckling interaction between sow and offspring consists of acoustic signals from mother to offspring, or vice versa (Algers, 1989). The gruntings of the sow act as three different signals. The first, a constant grunt rate, is probably a signal to the piglets to massage the udder. The second, an increase in grunt rate, is perhaps a signal to the piglets to shift to sucking. The third, the peak, may be a signal to the piglets to prepare for the sow's milk let-down. The piglet vocalizations seem to consist of five discrete classes, i.e. croaking, deep grunt, high grunt, scream and squeak. Some of the piglet vocalizations are probably signals to the sow, some to the littermates (Jensen and Algers, 1982).

2.12 Senses of Smell and Taste, Olfaction

Olfactory abilities are extremely well developed and match those of dogs. Pigs use their sense of smell to search for and examine food and to recognize group members. They can scent humans over hundreds of metres. Pigs can recognize and remember at least 30 individuals within a group (McLeman *et al.*, 2008; Spinka, 2009).

In preference tests, it has been shown that pigs respond to some sweet tastes but that there are individual variations to this. They respond strongly to sucrose solutions, but their preference for glucose and lactose is more modest. They do not show preference for sodium cyclamate, which is sweet for humans (Kare *et al.*, 1965).

The odour of a boar acts as an olfactory stimulus and increases the sexual receptivity of sows, and thus a boar is often kept in herds using artificial insemination.

2.13 Tactile Sense, Sense of Feeling

Tactile touch

Pigs do not groom each other like cattle and horses. However, individuals who have no fear of humans seem to appreciate being scratched behind the ears or on their back.

Calm and methodical rubbing of the udder is an old treatment of sows which have difficulty in nursing and letting down milk during the first days after parturition. By rubbing on the udder, sows that are lying on their belly instead of offering the piglets their udder can be brought to display the udder and, after some more rubbing stimulation, also to let down their milk. Even sows which are standing instead of offering the piglets the udder can be brought to lie down, display their udder and let down their milk by rubbing the udder.

Sense of pain

Studies of piglet physiological and behavioural responses to castration, tooth grinding and clipping have revealed that they feel pain during and often after the operation. In the case of castration, the pain, demonstrated in changed behaviour, remains for several days after the operation (Hay *et al.*, 2003).

In the case of teeth grinding or clipping, not only the operation but also the subsequent lesions, e.g. pulp cavity opening, fracture, haemorrhage, inflammation or abscess, and osteodentine formation presumably can cause suffering, as these are known to cause severe pain in humans (Hay *et al.*, 2004).

2.14 Perception of Electric and Magnetic Fields

In a survey, Hultgren (1990a) concluded that sows showed signs of discomfort and were reluctant to drink when subjected to 0.5 V AC between the watering system and the concrete floor. Currents of less than 1 mA seem to alter drinking behaviour in pigs, and 4 mA may even reduce their water consumption.

2.15 Heat and Mating Behaviour, Pregnancy

The wild boar

Wild boars exhibit a seasonal breeding pattern. The main breeding season is in the late autumn and farrowing occurs in late winter/early spring, with most sows producing just one litter a year. However, a second breeding season may occur in April, resulting in occasional litters in August. Wild boar sows are anoestrous between July and September. Compared to the domestic pig, they have a slightly longer gestation length (Mauget, 1982).

Females are in heat for about 72 h and actively seek boars at this time. The boar produces an odour, a sound, the 'chant de coeur', and behaviour which stimulate the female. Wild boar gilts mate for the first time at 8–10 months old (Gundlach, 1968). Pregnancy lasts about 115 days in wild, feral and domestic pigs.

The domesticated pig

Under normal feeding regimes, domestic gilts reach sexual maturity at about 6 months of age. Gilts' skeletal joints mature with age rather than with weight, which explains why rapidly growing gilts fed *ad libitum* experience joint stress due to their weight (Dewey, 1999). Whereas breeding is seasonal in wild boars, it is aseasonal in domestic pigs. Breeding during the domestication process has entailed year-round reproduction (Spinka, 2009).

In the first weeks after fertilization, stressors – for example high environmental temperature and high animal density causing a lot of aggression – can result in embryonic loss.

2.16 Before and After Parturition

The maternal behaviour of swine appears to pass through six different phases: nest-site seeking, nest building, farrowing, nest occupation, social integration and weaning (Jensen and Ekesbo, 1986).

Behaviour before and during parturition

The sows show a substantial change in behaviour in the 1 or 2 days before parturition. Studies of domestic sows and litters under semi-natural conditions (Jensen, 1988c; Stolba and Wood-Gush, 1989) and under traditional housing conditions (Jensen, 2002) have revealed that the behaviour shown by wild boars also exists in domesticated swine. One or 2 days before parturition, the sows leave the herd and can walk 2.5–6 km outside the home range until they find one or more suitable nesting sites and start building a bed or a nest. They usually select a place as dry and protected from rain – and snowfall – as possible, for example on a slope and under a tree, and protected from sight.

The sow starts 'bedding', i.e. gathering material, for the nest (Fig. 2.16). It may be that the sow moves the bed or starts building two or more beds before the final one is selected and completed. By rooting, pawing, carrying preferably grass but also branches, leaves, etc., a hollowed-out nest is formed (Fig. 2.17) which may include large quantities of material (Jensen, 1988a,c). The sows often try to treat and form the material by chewing it. The nest is usually finished 2–4 h before parturition. Thus, the sow performs a considerable locomotor activity during the day or days before farrowing.

Even gilts and sows kept indoors show nest-building behaviour before parturition, whether or not nest-building material, like straw, is available. In the absence of suitable materials or substratum, many elements of this behaviour, including restlessness, occur in sows at this time. This is a behavioural need which cannot be satisfied with the supply of a ready-made bed (Jensen, 1993). Swine are the only ungulates to perform nesting behaviour (Jensen, 2002).

Before the use of the combine harvester, European and North American farms usually had one or more huge straw-stacks outside the farmhouse. If, in the summer, the sows were kept in enclosures with such stacks available, they chose without exception their lying areas in excavations within the stacks. It happened that

Fig. 2.16. Sow gathering material for the bed or nest before parturition. This behaviour occurs in domesticated sows as well as in the wild boar (image courtesy of Per Jensen).

Fig. 2.17. Sow with piglets in the bed or nest. Notice the snow in the background. In the nest built by sows during winter, with the ambient temperatures a bit below zero, temperatures of +28°C are registered close to the piglets (image courtesy of Per Jensen).

highly pregnant sows disappeared into the stacks, had deliveries there and that the farmers did not see the piglets until they came out after maybe 1 or more weeks.

The sow normally lies down on her side during parturition and, if not disturbed, will remain in this position until parturition is complete. Nervous sows often get up one or more times during farrowing. Each such movement means a great risk of piglets being trampled by the sow.

The labour pains can be seen as a slight shiver or wave on the side of the belly and, just before expulsion, the sow mostly makes a characteristic jerk with the tail. The usual presentation of the piglet is the nose first, followed by the shoulders, with the forelimbs beside the trunk, followed by the hindlimbs extended backwards. During the actual delivery, the sow pays very little attention to the piglets.

Normally, parturition takes 2–4 h. It is significantly more common for confined sows to have a farrowing time exceeding 8 h than for sows kept loose in pens (Bäckström, 1973). Oliviero *et al.* (2010) found the average total duration of farrowing to be 4.5 h in sows kept in cages and 3.5 h in sows kept loose in pens.

The afterbirths are usually expelled at the end of the delivery. When the sows get up after the delivery, they usually eat the afterbirth.

Litter size

The wild boar

For the wild boar, a variation in the number of liveborn piglets of between 4 and 13 has been reported (Mohr, 1960). The average litter size in wild boars is 5, but the figure is uncertain as it is difficult to estimate stillbirths and neonatal deaths (Spitz, 1986). Graves (1984) reports observing an average of 5.1 piglets in litters accompanying sows in one-sow groups and 7.0 in litters accompanying sows in groups with two or more sows.

The domesticated pig

The average number of piglets born to domesticated sows varies between the extremes of 3–5 and more than 16–17. Litter size increases with parity (Spicer *et al.*, 1986), with an average of about 9.5 for primiparous sows to about 12 for older sows (Bäckström, 1973).

However, in Sweden there was an increase in litter size from 1930 to 1980 of 0.3 piglets and the total number of piglets born per litter in 2003 was 11.5. The number alive at 3 weeks was 9.0–9.1 that year, and the number of piglets weaned per sow and year (PW/S/Y) was 18.7 (Jordbruksverket, 2004). Average litter size for landrace and Yorkshire breeds in Sweden in 2003 for primiparous sows was 9.3–9.5 and for sows at their fourth to fifth farrowing 12.0–12.7, but as a result of lower piglet mortality in the primiparous animals, the difference in litter size was reduced to 1.5 piglets by 5 weeks of age and good farrowing results were demonstrated for sows up to the tenth litter (Jordbruksverket, 2004). For domestic sows in the UK, the number born in 1986 was 10.48 and 10.84 in 1995 (Cutler et al., 1999).

According to Ashworth and Pickard (1998), the average litter size has not changed since the early 1900s. However, in 1990, the total number born per litter in the USA was 10.37 (alive 9.47) and 10.75 in 2000 (alive 9.94) (USDA, 2005, 2007a,b,c). The slight increase in litter size seems mainly to be a result of less mortality, even though some of it may be an effect of genetic selection. In US herds with fewer than 100 sows, litters produced from July 2006 through to June 2007 contained an average of 8.8 piglets, of which 8.0 were born alive and 7.3 were weaned (USDA, 2009).

The number of mammary glands in 95% of all sows varies between 10 and 14, with 12 being the most common. However, it can vary between 8 and 18 (Schmidt, 1971). The teats are short, and present on their apexes the 6–12 small orifices of the excretory ducts (Sisson and Grossman, 1948).

After the birth

The wild boar

As soon as a piglet is born, it will try to reach the sow's udder and find a teat. When the sow gets up after parturition, she will sniff the young but does not lick them or assist them to free themselves of the fetal membranes.

Studies of wild boars (Gundlach, 1968) show that the piglets usually stay in the bed or nest during the time the sow leaves it during the first days. However, they leave it for excretory behaviour. Piglets thus are 'hiders', like calves and kids. Wild boar piglets stay in or close to the nest for 7–14 days (Gundlach, 1968) and thereafter the nests are suddenly abandoned and the piglets change from hiders to followers (Jensen and Redbo, 1987; Jensen, 1988c).

The domesticated pig

The newborn piglet gets on its feet within 1 or a few minutes after birth and finds its way to the sow's udder. After a short time, each piglet has chosen a teat. If they fail to obtain colostrum within 20 h, they usually die. Each piglet establishes itself on a nipple and will always feed from this one and defend it against other piglets. When parturition is complete, the sow often utters a characteristic sound, thus calling the piglets to suck.

Newborn piglets suck independently of each other during the first hours after birth (Castrén et al., 1989a,b) and they suck on average 2–3 times per hour; they suck about 15 times during their first 12 h and take in about 15 ml each feeding (Wehrhahn et al., 1981). The piglets perform cyclic and synchronized sucking about 8–10 h after the onset of farrowing (Lewis and Hurnik, 1986).

After the first few hours, milk ejection occurs for about 20 s every 40–60 min. Suckling starts with a deep pitched, distinct grunt by the sow, which is then followed by rhythmic grunting, which encourages the sucking behaviour of piglets. The piglets gather at the udder, find their own teats and start massaging the udder by up-and-down movements with their snouts. About 20–25 s after the grunt, peak rate milk flow starts. The milk flow time is about 20 s. Hence, piglets in a litter must suck simultaneously and those which do not find an available teat will go short and may eventually die. Piglets resume massaging and continue with this for 10–15 min to promote later milk production (Fraser, 1980; Algers, 1989).

Algers and Jensen (1991) reported an average milk intake of about 25 g when studying piglets from birth to 3 days of age. The number of sucking periods then decreases from day 2 successively to 6–8 times per 24 h when the piglets are 6 weeks old. Unused teats become dry.

To induce a let-down, the anterior teats of the sow have to be stimulated (Fraser, 1973). The piglets' attraction to the anterior teats is perhaps only to ensure that the anterior teats of the sow will be stimulated when suckling is started. The duration and intensity of teat stimulation given to a teat influences the production of that teat during the first days of lactation (Algers, 1989).

During the first day, most of the suckling is initiated by the sow (Algers and Jensen, 1985). At the piglet age of 10 days, the sow initiates 40% and the piglets terminate 55% of the suckling (Jensen et al., 1991). Although most of the suckling during the first 3 days is still initiated by the sow, the number initiated by the piglets increases and about 30% of these are unsuccessful (Castrén et al., 1989a,b).

The nursing–suckling interaction between sow and offspring consists of a number of functionally important elements, many of them acting as signals from mother to offspring, or vice versa (Algers, 1989).

Fighting over the same teat seems to be more common among pigs using teats in the front part of the udder. Fighting is frequently reported to peak during the second hour of age (Hartsock and Graves, 1976), but it sometimes persists and the two piglets can go on fighting until weaning. In rare cases, there are fights between more than two piglets in a litter – such fights can go on for weeks. Bite wounds to the face and ears resulting from such fights can sometimes lead to serious infections. Fighting pigs can disturb nervous sows and cause them to interrupt suckling.

Studies of domesticated sows kept in semi-natural enclosures (Jensen, 1986, 1988a,c, 1989) show that the sow lies in the nest or close to it for most of the time for up to 10 days. The piglets stay in the bed or nest when the sow leaves it during the first few days. However, they leave it for excretory behaviour. After about 10 days, the nests are suddenly abandoned (Jensen, 1988a) and the sow and the litter join the herd, but the litter remains as an intact social unit throughout lactation and mixes with other piglets only occasionally during resting. At this time, there is seldom any damaging fighting but, at later ages, mixing poses serious problems for piglets.

Free-ranging domestic piglets and those offered suitable pen material may, by rooting, investigate potential solid food when 1 day old, but only from 4 or 5 weeks will they eat substantial amounts. Also, they continue to suck as long as they have the opportunity – for domestic piglets kept in a semi-natural enclosure, this lasts until natural weaning at about 17 weeks (Jensen, 2002).

Some sows show very protective behaviour towards their piglets but a furious aggressiveness against humans, especially in the first days after delivery. Nervous sows might, as a rarity, show flight reaction at the sight of the newborn piglets, others can show great aggressiveness against the piglets and try to catch and bite them, so-called farrowing hysteria.

For sows with tendencies to bite the piglets or attack the stockman violently during or after parturition, some farmers in Sweden used to make individual wooden cages in which the sows were kept until this behaviour disappeared one or several days after farrowing. It was estimated that this was used for less than 5% of sows. It seems as if, during recent decades, the percentage of violent piglet-protecting sows has decreased, maybe as a result of breeding. It cannot be ruled out, however, that this has resulted simultaneously in a decrease in the percentage of sows producing large numbers of weaned piglets efficiently without human assistance.

During the 2000s, the most advanced pig husbandry system, from the animal health and welfare point of view, is probably represented by piglets born in one of the secluded birth pens, in a straw-bedded collective pen, or born to a loose-kept sow in a traditional pen – in all cases, being kept there until abrupt weaning occurs at about 5–7 weeks by the sows being transferred to a mating section. However, this takes up more space and may even be somewhat more labour-intensive. Alternatively, the sow and its litter, together with other sows and their litters, are transferred to a large collective straw-bedded pen at the age of 2 weeks, weaned abruptly at 5–7 weeks by removal of the sows and left with the other piglets in that pen until they are transported to slaughter. However, in most countries in the world, the majority of piglet litters are still born in small pens, without bedding or other material for rooting and partly or totally furnished with a perforated floor, by sows confined in cages without the possibility to do more than stand up and lie down and forced to defecate and urinate on the spot, and being kept there until weaning.

Play behaviour

Piglet play occurs from an early age and develops notably until their third week; thereafter it declines, although fresh straw can even stimulate adult sows to play. Play is a very prominent feature of the piglets' total behaviour (Broom and Fraser, 2007; Spinka, 2009). In young piglets, play fights are the most typical behaviour and, when they are a few weeks older, chasing in short rushes and gambols begin to be seen.

Playing and fighting seem to be correlated and related to weight gain during the first 5 days after weaning when performed between 33 and 50 days of age, suggesting that these two behavioural categories may be part of one continuum. This playing/fighting behaviour is highly variable between individual farms, independent of whether they employ group or individual housing of lactating sows and litters (Silerova et al., 2010).

Weaning

The wild boar

Young wild pigs from about 2 weeks of age, together with their dams, gradually join individuals in the main herd and are eventually weaned at about 14–17 weeks. They live in a complex and rich environment. Weaning under natural conditions is a slow and gradual process, where suckling frequencies decrease almost linearly from soon after birth up to the final milk intake (Jensen, 2002).

The domesticated pig

Under semi-natural conditions, domestic piglets are weaned between 14 and 17 weeks of age (Jensen, 1988a,c; Jensen and Recén, 1989). Domestic pigs were, in traditional animal husbandry systems, weaned at 6–8 weeks of age. During the 2000s, however, weaning more often occurs at about 5–7 weeks. Since the 1970s, early weaning has often been practised, with weaning ages varying from 3–4 weeks in some regimes to as early as 7–10 days in others.

2.17 The Young Pig

The wild boar

From weaning, young wild boars grow up naturally in groups of animals of different ages, and thus are in a complex and rich environment.

The domesticated pig

A few weeks after weaning, most domestic piglets are transferred to a fattening pig unit, where they are kept together with eight to ten others in pens with a feeding trough, a dunging area and a resting area, usually without any material for rooting. In a few countries, animal welfare legislation demands that they are provided daily with at least 100 g straw or similar material per pig. In several countries, the total floor area is still slatted or otherwise perforated and no bedding for rooting or any other enrichment of the environment is provided.

2.18 Assessment of Pig Health and Welfare

In order to assess whether or not a pig is healthy, special attention should be paid to bodily condition, condition of the skin, eyes, ears, tail, legs and feet; sounds, activity, movements and postures appropriate to its age, sex, breed or physiological condition; and to its behaviour in general.

The relatively short neck in the pig limits the potential vertical movement of the head – an important indicator of lameness in many species. As pigs have a stilted locomotion and their natural response to disturbance is a short rapid locomotion rather than steady walking or trotting, there are difficulties in observing and defining lameness in pigs.

The healthy pig

The healthy pig makes sounds, including grunts and squeals, and has activity, movements and posture appropriate to its age, sex, breed or physiological condition, for example respiratory characteristics, stage of sexual development or reproductive condition. It has clean and bright eyes, clean hair, skin free from obvious lesions or damage, normal legs and feet; normal feeding, drinking, defecating and urinating, sucking or suckling behaviour; normal locomotor and exploratory behaviour; and uses separate resting and dunging areas if available.

The sick pig

Each deviation from the signs of health mentioned above, including normal behaviour patterns, must be regarded as a symptom of disorder and may be a symptom of disease. The sick pig leaves the group, ceases eating, shows lack of activity and lies down and, if possible, often buries itself into the bedding. Other symptoms might be shivering or vomiting.

Abnormal behaviour and stereotypies

THE WILD BOAR Abnormal behaviour and stereotypies have not been described for wild boars kept in nature.

THE DOMESTICATED PIG *Examples of abnormal behaviour and stereotypies.* Tethered or otherwise restricted or confined sows frequently perform a number of stereotypies and vacuum chewing activities (Fig. 2.18) very rarely seen in sows kept loose in pens (Ekesbo *et al.*, 1979;

Vestergaard and Hansen, 1984; Jensen, 1988b; Arellano et al., 1992; Goossens et al., 2008). Sows in crates show higher agonistic behaviour and lack several behaviours performed by sows loose in pens (Ekesbo et al., 1979; Jensen, 1984).

Abnormal behaviour, 'belly nosing' (Fig. 2.19), ear and tail biting usually occur in early-weaned piglets, especially if kept in barren environments (e.g. Algers, 1984b). Such behaviour is very seldom seen in healthy piglets weaned after 5 weeks of age and kept loose in pens with access to straw.

Shoulder lesions are common in sows kept in crates (Fig. 2.20) especially if no or only little bedding is used. Slatted or perforated flooring as a lying area is a risk factor for such injuries in sows kept loose in pens (e.g. Holmgren and Lundeheim, 2010).

Sows confined in closed unstrawed stalls show higher amounts of stereotyping incidence and aggression than group-housed sows in strawed pens. These differences increase from the first until the fourth pregnancy (Broom et al., 1995). Provision of straw to the sow before parturition has a positive effect on the behaviour of the sow both during and after parturition, which affects the survival of the piglets (Westin and Algers, 2006). Piglet creep areas without any bedding are associated with a higher mortality rate than those with bedding (O'Reilly et al., 2006).

When comparing sows kept in strawed pens with individual feeding stalls with sows in strawed yards with an electronic sow feeder, the latter show more fighting but fewer total agonistic interactions than sows in groups with stall feeding (Broom et al., 1995). The provision of straw of any length reduces the occurrence of behaviours such as nosing other pigs, aggression and tail biting, compared with pigs with no access to straw. Chopped straw increases the prevalence of behaviours such as licking and decreases the prevalence of behaviours such as picking, suggesting that pigs are not able to manipulate the chopped straw in the same way as full-length or half-chopped straw. In addition, levels of tail biting are higher in groups that are provided with chopped straw compared to groups with full-length or half-chopped straw (Day et al., 2008).

In a Spanish study, pigs kept outdoors all year without any supplementary feed used 6–7 h per day for foraging (Rodriguez-Estevez et al., 2009). Pigs kept indoors in commercial farms might use about 30 min daily for feeding, and even if their nutritional needs are fully covered, they attempt to meet their behavioural need to forage further for food. This need might be strengthened by hunger if restricted feeding is applied, which often is the case in modern pig production. If straw or other edible material is used as bedding, these behavioural needs may be met by rooting and manipulating the

Fig. 2.18. Sows confined in crates performing stereotypies and vacuum chewing activities, behaviours which are rare in sows kept loose in pens (image I. Ekesbo).

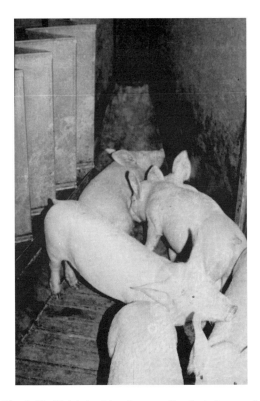

Fig. 2.19. Piglets kept in a house without windows and in a pen with only slatted flooring. They were fed twice daily from the automatic feeding device to the left. Such barren environments usually give rise to sterotypies, vacuum chewing activities or abnormal behaviour. Here, one pig is performing 'belly nosing', a behaviour imitating sucking behaviour (image I. Ekesbo).

Fig. 2.20. Sows confined in crates very often get injuries by lying on their side on a concrete or perforated floor without bedding. Note the swelling under the wound, the remains of an earlier abscess originating from a similar pressure injury (image I. Ekesbo).

bedding material. If such material is not available to them, stereotyped behaviours will develop.

The allocation of straw stimulates exploratory behaviour and reduces the amount of abnormal exploratory behaviour redirected towards pen mates. The more straw available, the more exploratory the behaviour directed towards the straw (Studnitz *et al.*, 2007).

Piglets, especially if weaned early, kept in a barren environment usually perform sterotypies and vacuum chewing activities very rarely seen in piglets kept loose in pens with access to straw.

Nose ringing has been used in order to prevent rooting of sows kept on pasture. When sows kept indoors are prevented from rooting through nose ringing, they try to chew, sniff and manipulate the available rooting materials, while unringed sows also root in them (Studnitz *et al.*, 2007).

Space allowance has a greater effect on behavioural patterns than group size (Randolph *et al.*, 1981).

Of four groups of sows that were tethered, confined in stalls and group housed respectively indoors and outdoors, the sows in stalls showed 46% more manipulation of drinkers, bar biting, etc., compared with the other groups (Barnett *et al.*, 1985). Sows confined in stalls require a further 2.5 megajoules of digestible energy (MJDE) per day for each 1°C below the LCT (lower critical temperature) in order to allow for the higher activity levels associated with stress response and stereotypies (Cronin *et al.*, 1986).

High growth rate is reported to be associated with aggressive behaviour (Rydhmer *et al.*, 2006).

Examples of injuries and diseases caused by factors in housing and management, including breeding. The abnormal behaviour seen in sows confined in crates indicates that they are in a stressful situation, which increases risk of disease. A comprehensive study by Bäckström (1973) indicates this. Sows in crates (*n* = 1283) showed 11.2% mastitis–metritis–agalactia syndrome (MMA), 5.4% farrowing > 8 h and 24.1% total morbidity, whereas the corresponding figures for sows kept free in pens (*n* = 654) were 6.7%, 2.3% and 12.8%. Sows kept confined in crates also have a prolonged parturition time compared to sows kept free in pens (Bäckström, 1973; Oliviero *et al.*, 2010).

Environmental disturbance like transferring sows into crates can disrupt parturition in the pig; this disruption is accompanied by depression of circulating oxytocin (Lawrence *et al.*, 1992).

The presence of MMA may lead to other diseases or injuries. Thus, the prevalence of shoulder lesions increases due to the presence of MMA (Ivarsson *et al.*, 2009). There is a relation between the duration of long, uninterrupted lying bout times after farrowing and the occurrence of shoulder lesions, even in well-conditioned sows provided with a small amount of straw present at the time of farrowing (Rolandsdotter *et al.*, 2009). Injuries caused by lying on the side on a concrete or perforated floor without bedding (Fig. 2.19) might become infected and result in purulent abscesses.

Bad floor quality of the lying area can cause severe claw injuries to newborn piglets (Westin and Algers, 2006).

A higher proportion of skin lesions caused by biting are observed in group-housed sows than in sows housed individually (Goossens *et al.*, 2008).

A total summary of study results thus indicates that keeping sows free in groups or in single pens with access to bedding for rooting means less risk of disease and injuries.

Early-weaned piglets confined in cages without bedding or other stimuli usually perform tail biting (Algers, 1984a). Tail docking at an early age does not prevent the behaviour. Tail-biting behaviour usually starts early after weaning and continues until slaughter. However, tail biting also often starts in piglets weaned after 5 weeks of age if kept in barren environments, e.g. pens with slatted or perforated flooring. The wounds (Fig. 2.21) are often sites of infection that may end in purulent abscesses. Tail-biting behaviour that has started early after weaning usually continues until slaughter.

Scour is a common disease problem after weaning, especially if, at the same time, the piglets are brought to a new pen and mixed with other litters. The risk of scour is reduced if the piglets are kept in the same pen with access to straw or other suitable material for rooting before and after weaning. Mixing sows and litters in group pens (Fig. 2.10) and letting the piglets stay in them after weaning seems to reduce the risk of scour even more.

Studies of the impact of space allowance show that weaned pigs in large and small groups are similarly affected negatively by crowding. Highly crowded pigs show more leg lesions than uncrowded pigs. Greater leg lesions have been recorded among pigs housed in large groups, as have greater lameness scores (Main *et al.*, 2000; Street and Gonyou, 2008). The risk of tail biting is higher in cases of reduced levels of floor space per pig and ear-biting

behaviour occurs more often when tails are docked short (Goossens *et al.*, 2008).

When unacquainted pigs are mixed together, this means stress, measurable via physiological parameters, and aggression, measurable via ethological methods (Bradshaw and Hall, 1996; Geverink *et al.*, 1996, 1998). In pigs kept some time at the abattoir before slaughter, it has been shown that the stress caused by mixing them the day before slaughter, during transport or in the abattoir implies an increased risk of gastric ulcers. If the pigs are slaughtered directly after having been mixed during transport, the changes in the mucous membrane of the stomach are fewer (Muggenburg *et al.*, 1967).

The correlation between post-weaning growth and playing/fighting behaviour supports the fact that playing/fighting behaviour can be used as an indicator of how piglets are coping with the stress of weaning (Silerova *et al.*, 2010).

A great problem in many pig herds is bad air quality. Airborne dust and toxic gases are two major concerns. Respirable dust, particles smaller than 5–10 microns, can be inhaled into the respiratory system. Very fine particles, smaller than 1 micron, can penetrate the lung tissue and cause permanent lung damage (Hartigan, 2004). Toxic gases, like ammonia, pave the way for respiratory infections. Most dangerous is hydrogen sulfide, which is formed in stored liquid manure. Even in small amounts, hydrogen sulfide causes haemorrhage, e.g. in the hooves (Bengtsson *et al.*, 1965; Ekesbo and Högsved, 1976). More than 20 ppm hydrogen sulfide in the air should be regarded as a danger to life. The genetic selection for increased growth capacity in domestic

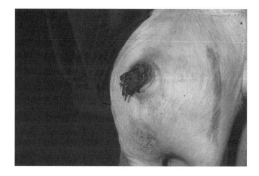

Fig. 2.21. Early-weaned piglets confined in cages without bedding or other stimuli usually perform tail biting. Tail docking at an early age does not prevent this behaviour, which usually starts early after weaning and continues until slaughter (image courtesy of Bo Algers).

pigs has caused skeleton diseases, for example osteo-chondrosis, with lameness and even fractures in sows between 6 and 18 months of age (Reiland, 1975; Dewey, 1999). Pigs reach sexual maturity at 5–6 months of age but do not have a mature skeleton until 18 months of age (Dewey, 1999).

Symptoms of pain. Pigs do not show any distinct signs of pain in the way they show signs of fear, e.g. by screaming. Even disorders which cause consider-able pain in humans, intestine torsion for example, seemingly cause no signs other than the pig leav-ing the group and lying down and, if enough straw bedding is available, often totally nestling down into the bedding.

2.19   Capturing, Fixation, Handling

Swine react to capturing and fixation by furious attempts to free themselves and by loud screaming. To carry out a number of essential tasks, such as blood sampling, pigs need to be restrained at times. Two common procedures for pigs of more than about 6 months of age are the snout snare and the stanchion. Pigs that are physically restrained for more than short lengths of time show evidence of an acute physiological stress response (Johansson and Jönsson, 1977, 1979; Jönsson and Johansson, 1979; Kelley, 1985).

When handling pigs, their characteristics as gre-garious animals must be considered. By patient and calm handling of the single pig or a group of pigs, their tendency to be panic-stricken can be largely avoided.

2.20   The Importance of the Stockman

The behaviour of humans and a pig's early experi-ence of humans are factors that influence pigs greatly. The important role of humans as a positive stimulus should not be underestimated. In a study of tethered sows, a positive human–animal rela-tionship obviated some of the negative effects of being tethered by lowering the physiological stress level and strengthening the immunity system (Pedersen *et al.*, 1998). A good stockman's behav-iour is reported to reduce neonatal mortality (Seabrook and Mount, 1995b; Prime *et al.*, 1999). Hemsworth (e.g. Hemsworth *et al.*, 1981, 1986a,b; Hemsworth and Barnett, 1987; Hemsworth, 2007) has performed several experiments on pigs showing, for example, the effects on pig behaviour of pleas-ant, unpleasant and inconsistent handling, often combined with measuring stress response by corti-costeroid concentrations, and also the consequences of different human behaviour on pig-production traits. Thus, pigs approached the experimenter sig-nificantly more when he did not approach, squat, have bare hands or did not initiate interactions com-pared with when he did approach, stood erect, had gloved hands or initiated interactions with the pig. Unpleasant and inconsistent treatment often resulted in a chronic stress response, with consequent adverse effects on growth performance. The early handling of pigs will influence their behavioural response to humans later in life (Hemsworth *et al.*, 1986a; Tanida *et al.*, 1994).

Barnett and Hemsworth (1986), by studying the influence of handling, showed that gilts treated kindly had a significantly higher pregnancy rate (87.5%) than those treated unkindly (33.3%). The non-treated controls had a 55.6% pregnancy rate. Blood samples indicated that the unpleasantly treated gilts had a chronic elevation of free corti-costeroid levels.

Seabrook (1995) considers the interaction between humans and animals as three interrelated components; hand and arm interaction, vocal inter-action and holistic empathetic interaction.

The experienced stockman and veterinary practi-tioner know that swine are individuals, where one reacts in one manner and another reacts otherwise to identical human behaviour. These experiences are verified in ethological experiments where tele-metric heart rate measurements are also used (Seabrook and Mount, 1995a).

Regrouping has a deleterious effect on average daily weight gain (Stookey and Gonyou, 1994; Ekkel *et al.*, 1995). The setback is attributed to both the physical stress of fighting during the first 24h and the social stress that persists beyond the first 24h period.

3 Rabbits (*Oryctolagus cuniculus*)

3.1 Domestication, Changes in the Animals, their Environment and Management

After the last glaciation, the wild European rabbit, *Oryctolagus cuniculus*, was restricted to the Iberian Peninsula. Fossils indicate that from there it spread to the south of France, due to natural migration. The rabbit was first described by the Phoenicians when they reached the Iberian shores about 1100 BC and by Pliny in Spain during the 1st century AD. The Phoenicians termed the Iberian Peninsula 'i-shephan-im' (literally, 'the land of the rabbit'), which the Romans converted to the Latin form, Hispania, and hence the modern word, Spain. The rabbit has long been kept and reared in captivity and appears to have been introduced to Italy in the 1st century BC, when it is described by Varro in his book on agriculture. There they were kept in large gardens, leporaria, containing cover and surrounded by a whitewashed wall. Actual domestication was probably accomplished by medieval monks (Zeuner, 1963).

The rabbit was introduced into Britain in the 11th century and kept in captive warrens as a food and fur resource (Surridge *et al.*, 1999). During the 16th century, it was introduced in Sweden as a delicacy in superior circles, but later became the poor man's food. By the middle of the 17th century, domestic rabbit breeding was in full swing in England (Zeuner, 1963). The rabbit's current distribution results from extensive introduction by humans, so that the animal now occurs on every continent and on oceanic islands ranging from Alaska to Macquarie (Bell, 1999). They were brought to oceanic islands by sailors to provide a store of fresh meat available to passing ships (Clutton-Brock, 1999). Their propensity for burrowing has meant that rabbits have always been difficult to keep enclosed and they have escaped to breed in larger numbers wherever they have been taken. During the two world wars, many families who had the possibility to do so kept rabbits in cages for food, as other sources of food were scarce in European countries (O'Malley, 2005).

Rabbits are territorial, social, burrowing herbivores that prefer a sandy, hilly terrain with shrubs and woody plants and are not found at altitudes above 600 m. European wild rabbits live in small, stable, territorial breeding groups. This is the social unit and consists of 1–4 males and 1–9 females, but different breeding groups may comprise colonies of up to 70 rabbits. In the wild, they live loosely in large warrens that can contain up to 60–70 individuals. The home range is rarely larger than 20 ha. They mark out their territory with the scent from the chin glands, the anal glands and the inguinal glands, and by way of urination and deposition of faecal pellets. They dig burrows or warrens that can be 3 m deep and 45 m long. The tunnels are about 15 cm in diameter and the living chambers 30–60 cm high. Each warren has several entrances which allow quick escape from predators. Does primarily dig burrows, and only pregnant or pseudopregnant females attempt to dig very deep tunnels. As a prey species, rabbits have evolved not to be florid or overt, constantly vigilant, lightweight and fast-moving, with a highly efficient digestive system that enables them to spend the minimum time possible above ground and in danger of capture (Cowan, 1987; Harcourt-Brown, 2002; Donnelly, 2004; Meredith, 2006; Fraser and Girling, 2009).

Rabbits have a thin skin and dense fur that consists of a soft undercoat and stiff guard hairs. Most rabbits moult approximately twice a year (spring and autumn), but this can vary. Moulting starts at the head and proceeds caudally. The tail is short. The skin on the neck is loose and pendulous and forms a pronounced dewlap in the females of some breeds. Pregnant or pseudopregnant does undergo a loosening of the hairs on the belly, thighs and chest, which are then easily plucked to line their nests before kindling. This exposes the nipples. The only hairless areas are the tip of the nose and the inguinal folds and, in the bucks, the scrotal sacs (Harcourt-Brown, 2002; Donnelly, 2004; Meredith, 2006).

The forelimbs are short and fine with five digits, in contrast to the long and powerful hindlimbs, with four digits. The digits, metacarpal and metatarsal areas are completely covered with hair and there are no footpads. The claws of rabbits are very sharp. The spine is naturally curved. Compared with other animals, the rabbit has a fragile skeleton. The skeleton represents only 7–8% of the body weight, whereas the skeletal muscle comprises more than 50% of the body weight. In comparison, the skeleton of a cat constitutes 12–13% of the body weight (Donnelly, 1997; O'Malley, 2005; Meredith, 2006).

The upper lip of the rabbit is cleft (Fig. 3.1). The permanent set of teeth erupts during the first 5 weeks of life. Rabbits, and hares, differ from other mammals in that the incisor teeth are arranged peculiarly in having the second upper incisors (peg teeth) placed behind (posterior to) the first, instead of beside them. There is a large diastema between the incisors and premolars. All the teeth are open-rooted and grow continuously. The incisors are adapted to cut through vegetation. Enamel is deposited on the front side of the anterior incisors only. The back side is composed of dentin, whereby the front surface wears down more slowly. By this, the

incisors remain permanently sharp from gnawing. A deciduous set is present in fetal rabbits and is shed just before or just after birth. In adulthood, males usually can be distinguished from females by a more rounded, broader appearance to the front of the head compared to a more pointed area in the female (Donnelly, 1997; Bell, 1999; Harcourt-Brown, 2002; Capello and Gracis, 2005).

In the male rabbit, the testicles are found in hairless scrotal sacs on either side of and cranial to the penis. The inguinal canal remains open and the testicles can be retracted into the abdomen. Retraction occurs during periods of sexual inactivity or during periods of insufficient food. Does possess four to five pairs of nipples on the ventrum. Male rabbits have no or rudimentary nipples (Harcourt-Brown, 2002; Meredith, 2006).

Both sexes have three pairs of scent glands: the chin glands, the inguinal glands and the anal glands (Harcourt-Brown, 2002; O'Malley, 2005).

Depending on breed, the weight varies between 1 and 7 kg, the males usually somewhat smaller than the females (Donelly, 1997; O'Malley, 2005).

In the wild, rabbits can live until 8 years old, although predation and road casualties often result in the death of rabbits at an age much younger than this (Fraser and Girling, 2009). The reproductive life of a domestic rabbit depends on its breed but is about 5–6 years for the buck and up to 3 years for the doe (Donnelly, 1997). Maximum ages for parkland

Fig. 3.1. The rabbit grazes grass and vegetation and is a selective feeder, with a wide food range. The head is raised at intervals to survey the surroundings (image courtesy of Eva Ekesbo).

rabbits are 8 years and 9 years for males and females, respectively (Bell, 1999). Life expectancy of the pet rabbit is generally 5–10 years, but some individuals live to 12 years or more (Meredith, 2006).

The rabbit is sensitive to stress. Examples of stress situations are imminent threat from predators, the presence of a rival, when it loses contact with the group or when it is put in an unsuitable environment (Drescher, 1994; Deeb, 2000).

There are more than 70 varieties of domestic rabbits, all descended from the wild European rabbit. They are raised for their meat and fur, for use as laboratory animals and for pets. They vary greatly in colour and may weigh from 1 to 7 kg. Some have small, erect ears; others have long, hanging ears (O'Malley, 2005; Meredith, 2006; Fraser and Girling, 2009).

The behaviour of wild and domestic rabbits is very similar. The modern rabbit retains many of the characteristics of its wild counterparts, despite changes in size, colour, coat texture and temperament. The major difference is their response to confinement: wild rabbits do not adapt well to cages, often fail to breed and exhibit abnormal behaviour not seen under natural conditions. Other wild lagomorphs, e.g. hares, resist domestication and cannot be raised successfully in cages. In contrast, domestication of rabbits has resulted in an animal that is less stressed by confinement and has a more placid disposition toward humans while retaining most of the behavioural repertoires of its wild ancestors. The cage housing and management systems used in commercial rabbit husbandry often conflict with basic rabbit biological needs. Several attempts have been made to improve housing systems, e.g. the use of group housing. Thus, keeping rabbits in group housing instead of cages and with a natural reproduction rhythm instead of artificial insemination leads to a considerable increase in litter sizes. However, a great number of does in a group-housing system get skin injuries through increased agonistic behaviour, which arises when several rabbits are kept in limited areas (Walshaw, 2000; Harcourt-Brown, 2002; Donnelly, 2004; Hoy *et al.*, 2006; Rommers *et al.*, 2006).

Rabbits, like hares, belong to the order Lagomorphs. Rabbits and hares look a lot alike and are often mistaken for one another. But the two animals differ in a few important ways. In rabbits, the young are born naked and blind, while newborn hares have fur and their eyes are open. Newborn hares are able to hop almost immediately

and leave the place of delivery. Hares are usually bigger, have longer ears, longer hind feet and a somewhat longer tail than rabbits. They usually live alone, not in groups, and do not dig burrows. When a rabbit senses danger, it hops for cover. It tries to run and hide from a predator. But a hare will leap long distances across an open field. It attempts to outrun its enemy.

The female rabbit is called a 'doe', the male a 'buck' and the offspring 'kits'.

3.2 Innate or Learned Behaviour

The locomotion of rabbits is an example of innate behaviour.

Mammary odour can boost learning. Newborn rabbits discriminate between different categories of adult conspecifics on the basis of their abdominal odour cues. Kits seem able to detect individual maternal odours. In fact, they appear to react to both species-specific cues and individual cues that they have learned. The mammary pheromone by itself may act as both a releasing and a reinforcing signal in these early socially oriented behaviours. These mammary signals and cues confer success in an offspring's approach and exploration of the maternal body surface, and ensuing effective initial feeds and rapid learning of maternal identity (Patris *et al.*, 2008; Schaal *et al.*, 2009).

3.3 Different Types of Social Behaviour

Social behaviour

Rabbits are gregarious animals preferring to live in the company of others in stable, territorial breeding groups with a defined social hierarchy. Within this main grouping, rabbits will live in closer, smaller groups – either in a male/female pair or, more commonly, in groups of between 2 and 8 individuals. Within the group, a hierarchy exists where male animals will not tolerate the presence of other males and older males will thus drive out younger ones. Dominance hierarchies are formed within each sex for each breeding group, and they are stable over time (Bell, 1983). Generally, a dominant male rabbit (or 'buck') will choose a female rabbit (or 'doe') as a permanent mate. The more dominant males will also mate with one or more additional does as a sort of harem. Young bucks normally move to a new social group before starting their first breeding season, while young does stay on to breed in their natal group (Parer, 1982; Bell, 1983, 1999; Webb *et al.*, 1995; Donnelly, 1997; Harcourt-Brown, 2002; Fraser and Girling, 2009).

Rabbits engage in 'amicable' activities such as lying together, grooming and nuzzling. These activities may occupy a considerable portion of each day (Donnelly, 1997).

Communication

Rabbits communicate with each other by thumping their hind legs on the ground as a warning when danger is near (Fraser and Girling, 2009).

Behavioural synchronization

Rabbits show no apparent behavioural synchronization as shown by cattle or sheep.

Dominance features, agonistic behaviour

Male rabbits within the group will establish a dominance hierarchy with the older heavier males at the top. Aged males that have been usurped by younger, fitter rabbits are driven from the group to become solitary satellite males. Young male rabbits are also often driven from the group when they reach puberty, either to join another warren or to solitary lives in hedgerows. The females tend to remain within the original group. Bucks scent mark does and young rabbits of their breeding group by spraying urine on them (Mykytowycz, 1968; Donnelly, 2004).

Female rabbits are less aggressive towards each other than males, but will defend a chosen nesting site ferociously. They are more aggressive towards juvenile does than towards juvenile bucks. Under natural conditions, both bucks and does on their own territory, surrounded by their own odour and that of their clan, win two-thirds of all aggressive encounters. Mature males (this can be as early as 4 months of age) are territorial and will fight, often even if they have been kept together since kits. This will be exacerbated in the presence of a sexually active female. During territorial disputes, rabbits will sometimes 'box' using their front limbs. Males mark territory more intensively and frequently than do females; dominants of both sexes mark more frequently than subordinates; and dominants mark most in the presence of subordinate rivals. Does scent mark their young, attack other young within the same breeding group and may chase, and even kill,

young from other breeding groups (Mykytowycz, 1968). They may attack even their own young if they have been smeared with foreign urine (Mykytowycz, 1968; Cowan, 1987; Harcourt-Brown, 2002; Donnelly, 2004; Fraser and Girling, 2009).

3.4 Behaviour in Case of Danger or Peril

Rabbits are prey species and therefore spend large amounts of their time watching for attack from predators. If the rabbit is scared at any time, then it will demonstrate a variety of responses. This may be crouching low to the ground, ears flat to the head and not moving, or demonstrating a fight-or-flight response. If the rabbit does not run away to hide in a hole, then it may stamp and thump with its hind feet and show its scut, as warning signals to other rabbits to alert them, or rarely may vocalize, grunting or squeaking, and bite or attack the problem with its front feet. When fleeing from a predator, it hops and bolts for hide and cover in quick, irregular movements, designed more to evade and confuse than to outdistance a pursuer. If attacked directly, a rabbit will scream a blood-curdling sound. As part of the fear response, rabbits can go into a trancelike state, a stress response, resulting in physiological changes associated with fear (Harcourt-Brown, 2002; Fraser and Girling, 2009).

3.5 Some Normal Physiological Frequency Values of Interest at Clinical Examination

The normal body temperature is 38.5–39.5°C, the pulse frequency is 150–250 beats per min and the respiration frequency is 30–60 breaths per min (O'Malley, 2005). Rabbits are nose breathers. The nose moves up and down ('twitching') in a normal rabbit 20–120 times per min, but this will stop when the rabbit is very relaxed (Meredith, 2006).

3.6 Active and Resting Behaviour Patterns

Activity pattern, circadian rhythm

Rabbits are essentially nocturnal, leaving the burrow in the early evening and returning in the morning, although they can be seen grazing or basking during the day. Although rabbits use burrows that serve as a primary haven and restrict their home range, they are, if no predators are present, active above ground, moving around, hopping, running, chasing and playing. Members of neighbouring groups may move out into communal grazing areas during peak dawn/dusk feeding periods. Rabbits' activities such as lying together, grooming and nuzzling occupy a considerable portion of each day. When they are above ground, they spend about 44% of their time eating, 33% inactive, 13% moving and 10% on other activities (Gibb, 1993; Donnelly, 2004). They will therefore often remain underground re-eating caecotrophs and sleeping during the day. The primary feeding times for rabbits are in the early morning and at night, with coprophagy commencing 3–8 h after eating. Where few predators exist, rabbits will demonstrate diurnal behaviour, coming out of the tunnels to feed during daylight hours, but this behaviour is less common. If given the chance, then rabbits can be observed basking in the morning sun (Donnelly, 1997; Bell, 1999; Harcourt-Brown, 2002; Fraser and Girling, 2009).

Exploratory behaviour

The exploratory behaviour of objects is sniffing, of the environment is standing on the hind legs looking and listening.

Diet, food searching, eating and eating postures (feeding behaviour)

When wild rabbits emerge from their burrows at dusk, they begin to feed. Initially, they graze grass and vegetation, raising their heads at intervals to survey the surroundings. Field studies of wild rabbits indicate that they are selective feeders with a wide food range. Unlike the grazing horse or cow, which eats the entire plant, the rabbit selects the most nutritious part of the plant, favouring young, succulent plants over mature, coarse growth. Rabbits chew their food thoroughly. They ingest coarse fibre only to stimulate gut motility and excrete it rapidly, unlike horses, which carry fibre for up to 3 days. The rabbit caecum is the largest of all animals, relative to size, with ten times the capacity of the stomach and containing 40% of the intestinal content. Compared with other animals, rabbits have a high water intake. A rabbit's average daily water intake is 50–150 ml per kg of body weight; a 2 kg rabbit drinks about as much water daily as does a 10 kg dog (Harcourt-Brown, 2002; Donnelly, 2004; O'Malley, 2005).

Pellets of soft caecal contents (caecotrophs) are expelled periodically from the anus and reingested as a source of nutrients. The caecotrophs have an outer greenish membrane of mucus that encloses semi-liquid caecal ingesta; they contain high levels of vitamins B and K and twice the protein and half the fibre of hard faeces. The caecotrophs are swallowed intact and resemble a bunch of light green, sheeny peas in the rabbit's stomach. This digestive strategy utilizes bacterial fermentation to synthesize nutrients and avoids the need to store large volumes of food in the digestive tract. Vegetation can be digested efficiently below ground without the need to spend long periods grazing and exposed to predators (Harcourt-Brown, 2002; Brooks, 2004).

Rabbits have so-called milk oil, an antimicrobial fatty acid distinct from that of secreted gastric acids. This antimicrobial fatty acid is produced from an enzymatic reaction with a substrate in the doe's milk in the suckling rabbit's stomach. This special physiological adaptation controls the gastrointestinal microbial contents of young rabbits. Rabbits fed milk from other species do not develop this antimicrobial factor and are more susceptible to infections (Brooks, 2004).

Selection of lying areas, lying down and getting up behaviour, resting postures, sleep

A healthy rabbit typically will rest compactly on all four limbs with regular twitching of the nostrils. The plantar surface of the hindlimbs from the tarsus distally is in contact with the ground at rest. When a rabbit is sitting, undisturbed, the plantar surface of the lower hindlimbs, from the toes to the hock, is in contact with the ground (Donnelly, 1997; Huerkamp, 2003; Meredith, 2006).

Locomotion (walking, running)

Rabbit movement consists of hopping, crawling and intensive locomotion. Hopping is used to travel longer distances, whereas crawling is performed when feeding on grass or exploring on the spot and during social encounters (Kraft, 1979; Lehmann, 1991).

Rabbits move about on the tips of the digits in a fashion known as digitigrade locomotion. When browsing, rabbits move slowly using all four legs. In case of danger, the powerful hind legs enable them to reach a high speed when seeking protection in the burrows (Donnelly, 1997; O'Malley, 2005).

Swimming

Rabbits do not swim.

3.7 Behaviour at Defecation and Urination (Eliminative Behaviour)

Hard faecal pellets are always voided above ground, never in the burrow (Harcourt-Brown, 2002).

Caecotrophs are produced during the day in wild rabbits. They are produced several hours after feeding, during a quiet undisturbed period, which is during the day for a wild rabbit in its burrow but can be during the night or early morning for a domestic or laboratory rabbit in a cage. In healthy rabbits, caecotrophs are consumed straight from the anus and are swallowed whole. Soft caecotrophs are usually consumed during periods of rest underground, although occasionally rabbits exhibit this behaviour above ground. When food is scarce, all caecotrophs are consumed. When food is available *ad libitum*, the protein and fibre content of the ration influence the amount of caecotrophs consumed. Increased levels of fibre increase caecotrophy, whereas high protein levels reduce it (Harcourt-Brown, 2002).

When kept confined, rabbits create a specific toilet area for faeces and urine (Fraser and Girling, 2009).

3.8 Body Care, Cleanliness

The wild rabbit grooms its coat regularly, using the incisors to pull out dead hair. The coat is short and dense and does not knot and mat easily like the fluffy coat of many pet rabbits. In wild rabbits, mutual grooming is an important part of social behaviour. During periods of rest, they lie together and groom each other, especially around the face and head (Harcourt-Brown, 2002). The rabbit licks its forepaws to clean its ears and face thoroughly and uses its tongue to preen its entire hair coat (Walshaw, 2006).

3.9 Temperature Regulation and Climate Requirements

Temperature regulation

Rabbits tolerate cold better than heat. Because they do not possess brown fat, they shiver when exposed to cold. Shivering works well in the short term, and rabbits can tolerate cold weather if properly acclimatized and sheltered. The pinnae are highly vascular and in cold weather they retain body heat by shunting warm blood from the ears to warm the body core and then largely shutting down the circulation of blood in the ears. In cold weather, rabbits adopt a hunched posture and huddle together to decrease the collective surface area. They are unusually sensitive to temperatures higher than 28°C and have little protection against high ambient temperatures. The pulse rate of rabbits increases only mildly in response to an increase in body temperature. The pinnae represent a large portion of the total body surface in rabbits, approximately 12% in the New Zealand breed. The pinnae have the largest arteriovenous shunts in the body. Directly cooling the ears causes a drop in body core temperature, and vice versa. High temperatures inhibit drinking and panting, which can hasten dehydration and be fatal. Rabbits cannot sweat, except through sweat glands confined to the lips; also, they pant ineffectively, and, when sufficiently dehydrated, they stop panting. Rabbits do not increase water intake when the ambient temperature becomes high; heat actually seems to inhibit drinking. They can tolerate a water loss equivalent to 48% of their body weight. In contrast, when water loss reaches 11–14% of body weight in dogs, circulatory failure occurs. Rabbits are sensitive to low humidity. High humidity levels are not a problem as underground burrows are naturally quite humid (Donnelly, 2004; Marai *et al.*, 2004; O'Malley, 2005).

Climate requirements

Although rabbits use their ears to dissipate heat, they actively seek shade and burrow to conserve water and to shelter themselves from heat. Shelter from direct sunlight is essential in the design of any rabbit housing. Because of their extreme sensitivity to heat, rabbits should be housed at between about 10 and 20°C. In the wild, they cool down by seeking shade in their burrows or by stretching out to increase body surface area. Domestic rabbits can be housed outdoors if protected from adverse weather cold below just above 0°C. The response of rabbits to high and low ambient temperatures is also important during transportation. The risk of mortality is greater in hot weather than in cold weather (Huerkamp, 2003; Donnelly, 2004; O'Malley, 2005).

3.10 Vision, Behaviour in Light and Darkness

The rabbit has a wide visual field that allows it to watch for predators while it is grazing. This is facilitated by their laterally positioned eyes, which give an almost 360° field of view. The visual field does not include the area immediately under the nose. Their accommodation is poor. They have good nocturnal vision (Harcourt-Brown, 2002; Fraser and Girling, 2009).

Exposure of domestic rabbits to a long daylight regime affects their health adversely (Marai *et al.*, 2004).

It is believed that rabbits are able to distinguish by eyesight people known to them from people they are unfamiliar with (Meredith, 2000).

3.11 Acoustic Communication

Hearing

Rabbits have good hearing (Meredith, 2000). Humans are able to localize sounds within 1–2°. In contrast, many animals localize sounds with less precision than humans. Rabbits, for example, have sound localization thresholds of approximately 22° (Ebert *et al.*, 2008).

Vocalization and acoustic communication

Rabbits are naturally very quiet animals. They communicate with each other by thumping their hind legs on the ground as a warning when danger is near. The only vocal sounds that are made are a loud, high-pitched scream of terror or a range of growls and hums that denote pleasure or defence. Humming sounds may be emitted by bucks as part of mating behaviour (Harcourt-Brown, 2002; Fraser and Girling, 2009).

3.12 Senses of Smell and Taste, Olfaction

Rabbits rely heavily on scent. Food selection and ingestion are based on smell and on tactile information gained from the sensitive vibrissae around the nose and lips. Smell and taste are more important for identifying members in their own group than is sight. The outermost sensitive vibrissae facilitate seeking feed as well as orienting the rabbit in aisles and caves (Harcourt-Brown, 2002; Meredith, 2006).

3.13 Tactile Sense, Sense of Feeling

Tactile touch

The tactile information gained from the large numbers of tactile vibrissae around the nose and lips is used to help locate when underground (Harcourt-Brown, 2002; Meredith, 2006).

Sense of pain

As a prey species, rabbits show few clinical signs of pain or discomfort, i.e. anything that would draw attention to their weakness. Their natural response is to become quieter, less active and often to reduce or stop eating and performing caecotrophy. These are subtle signs, often in the face of quite serious pain, but it is a survival tactic. A wild rabbit which is obviously injured or in pain would be easy prey for a predator and likely to be selected out of a group of others. A rabbit presenting with severe pain will show a hunched posture and immobility and may grind its teeth. Only by monitoring carefully features such as food intake, activity levels and faecal pellet/caecotroph output is it often possible to detect the signs of discomfort and pain. Very occasionally, some rabbits will vocalize when in pain (Huerkamp, 2003; Fraser and Girling, 2009).

3.14 Perception of Electric and Magnetic Fields

This does not seem to be described in rabbits.

3.15 Heat and Mating Behaviour, Pregnancy

Rabbits are well known for their ability to reproduce quickly. Like the cat and the ferret, rabbits are induced ovulators. Ovulation occurs between 10 and 13 h after copulation. Does have no uterine body, the two separate uterine horns and two cervices open into the vagina. Although they do not show a regular oestrus cycle, they do vary in receptivity, and a cyclic rhythm exists. Does can be mated soon after giving birth and may be lactating and pregnant at the same time. They will often circle the male when in heat. Sexual receptivity in a doe is characterized by lordosis, which is where she lies in sternal recumbency with the pelvis raised and the mid-lumbar region dipped ventrally, or presenting of the perineum in response to a buck's attempts to mount. Similar behaviour is seen in cats.

Generally, when in heat, female rabbits are hyperactive and brace themselves when touched. When not receptive, they do not allow males to mount. In does kept in cages, depending on cage space, non-receptive behaviour often takes the form of running away, cornering, biting and vocalizing. Pseudopregnancy occurs especially after stress in farmed rabbits and lasts 16–18 days (Donnelly, 1997; Harcourt-Brown, 2002; Meredith, 2006; Walshaw, 2006; Fraser and Girling, 2009).

Bucks show a constant libido after puberty. Initiation of copulation in rabbits is confined to basic patterns such as sniffing, licking, nuzzling, reciprocal grooming and following the doe. Males may also circle the object of their affections and may make a low humming noise while they do so. Bucks may also exhibit tail flagging and enuriation, the emission of a jet of urine at a partner during a display of courtship. Experienced males usually initiate copulation within minutes or even seconds after a receptive female is introduced. Inexperienced males generally require longer. Bucks rapidly mount receptive females and accomplish intromission after a series of rapid copulatory movements. Reflex ejaculation follows immediately on intromission. The copulatory thrust is generally so vigorous that the buck falls backwards or sideways and may emit a characteristic cry. Vigorous bucks may attempt to copulate again within 2–3 min (Donnelly, 1997).

Gestation in European wild rabbits lasts 28–30 days according to Bell (1999) and in domesticated rabbits 28–34 days, with most litters being born on day 31, according to Bennett (2001). According to Donnelly (2004), the gestation period varies with the breed but is approximately 30–32 days; according to Fraser and Girling (2009), it lasts from 29 to 35 days.

Comparisons between natural mating and artificial insemination (AI) show that AI leads to a lower kindling rate and a lower kit weight at weaning (Rommers *et al.*, 2006).

3.16   Before and After Parturition

Behaviour before and during parturition

Pregnant does dig tunnels to provide nests for parturition. A few hours to days before kindling, rabbits pull fur from their abdomen, sides and dewlap to make a nest. Although the underlying skin can look inflamed, this is normal behaviour.

High-ranking does often give birth in a special breeding chamber dug as an extension to the warren, whereas some of the subordinate females are chased away from the warren and forced to give birth in isolated breeding stops (Mykytowycz, 1968).

Does usually give birth in the early morning. Total parturition time would be 10–14 min for a large litter and 5–7 min for an average litter. Parturition is rarely difficult, with dystocia being uncommon.

Number of kits born

The size of the litter is 5–6 young for European wild rabbits, but for domesticated rabbits litter size depends on parity and breed. Primiparous animals tend to produce smaller litters. Small breeds produce litters of 4–5 kits, whereas larger breeds produce litters of 8–12 kits (Donnelly, 2004).

It is possible for a doe to have six litters in a year and to produce 40–50 offspring (Harcourt-Brown, 2002).

After the birth

Blind, deaf and hairless at birth, rabbit newborns need, as all mammal newborns, to interact rapidly with the mother to find the nipples and suck. In the wild, newly born rabbits, or 'kits', are cleaned and nursed by the doe before she leaves the nest and blocks the entrance. This behavioural pattern serves to protect the kits from predators and is a highly evolutionary adaptive behaviour. She will stay in the vicinity of the nest but usually returns only once daily to feed the kits for a period of 3–5 min, during which time a baby rabbit can consume 20% of its body weight. Newborn rabbits are directed to their mother's nipples by specialized odour cues. Olfactory cues are critical during the nursing period: a gland in the region of the nipple produces a pheromone that attracts kits. Efficient odour cues are released only from the nipples. The eyes of the kits open at 10–11 days, and hearing develops at the same time. Rabbit milk is concentrated, containing 13–15% protein, 10–12% fat and 2% carbohydrate. Most passive immunity is obtained before birth, although some antibodies are present in the colostrum. During the rapid development of the young, lactating rabbit females and kits have developed some sensory, physiological and behavioural adaptations, allowing them to communicate and allowing the young to ingest milk then solid food efficiently. After days 10–15, the mother–young interactions change progressively. After 21 days, the doe ceases closing the burrow and the young rabbits emerge from the nest at about 18–21 days, start nibbling grass or hay at 3 weeks and are weaned at about 25 days old. Does mark kits with chin and inguinal secretions, and they are openly hostile to young that are not their own. Recognition of the mother by neonates is assumed to emerge only once the dam and her young leave the nest and encounter, for the first time, other family units (Kersten *et al.*, 1989; Bell, 1999; Harcourt-Brown, 2002; Marai and Rashwan, 2003; Donnelly, 2004; Baumann *et al.*, 2005; Moncomble *et al.*, 2005; Coureaud *et al.*, 2008; Val-Laillet and Nowak, 2008; Fraser and Girling, 2009).

Play behaviour

Rabbits are not particularly playful (Huerkamp, 2003).

Weaning

Rabbits are weaned at about 4–6 weeks of age (Meredith, 2006).

3.17   The Kit and the Young Rabbit

Small rabbit breeds develop more rapidly and are sexually mature at 4–5 months of age. Medium-size breeds mature at 4–6 months and large breeds at 5–8 months of age. Does mature earlier than bucks (Donnelly, 2004).

Young females stay on to breed in their warren-group of birth, while young males move to a new social group before the start of their first breeding season (Bell, 1999).

3.18   Assessment of Rabbit Health and Welfare

The healthy rabbit

The healthy rabbit has activity, movements and posture appropriate to its age, sex, breed or physiological condition, stage of sexual development or reproductive condition. It has clean and bright eyes, clean coat free from obvious lesions or damage, normal legs and feet; normal feeding, drinking, defecating and urinating, normal locomotor and exploratory behaviour; and uses separate resting

and dunging areas if available. The caecotrophs and the hard faeces have normal amount, consistency, colour and smell.

The sick rabbit

Symptoms

Mouth breathing is a very poor prognostic sign. Immobility may indicate pain or fear. Respiration via the mouth is a serious disease symptom. Abnormal gnawing usually indicates understimulation or wrong diet composition. Assessment of the teeth should always be an essential part of the clinical examination (Donnelly, 2004; Meredith, 2006).

Examples of abnormal behaviour and stereotypies

Giving the rabbits a protected environment where they are not placed under any stress and are allowed to exhibit their natural behaviour will prevent the development of many behavioural problems that are presented in practice (Fraser and Girling, 2009).

Cannibalism practised by wild rabbit does in captivity is a manifestation of the failure of maternal behaviour as a consequence of the stress they experience in captivity (Gonzalez-Redondo and Zamora-Lozano, 2008).

Hay is a more effective object than gnawing sticks and other objects in reducing abnormal behaviour and giving the individually housed rabbits some alternative occupation (Lidfors, 1997).

Examples of injuries and diseases caused by factors in housing and management, including breeding

Heat stress affects adversely the respiration rate and ear temperature in does, litter size at birth, litter weight at birth, litter size at weaning and gestation length (Marai *et al.*, 2004; Kumar *et al.*, 2005).

Rabbits kept in a conventional cage system, especially the females, showed more restlessness, excessive grooming, bar gnawing and timidity than rabbits kept in an enriched cage system (Hansen and Berthelsen, 2000).

The isolated domestic rabbit is deprived of the protection of burrows and of a social hierarchy and often lives in a limited space, which is a fraction of a metre of a hard sanitized space, removed from odours, markers and social interaction (Marai and Rashwan, 2004). Despite their gregarious nature, rabbits used for research are often housed individually, due to concerns about aggression and disease transmission. However, conventional laboratory cages restrict movement and rabbits housed singly in these cages often perform abnormal behaviours, an indication of compromised welfare (Kersten, 1995; Chu et al., 2004).

Under commercial husbandry conditions, the entrance to the nest box is not closed, as under natural conditions, but stays permanently open, potentially counteracting a doe's behavioural goal of a closed nest. Due to the non-manipulable floor and the absence of roughage or other appropriate materials, a doe will fail to achieve the feedback of a successful removal of nest stimuli, in spite of conducting the appropriate behavioural patterns. This leads to repeated nest contacts, nest visits and nest closing attempts and can increase kit mortality due to the crushing of kits, of out-of-time kit activation or sucking and the disturbance of their energy-saving strategy of resting deep inside the insulating nest material between nursing visits (Coureaud et al., 2000; Baumann et al., 2005).

Crowding causes rabbits to become aggressive and bite one another during the first few days of nest sharing. When rabbits are exposed to noise, it causes adverse effects, including nervous and behavioural abnormalities and can cause a startled response and traumatic injuries to limbs and back. Keeping rabbits in a hot climate is the main cause of abnormal maternal and sexual behaviour (Marai and Rashwan, 2004).

Keeping several does together in group pens, transport and other stress situations in farmed rabbits may trigger pseudopregnancy (Donnelly, 1997; Rommers et al., 2006).

Skin disease is common in pet rabbits, but apparently rare in their wild counterparts (Harcourt-Brown, 2002).

Culling and mortality in adult breeding rabbits were investigated on 130 commercial farms with 82,352 does, 50,834 of which were culled or euthanized in Spain during 2000–2005. The highest causes, calculated using the median of the monthly cumulative incidence (MCI), were 1.3% low productivity (0.9% infertility and 0.4% other causes), 0.5% mastitis, 0.5% poor body condition and 0.3% sore hocks. The causes of culling males were estimated based on a population of 6514 males, 5313 of which were culled. The highest causes were 2.0% MCI low productivity (0.9% infertility and 0.5% no libido), 0.3% abscesses, 0.3% sore hocks and 0.2% poor body condition (Rosell and de la Fuente, 2009).

Heavy rabbits housed on wire floors often develop an ulcerative pododermatitis (sore hocks) of the plantar surface of the lower hindlimbs. Pododermatitis is a common injury in rabbits kept in cages with a wire floor (Drescher and Schlender-Bobbis, 1996; Donnelly, 2004).

Symptoms of pain

Rabbits are very poor at demonstrating pain. Their natural response is to become quieter, less active and often to reduce or stop eating and performing caecotrophy. These are subtle signs, often in the face of quite serious pain, but it is a survival tactic as rabbits are a prey species. A wild rabbit which is obviously injured or in pain could be easy prey for a predator and likely to be selected out of a group of others. Covering up pain makes this less likely, but also makes it very difficult for the nurse and vet to spot when a rabbit is actually suffering. Only by monitoring carefully the features above, such as food intake, activity levels and faecal pellet/caecotroph output, is it often possible to detect the signs of discomfort and pain. Very occasionally, some rabbits will vocalize when in pain. This is often associated with intense visceral pain, such as pyelonephritis/renoliths or gastrointestinal tract bloat. Teeth grinding (bruxism) can also be another inconsistent clinical sign of discomfort (Fraser and Girling, 2009).

3.19 Capturing, Fixation, Handling

Rabbits are at significant risk of injury from improper handling. They should never be lifted by the ears. If the hind legs are not restrained, sudden, straight-out kicks risk the fracturing of long bones or the spine, usually at the seventh lumbar vertebra. The proper manner to carry a rabbit is to grasp the scruff of the neck with one hand and support the abdomen and the hindquarters with the other. To traverse a longer distance, the handler should tuck the head of the rabbit into the crook of the elbow of his or her arm (usually the left) and use that arm to support the body weight (Mader, 1997; Huerkamp, 2003).

As part of the fear response, rabbits can go into a trancelike state. 'Trancing' is sometimes described as a method of restraining rabbits for examination. By placing the rabbit on its back, it will freeze. This method should not be used as the reaction is a stress response, resulting in physiological changes associated with fear. The person restraining a rabbit on a table should stand facing the flank of the rabbit, placing one hand gently over the thorax and the other on the hindquarters (Huerkamp, 2003; Fraser and Girling, 2009).

3.20 The Importance of the Stockman

Rabbits show some similarities to other species in that they should be socialized when young. The sensitive period has been shown to be 10–20 days (Kersten *et al.*, 1989). Rabbits introduced to humans and handling at this time will be much more responsive than rabbits that are not handled (Fraser and Girling, 2009).

4 Cattle (*Bos taurus*)

4.1 Domestication, Changes in the Animals, their Environment and Management

The aurochs, *Bos primigenius*, the progenitor of current domestic cattle, was a wild ox that once lived in most parts of Europe, Asia and North Africa, with the exception of Siberia, north Sweden and Finland. The species was characterized by a long, narrow head, with powerful horns bent forward and inwards (Fig. 4.1). The bull was brown black to black, with a light streak over its back; the cow was reddish-brown or dark brown and the calf reddish-brown. In all probability, the head was much darker than the body, a type of coloration which still occurs in certain domesticated breeds. The bulls were big, weighing about 1 t, while the cows were smaller. This difference still remains in cattle (Fig. 4.2). The earliest findings of aurochs are more than 300,000 years old. Most died out about 2000 years ago. However, some individuals survived in a game preserve in Poland until the 1600s and the last individual, a cow, died in 1627 in the Jaktorow forest, south-west of Warsaw in Poland (Clutton-Brock, 1999).

The domestication of cattle (Fig. 4.3) probably began about 9000 or even 10,000 years ago (Clutton-Brock, 1999; Geigl, 2008). Today's European cattle seem to be descended from cattle brought from the Near East by the first farmers (Troy *et al.*, 2001; Hall, 2002).

At the beginning of the 2000s, there existed more than 1000 breeds of cattle (Hall, 2002). During domestication, breeds were developed of different sizes, colours and ability to cope with different climates and types of vegetation. Some beef breeds, e.g. the Belgian Blue, were developed genetically for extreme meatiness, others, e.g. Holsteins, for extremely high milk production. Until about 1800, the average annual milk yield was about 1000 kg. In the next 150–160 years, it increased to about 3000–4000 kg. In the 1950s, dairy cows in most countries yielded about 3000 kg per year; in the 2000s, herds in many countries have cows producing annually far more than 10,000 kg of milk. In Sweden, the average annual milk yield increase per cow between 1800 and 1900 was less than 20 kg. Between 1906 and 1955, it was about 25 kg; between 1956 and 1980 it was about 85 kg; and between 1981 and 2005 slightly more than 140 kg (Ekesbo, 2006). Thus, most replacements produce more milk than their mothers.

From the start of domestication until the 1920s, when milking machines were first introduced (Fig. 4.4), cows were hand-milked mostly by women (Fig. 4.5). Hand milking involved a close intimate human–animal relationship. Milking was often performed three times a day. However, in most countries, milking machines generally were not used until the 1950s. Using the machine, one person could then milk five cows in the same time as hand-milking one cow. However, the time for building the human–animal bond day by day became reduced compared with hand milking. At the same time, mostly men took over the milking. In order to save labour, it was performed generally twice a day. The increased milk yield during the 2000s, in combination with the emergence of robot milking or, in very large herds, milking work in three shifts, has led to cows in some herds being milked three times a day or more. Thus, in 2008, in 8% of US dairy herds fresh cows were milked three times or more daily (USDA, 2008). The introduction of milking machines reduced the human–cow contact, as did the rapid increase in herd sizes that occurred in many countries from the last decades of the 20th century. The changeover from small or medium-size tied herds into loose-housed herds with milking parlours has reduced this contact even more, as has the introduction in the loose-housed herds of milking robots in the late 20th century (Figs 4.6–4.8). During milking in milking parlours or in robot rooms, the cow is often given some concentrate feed. In the robot rooms, each cow will not be

Fig. 4.1. A red cow from the cave of Lascaux (about 15,000 BC) with a black head and a white stripe along the back (image courtesy of photo archive HMH Department).

Fig. 4.2. A bull of the 20th century, Swedish Red Breed (image I. Ekesbo).

allowed to come back for milking until after about 6–8 h and an alarm system reports if more than 12 h have passed since the cow's last visit. In loose-housed herds, transponders, an identification box attached to the cow's neck strap or legs, are often used. In herds with robots, they are a necessity. The transponder emits the cow's identification and may record her activity, locomotion, feeding and resting at

Fig. 4.3. Hornless cow, with calf tied to foot, being milked from the side. In newly domesticated cattle, having the calf close to the mother's side triggers milk let-down. From the coffin of Kawit, Priestess of Het-Hert, Eleventh Dynasty, *c.*2050 BC (image courtesy of photo archive HMH Department).

Fig. 4.4. Milking machines came into use in some large herds during the 1920s. Here are two women with eight machines each (image courtesy of photo archive HMH Department).

Fig. 4.5. Until the 1930s, hand milking was, in most western countries, used in all herds except the largest. Here, as was usual for those times, it is performed by a woman (image courtesy of photo archive HMH Department).

different times. This information is gathered daily and supplemented with a recording from each milking opportunity regarding the amount of milk, etc. This information is treated in the herd computer. Besides production information, such information is also used for health control. Thus, if a cow transponder reports an abrupt decrease in milk yield or long resting or inactivity, this is a warning signal that the cow in question may have a disorder.

This reduction of human–animal contact means that a cow in the 21st century is in many ways less tame, less domesticated, than a cow from 1950 – i.e. the human–animal bond is less well developed in the modern dairy animal. The average herd size has in most countries changed from fewer than 10 in the 1950s to more than 40 in the beginning of the 2000s, in some countries, even more. Until the last decades of the 1950s, most dairy cows were kept on pasture during summer.

With increased herd sizes, from 200–300 cows per herd to more than 1000, it is more and more common to keep cows indoors all year round and bring the fodder in to them (so-called zero grazing). It has been shown that keeping them indoors year-round increases the risks of higher incidences of disease in dairy cows (Ekesbo, 1985; Bendixen *et al.*, 1986). This has led, for example, to Swedish animal welfare legislation stipulating that cows be kept outdoors, preferably on pasture, during summer.

Fig. 4.6. Part of an uninsulated loose-house building. In the forefront, three cows are queuing up on their way into the cold insulated milking room (a) with its milking robot that milks one cow at a time (image I. Ekesbo).

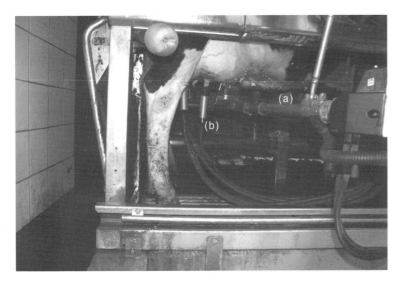

Fig. 4.7. The cow has, without any human contact, entered the milking room voluntarily and is restrained by the walls in the milking stall. The milking robot arm (a) – after having washed and dried the teats – by means of a laser beam, finds and carries one teat cup (b) up to the teat. Two teat cups are already in place (image I. Ekesbo).

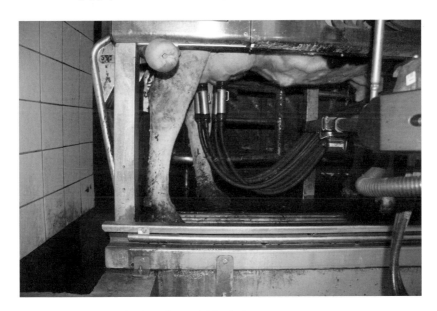

Fig. 4.8. Finally, all four teat cups are in place (image I. Ekesbo).

Even if domestication has produced a great number of cattle breeds with different morphological features, domestic animals apparently have retained much of the behavioural repertoire of their wild ancestors. There are not only morphological differences between breeds; scientific studies (e.g. Albright and Arave, 1997; Phillips, 2002) have endorsed the empirical experiences of farmers and veterinarians that there also exist great individual behavioural differences. Furthermore, within a population of animals in natural environments, there are differences between individual animals in the behavioural strategies adopted (Spinka, 2006). Due to the general use of artificial insemination, groups of commercial dairy cattle are nowadays almost totally female.

Cattle can reach an age of close to 20 years. Until the 1960s, cows were often kept for ten lactations or more. However, as a result of intense genetic selection for high milk production, most dairy cows are culled nowadays before having their fourth calf. This is because of genetically or disease-induced low milk yield, or because they do not fulfil the requirements for fertility. In 2000, the average interval between calvings in the USA varied from around 12.8 months for Jerseys to 13.4 months for Brown Swiss (Hare *et al.*, 2006). In Sweden, the average for all dairy breeds was 13.4 months in 2008. Cows with long intervals between calving are at risk of being culled. However, since the 1980s the calving interval has been increasing in all herds in most major dairy countries that select for higher milk production (Hare *et al.*, 2006).

From the end of the 1900s, hormonal methods have been developed to produce ova and embryos, and these can result in the birth of multiple viable offspring. These techniques may result in genetically similar cattle, which may also be used in experimental work (Ewbank, 2010).

Male cattle are described as 'bulls', the female as a 'cow' and the offspring as a 'calf'. Cattle of both sexes are defined as calves up to 6 months of age. The term 'heifer' is generally used for a female from 6 months old until just before her second calving. A 'bullock' is a castrated male older than 6 months. Adult bullocks, oxen, were used as draught animals until the early 1900s.

4.2 Innate or Learned Behaviour

In cattle, as in other animals, certain behaviours are innate, others learned.

Innate behaviours are, for example, the cow leaving the herd at the time of delivery; licking of the calf after delivery; the calf lying low or squatting on or close to the delivery place during the first few days after delivery; suckling of the calf; getting up and lying down behaviour; body positions at rest, urination, defecation, or in intense rain and wind; signs of anxiety if separated from the herd; and individual animals are mostly oriented in the same direction during grazing. Some innate behaviour, e.g. body positions, will usually be performed irrespective of external conditions. Other innate behaviours, e.g. synchronized grazing, are influenced by external conditions. Examples of learned behaviour are, for example: manipulating water bowls in order to get water; passing into milking parlours or

milking robot rooms; the use of cubicles, or seeking buildings as protection to keep out the weather. The impact of adverse weather on the behavioural and physiological responses of cattle is shown in several studies (e.g. Tucker *et al.*, 2007).

Cattle can learn discrimination tasks quickly, for example to discriminate between individual familiar conspecifics, and the speed of learning can be influenced by the identity of the stimulus individuals (Hagen and Broom, 2003).

4.3 Different Types of Social Behaviour

At pasture and when the animals move from and to the barn, one and the same cow is usually the leader cow. This cow is far from always the most dominant in the social hierarchy (Phillips, 1993). However, when the cow group moves, animals far down the social hierarchy are always among the last (Arave and Albright, 1981). When a cow group moves voluntarily, for example to the cow house or to another pasture, the animals usually go in some sort of near single file.

Knowledge of the memory function and memory capacity of cattle is limited. It is, however, assumed that cows are capable of remembering the appearance of 50–70 other individuals, which might explain why subgroups are created in large herds on pasture (Fraser and Broom, 1990). Cows seem to have the ability, by vision and probably also olfaction, to recognize their stockman. A cow recognizes her calf through olfactory, visual and auditory cues very soon after calving (Schloeth, 1958; Hafez, 1962).

Communication

Visual, but also acoustic and olfactory, signals are an important part of communication between cattle and thereby determine their social behaviour. If a cow is separated from the herd, has got outside the fence of the pasture for example, she will keep bellowing and often there are answers from her herd mates. Such acoustic communication can last for several hours.

Behavioural synchronization

Among domestic animals, cattle and sheep are the most pronounced group behaviour synchronized animals. Cows in a herd essentially perform resting

Fig. 4.9. During most of each grazing period, the majority or all of the cows are oriented in the same direction (image I. Ekesbo).

Fig. 4.10. During the resting period, it can be seen how single cows now and then rise, maybe defecate, possibly bite and swallow a small amount of grass and then lie down on their other side (image I. Ekesbo).

and grazing simultaneously (Figs 4.9 and 4.10). However, during the resting period, it can be seen how single cows now and then rise, maybe defecate, possibly rip some grass and then lie down on their other side. During the grazing period, the cows are mostly oriented in the same direction. Begall *et al.* (2008) demonstrate an explanation for this behaviour in their finding that domestic cattle

(8510 cattle in 308 pastures across the globe) align their body axes in roughly a north–south direction and that magnetic north is a better predictor than geographic north.

Dominance features, agonistic behaviour

The dominance features of cattle follow several patterns. A cow demonstrates dominance through lowering her head against another cow, or sometimes by just a throw of the head directed towards the other. Submission is demonstrated through bending the head sideways and, among cows kept loose, by moving away. Grooming behaviour can reveal a relationship of dominance. A higher-rank cow never licks a lower-rank cow, but low-rank cows often lick cows higher in rank. Cows of equal rank lick each other and often the same cow pair is seen licking. Age within the group gives, as a rule, a higher dominance than, for example, body weight and strength. Fights seldom occur among cows which have been kept together in the same group for a long time, as visual signals are sufficient in order to confirm the relationship of dominance between individuals. During the first hours after tied-up herds are let loose on pasture in spring, fights are normally seen, especially between young cows. Later, when the dominance order is established, these fights decline and finally cease. Such fights are not seen in loose-housed herds during their first day on pasture in the spring.

Aggression between individuals, however, occurs readily in narrow passages and other confined spaces, where the distance between individuals cannot be maintained and the social signals between them cannot be exchanged in a normal way.

Cows show aggression by lowering the head and sometimes pawing the ground. Bulls sometimes also scratch the ground with their horns. Before an attack, the bull usually bends the back vigorously. In bulls and cows, an attack occurs by a rapid rush with lowered head towards the adversary, trying to butt and gore. Attacks by bulls are performed with far greater force than attacks by cows.

4.4 Behaviour in Case of Danger or Peril

A cow, even a normally very friendly individual, with a newborn or a very young calf often may attack approaching people or animals. Therefore, caution should be used when approaching cows with newborn calves. Even the stockman may be attacked.

Just before the attack, the cow lowers its head and sometimes paws the ground with the front feet. An attack may take place with lightning rapidity by rushing at and trying to butt and gore the 'intruder'.

If a cow is attacked on the flank, by a dog for example, it tries to butt and gore sideways and at the same time it tries to kick the attacker with the hind legs. Cows often attack dogs even if they do not have calves at their sides.

When a herd of cows perceives a serious threat, such as from a dog, they often form a circle, with their hindquarters inwards and lowered heads directed outwards.

Swishing of the tail is seen when cattle are disturbed and seem to experience irritation, as when a dog is approaching or when a calf that the cow does not accept tries to suck. Tail swishing might be seen as a threat and is often followed by kicking.

4.5 Some Normal Physiological Frequency Values of Interest at Clinical Examination

The normal rectal temperature of adult animals is 38.5°C and up to 39°C in calves. The critical point in adult animals is > 39.0°C.

In adult animals, the pulse rate at rest is 60–65 per min; in calves it is higher, up to 100–120 per min. If a cow is anxious at the examination, the pulse rate will easily reach 80–85, even in a healthy animal (Andrews *et al.*, 2004).

In adult animals at rest, the respiratory rate is 10–12 breaths per min. In newborn calves, it is 30–35 breaths per min and in calves of about 3 weeks it is lower, at 20–22 breaths per min (Reece, 2004).

The number of rumen contractions is about 8–10 every 5 min.

4.6 Active and Resting Behaviour Patterns

Activity pattern, circadian rhythm

Cattle divide their time on pasture in principle between grazing, ruminating and resting. Their natural day and night rhythm is considered to include two grazing periods, one in the afternoon before sunset and one early in the morning. Between these periods, the animals ruminate and rest. Cows spend the greater proportion of this wakeful period in a state of drowsiness (Ruckebusch, 1972). In high-yielding cows, these two periods are not that marked as their day and night is divided into additional short grazing periods. Studies in New Zealand (Hancock, 1950, 1954) indicate that the animals graze up to 85% during the light part of the day and night and the remainder during the dark part of the day and night. Rumination is performed at standing or lying for about 5–8 h out of a 24 h period, in about eight bouts of about 45 min each (Fraser and Broom, 1990; Phillips, 2002; Broom and Fraser, 2007). Adult animals rest lying for 10.1–11.6 h of a 24 h cycle, calves a longer time (Arave and Albright, 1981; Krohn and Munksgaard, 1993; Albright and Arave, 1997; Phillips, 2002). Maximum grazing times occur for about 10–12 h. High-yielding cows can only achieve long grazing times at the expense of lying and ruminating times (Phillips, 2002). When not feeding, resting or ruminating, cattle exhibit other behaviour, grooming themselves or other cattle or browsing over the pasture. Young calves are playful, now and then running and prancing about, with the tail mostly held high.

Exploratory behaviour

Cattle, like most animals, behave cautiously to everything new and strange. Thus, they stop short when confronted with, for example, changes of ground in unfamiliar areas, but pass unhampered over new grounds or flooring in areas familiar to them. However, a piece of paper in an alleyway or a frog or a snake on a path in the pasture might make a cow refuse to proceed. In unfamiliar alleyways, cows often refuse to cross very light or very dark areas.

Like other prey animals, they may stampede readily if frightened by something new or unknown, e.g. a fluttering article of clothing.

Diet, food searching, eating and eating postures (feeding behaviour)

At just 2–3 weeks of age, the young calf will start eating hay or, if on pasture, pick some grass. Cattle kept on natural pastures with different types of vegetation show a notably varying feeding behaviour. They may, after having grazed an area of clover, change to an area with a quite different flora, and, whenever an opportunity arises, they combine their diet with sprigs from deciduous trees like aspen, oak or birch.

According to dietary profiles compiled from the world literature, the average cow's diet by botanical composition (over all seasons) consists of 72% grass, 15% forb (herbs) and 13% shrub (Alcock, 1992).

Cattle, by nature, have a need to search for feed and spend about 12 h actively looking for it. This behavioural need exists even if a cow has its nutritional requirements satisfied by a very concentrated feed in, for example, 3 h. Therefore, cattle kept indoors, or outdoors without access to pasture, must always have access to roughage.

Cattle drink by lowering the muzzle into the water, but keep their nostrils above the surface of the water and suck it into the mouth. Newborn calves suckle their dam 5–10 times per day (Edwards, 1982; Lidfors, 1994). They suck by squeezing the milk out of the teat cistern by means of the mouth and tongue and by creating a vacuum around the teat. By this mechanism, the teat cistern is filled and emptied about 75 times per min (Phillips, 2002).

It is generally difficult to get newborn calves to drink from a bucket. This can often be remedied by slipping two fingers in the calf's mouth and at the same time moving its muzzle, but not its nostrils, carefully under the surface. The fingers in the mouth trigger the sucking reflex. If they have to drink milk directly from a bucket, young calves usually lower their muzzle and also the nostrils under the surface of the liquid, causing them to snort now and then to get rid of the liquid in their nostrils. When a calf drinks directly from a bucket, its head position is different from when sucking the udder and this is considered to counteract an effective closing of the reticular groove. This might result in milk getting into the rumen. Eager drinking from the bucket might also result in some milk being aspirated into the respiratory tract. Newborn calves prefer sucking milk from a teat-feeder to drinking it directly from the bucket, despite the fact that, in the latter case, they can get a larger amount per time unit. Drinking from a bucket does not satisfy their sucking behaviour, which can result in them trying to suck on different objects, i.e. other calves or pen fittings in their surroundings. For these reasons, buckets with an artificial teat are preferable for young calves, as they thus obtain the liquid by sucking and not by drinking.

The calf requires colostrum from its mother or, if this is not possible, first-day colostrum from another cow, during the first 3 days of life. It requires milk, or a milk substitute, until at least 6 weeks of age. Thereafter, the milk can be replaced by water and special feed. Calves must have free access to high-class hay from the beginning of their second week of life. Water requirement varies with age and, for cows, milk yield (Murphy, 1992). Adult cattle require daily about 50 l water, dairy cows require 100–150 l. High temperatures increase the need for water. For a cow producing 35 l milk per day, the requirement is about 100 l of water during winter but 115 l during summer. A cow yielding 50 l per day needs to drink 150 l. If cattle have a choice, they prefer a drinking water temperature of over 15°C. The amount of water consumed per minute is higher when given from an open water surface than from a water bowl.

Cattle graze herbage by collecting it into the mouth and compressing it against the upper palate with the tongue and lower incisors. Cattle do not have incisors on the upper palate. Instead they have a hard, ridged dental pad that is used as a grinding surface during rumination. The herbage is then severed from the plants by jerking the head upwards. This is repeated many times a minute, typically 30–70, and the animal moves its head from side to side as it slowly walks forward. When cattle eat roughage, e.g. hay or silage, the tongue is used to a greater extent to manipulate particles into the buccal cavity (Phillips, 2002).

By rumination, the feed is exposed to more efficient microbiological digestion in the rumen. It is performed in bouts of about 45 min and thus accounts for a substantial part of the day, 6–8 h. It is characterized by a regular pattern of mastication, normally 60–70 jaw movements per min for adult cattle (Hancock, 1950, 1954; Hafez and Bouissou, 1975; Phillips, 2002). Cattle ruminate most often when lying down, but rumination also occurs in the standing position. The animal needs to be relaxed and calm for rumination to start. Rumination is associated with reduced alertness and seems to induce a soporific or even hypnotic effect in the animal. A cow that is disturbed stops rumination. This means that cattle are very seldom seen ruminating during stressful procedures such as transportation.

When grazing, cattle always have one foreleg before the other (Fig. 4.11). They cannot graze normally when both forelegs are together. Therefore, the surface level of mangers and feed bunks always must be elevated over the ground surface of the stall or the feeding area.

Cattle never graze on spots contaminated by manure, behaviour that reasonably, to a certain extent, implies protection against transmission of infection. Such uneaten patches are characterized by abundant green grass and remain untouched, even on hard grazed land. In the same way, cattle

Fig. 4.11. A grazing cow on an Irish pasture (image I. Ekesbo).

try to avoid areas top-dressed with liquid manure. On such pastures, cows are seen roaming about trying to find uncontaminated areas – this limits their feed intake. However, if cows are put on a pasture a few days after it has been top-dressed by urine, they do not hesitate to graze.

In adult cows, full permanent dentition is in place by 3.5–4 years of age (Sisson, 1975). In the calf, the temporary set of milk teeth is complete at 4 weeks of age.

Selection of lying areas, lying down and getting up behaviour, resting postures, sleep

Lying down on pasture (Figs 4.12–4.14) is initiated by the animal examining the ground briefly. Cattle avoid dirty and wet lying surfaces (Keys *et al.*, 1976).

Lying down begins by the cow lowering her head, then she places one hind leg well in front of the other and sinks down on first one, then the other fore knee (carpus), after which she lets her hindquarters sink to the ground on the side where the hind leg was put forward. In this position, the cow is still resting on her fore knees. Thereafter, she sinks down on the breastbone and the elbows, lets the hindquarters sink further down on one side, to sink down finally from the elbows and rest by dispersing the body weight on the lower hind leg,

thigh and trunk and by support from the bent forelegs.

Unlike horses, cattle cannot rest standing. Normal resting positions are: resting on the sternum with both forelegs bent, with the head forward or with the head fully supported on the ground or tucked round against the thorax; or the head forward and one foreleg stretched forward, which happens about 15% of the lying time (Fraser and Broom, 1990); or lying flat on the side, which often is seen in calves and for short periods also in adult animals (Fig. 4.15). When lying down, cattle tend to ruminate, idle (i.e. doing nothing that is obvious to an observer) or sleep (Rushen *et al.*, 2008). At complete rest, 'sleeping', the head is tucked round and held against the thorax, 'the milk fever position', which occurs for 6–8h of a 24h cycle (Fraser and Broom, 1990; Broom and Fraser, 2007).

Cattle show a strong motivation to rest in a lying position. They are inclined to invest a lot of effort in order to be able to lie down (Jensen *et al.*, 2005). Despite this, lying time will be reduced if the lying surface is dirty or damp (Keys *et al.*, 1976). If there is a choice between a damp and a dry lying surface, cattle usually choose the dry one.

When on pasture, no adult animals rest quite close to one another; each animal keeps a certain individual distance. Exceptions are calves and twin heifers. The latter often rest close together until

Fig. 4.12. Cattle lying down behaviour: the cow has lowered the head, put one hind leg in front of the other and is resting on her fore knees (image courtesy of photo archive HMH Department).

Fig. 4.13. Cattle lying down behaviour: the hindquarters of the cow are sinking to the ground (image courtesy of photo archive HMH Department).

Fig. 4.14. Cattle lying down behaviour: finally, the resting position (image courtesy of photo archive HMH Department).

they get their first calf, after which their social ties seem to be dissolved. Young calves, as a rule, often lie close together. Preference studies between different bedding materials in concrete stalls indoors indicate that cattle prefer straw before other bedding materials (Jensen *et al.*, 1988). Some diseases causing pain, e.g. traumatic peritonitis, lead to cattle being reluctant to lie down.

Fig. 4.15. Common cattle resting positions keeping individual distances. Cow A is lying on her left side with both front legs bent at the carpus. Cow B has the left foreleg stretched forward and the right bent at the carpus. Cow C is lying flat on the left side with the head bent slightly upwards. Cow D is resting on her left side with the head tucked round against the thorax, the 'sleeping' position (image courtesy of photo archive HMH Department).

Sleep in cattle has been classified in two groups: quiet, which lasts during 13% of the 24h cycle, and active, which lasts 3%. Cattle spend between 200 and 300 min in SWS sleep and between 20 and 45 min per day in REM sleep (Ruckebusch, 1972, 1974, 1975).

When getting up from lying on her side, the cow first throws the head sideways and at the same time, by a jerk, places herself so that she is lying on her breastbone and forelegs (Figs 4.16 and 4.17). When getting up from the breast lying position, she lifts and, in a fast movement, moves her head, neck and body forward in a slightly oblique angle directed from the side she is lying on so that she will then be resting on her fore knees (carpi). This movement is followed immediately by her rearing on the hind legs at the same time as stretching the head forward along the ground and is ended by her standing on first one, then both forelegs. Getting up is often followed by the cow bending her back upwards (and, if possible, often taking one or two steps forward) and usually defecating, and maybe urinating.

If the dairy cows' lying down and getting up behaviour cannot be performed in a normal way in the stall, there is an increased risk that udder and teat injuries may occur.

Locomotion (walking, running)

Cattle on pastures spend 30–110min in locomotion and, in addition, 500–600min in mobile grazing, which means about 10h of a 24-h day for each cow (Fraser and Broom, 1990).

The gaits of cattle are walking, trotting or galloping, depending on the circumstances. Normally, they move at a relatively slow walking pace. They trot only when they have to move fast, for example if the cow's calf is in danger. If another animal, such as a dog or even occasionally a stranger, arouses the herd's attention, they often trot towards the unknown animal or person. When there is panic, the herd most often moves at a trot, but this movement might change into a gallop. At gallop, the tail is stretched backwards and kept almost on the same level as the back.

Swimming

Cattle are natural swimmers. They often cross sounds or watercourses in order to reach other pastures or to make contact with conspecifics that they can see or hear.

Fig. 4.16. The cow has first got on to her knees, i.e. resting on her carpi, and is now rearing on the hind legs (image courtesy of photo archive HMH Department).

Fig. 4.17. Once the cow has got up, it is followed by the back being bent upwards (image courtesy of photo archive HMH Department).

4.7 Behaviour at Defecation and Urination (Eliminative Behaviour)

When defecating, cattle raise the tail and the back is arched slightly. Usually, defecation occurs when standing, but it may be performed while walking or, at least indoors, when lying down. Defecation occurs about 16 times in a 24-h day in cows, less in younger animals, and is not influenced by milk yield or lactation period (Fuller, 1928; Aland *et al.*, 2002). Cattle, if nervous or frightened, defecate more often and usually with more liquid faeces.

When urinating, females stop any activity, raise the tail, arch the back and usually splay the hind legs before the process starts. Cows urinate on average 9 times in a 24-h day (Hancock, 1953; Aland *et al.*, 2002). Cattle usually start urinating when standing, eating or grazing, seldom when resting. Urination cannot be initiated or performed if the animal is disturbed or frightened. Thus, it does not occur during transport if the procedure is not carried out in a very calm manner.

Unlike horses and swine, cattle defecate and urinate randomly over their enclosure. When cattle get up after a resting period, defecation and often also urination occur.

4.8 Body Care, Cleanliness

Body care, grooming, in cattle is performed in different ways and makes up about 5% of all behaviours (Figs 4.18–4.20). Cattle lick, and thereby clean, every part of their bodies that they can reach and this behaviour requires a significant freedom of movement. It has been estimated that calves groom themselves on 152 occasions and scratch 28 times a day (Schake and Riggs, 1966).

Cattle rinse the nostrils with the tongue, they lick the parts of the body they can reach with the head, not only forelegs but also the hind part of the back, and they use their hind claws for scratching their heads or parts of the body. They also use solid structures, poles, trees, etc., for body care, i.e. rubbing the horns and horn bases. In loose-housed herds, cattle often make use of fixed brushes, if provided, for body care.

Besides grooming themselves, cattle groom each other by one animal licking the neck or head region of the other. When one animal grooms another, it is commonly found that the one engaged in the cleaning is slightly below the other in the social order (Fraser and Broom, 1990).

A cow tie, a stall, cubicle or a calf box, therefore, must be so designed that a cow or a calf can lick itself on the spine or scratch its head with the hind claw.

Fig. 4.18. Cattle grooming behaviour: the cow scratches herself with great precision with the hind claw at the horn base (image courtesy of photo archive HMH Department).

Fig. 4.19. Cattle grooming behaviour: allogrooming, one cow is licking another, a component of social behaviour (image courtesy of photo archive HMH Department).

Fig. 4.20. Cattle grooming behaviour: the cow is licking her back while lying (image courtesy of photo archive HMH Department).

The importance for the animals to be able to carry out body care behaviours has not been investigated fully. There is, however, reason to assume that animals that are prevented from carrying out body care behaviours are likely to show signs of stress. Tied cows that are forced to stand under electrical so-called cow trainers are prevented from carrying out many body care behaviours.

Body care also involves the animals avoiding lying down on manure or on otherwise soiled or wet surfaces. Sick cattle often lose their ability to keep their body clean and, in loose-housed herds, such animals are sometimes even seen to lie down on dirty surfaces. Intense self-licking may occur in animals subject to acute restraint (Fraser and Broom, 1990).

4.9 Temperature Regulation and Climate Requirements

Temperature regulation

Cattle can usually adapt to most climate variations. They regulate their body temperature via the skin and through sweating (Andersson, 1977). Panting accounts for about 40% and sweating for about 60% of the relative proportion of total evaporative heat loss in cattle (Robertshaw, 2004). Cattle spend more time standing and less time lying down in warm weather (Tucker *et al.*, 2008)

As for many animals, heat stress is generally much more of a problem for cattle, especially lactating cows, than is cold stress. The upper critical temperature (UCT) is 25–26°C for high-producing dairy cows, irrespective of previous acclimatization or of their milk production (Berman *et al.*, 1985). At ambient temperatures above 26°C, the cow reaches a point where she can no longer cool herself adequately and enters heat stress. Because animals can adapt to hot environmental conditions by gradual acclimatization, it appears logical to assume that high-producing dairy cows acclimatize to gradual warming-up during normal summer months. However, if hot conditions are sudden and prolonged, as is often the case in Mediterranean areas and in some regions of the USA, cows are less likely to acclimatize. Significant decreases in feed intake and resultant drops in milk production punctuate such heat stress conditions. In 100 investigated counties in the USA, the average dairy cow experienced 47 days when the maximum temperature was 32.2°C or higher (Kadzere *et al.*, 2002). High-producing cows are affected more than low-producing cows because the zone of thermal neutrality shifts to lower temperatures as milk production, feed intake and metabolic heat production increase (Coppock *et al.*, 1981). The reproductive performance of lac-tating cows is reduced greatly during thermal stress, but non-lactating heifers generally show no seasonal trend in reproductive performance, even in the humid south-eastern USA. Both expression of oestrus and fertility are reduced in heat-stressed lactating cows (Collier *et al.*, 2006).

At temperatures under the thermoneutral zone, changes in the blood circulation are not enough for maintaining heat balance. In this situation, cattle must increase their metabolism. Thyroid hormones are released when cattle are exposed to cold – this increases the feed desire (Westra and Christopherson, 1976; Young, 1981). Increased feed intake or increased muscle activity, for example by shivering, are ways of maintaining heat balance. Shivering implies muscle shakings with a frequency of 10 per s (Andersson and Jonasson, 1993).

Cattle adapt to cold, and this adaptation depends on for how long the exposure to cold takes place. Adaptation does not happen after short exposure to cold or at intermittent exposure over short peri-ods (Kennedy *et al.*, 2005). Cattle must be exposed to cold for at least 1 week in order to begin the adaptation process (Christopherson and Young, 1986). Thus, steers kept outdoors but taken into a climate chamber on two occasions shivered at –20°C in September but not in December (Gonyou *et al.*, 1979). Coat thickness increases notably if animals have had the opportunity to adapt to cold over some time (Webster *et al.*, 1970). This adapta-tion of animal coat to annual changes of ambient temperatures occurs in other species, horses for example (Fraser, 1992), and is studied systemati-cally in reindeer. The insulation capacity of the hair coat of reindeer at Svalbard is thus three times more efficient in winter than during summer (Cuyler and Øritsland, 1999).

A wet coat increases heat loss (e.g. Hillman *et al.*, 1989; Jiang *et al.*, 2005). While heat loss from a dry coat is negligible, from a wet coat it can amount to 200–300 W/m² (Cena and Monteith, 1975). Newborn and very young calves succumb easily to rain and temperatures near 0°C if dry and wind-protected lying areas are not available, as such conditions are necessary for their heat bal-ance. Calves born in snow at zero temperature and not licked by the cow immediately and not able to suckle within a short time may die of cold, or even drown in melted snow. Therefore, it is important that the calf gets on its feet soon after the birth and that the mother does not remain lying, as a result of weakness, and thus be unable to lick it (Ekesbo, 2009). Humidity from mist or light rain does not affect coat insulation, whereas a more heavy rain impairs the insulation effect considerably (Cuyler and Øritsland, 2004).

There are differences in coat structure within species. Coat structure differs, for example, between the Aberdeen Angus and the Hereford breeds. The two coat structures have, however, the same insulation capacity (Gilbert and Bailey, 1991).

The growth of the coat seems also to be governed by light and it grows when the days become shorter. How long the winter coat is retained in spring depends on the temperature and the animal's heat production. Underfed and lean cattle keep their coat longer than well-fed ones (Christopherson and Young, 1986).

Calves are born with about 1° higher rectal temperature than adults, but their temperature falls during the first 6 h because of heat loss from the wet coat, and then it rises again. The newborn calf compensates for the heat loss via its hair coat and exhalation of air by increasing its heat production through metabolism of the body fat, but this does not compensate entirely for the heat loss. Hypothermia of newborn calves occurs within 6 h unless the calf quickly becomes dry (Thompson and Clough, 1970). If newborn calves are licked dry by their mothers and have been able to suckle shortly thereafter, they are notably resistant against cold. They can manage temperatures outdoors down to −20°C if two prerequisites are fulfilled: they must have efficient protection against adverse weather and they must have a dry lying area (Ekesbo, 1963; Radostits et al., 1999). The brown fat in the subcutaneous region between the scapulae and in the region of the kidneys, as well as that occurring within the myocardium in calves for a few days after birth, is an important reservoir for heat production without shivering (Robertshaw, 2004). Via stimulation of the sympathetic nervous system, the metabolism in the brown fat will be increased, and thereby also the heat production from this tissue (Andersson and Jonasson, 1993). Calves in calf huts (hutches) with dry bedding changed each week have been shown to manage temperatures between −8°C and −30°C if given several warm milk feeds a day (Rawson et al., 1988, 1989a,b).

The lower critical temperature (LCT) is the ambient temperature below which the basal metabolic rate becomes insufficient to balance heat loss, resulting in falling body temperature. Calculations

of LCT for cattle have been made (e.g. Webster, 1970, 1974; Young, 1981; Clark and McArthur, 1994); for adult beef cows −13°C in calm weather and −3°C at a wind speed of 4.5 m per s. The estimate of LCT for cows producing 30 kg of FCM per day is given as a range from −16°C to −37°C (Hamada, 1971). For a newborn calf weighing 35 kg and with a coat 1.2 cm thick, it is estimated at +9°C; at 1 month of age with a coat 1.4 cm thick, it is 0°C; and, for a dairy Friesian cow weighing 500 kg and at top lactation, it is −30°C (Webster, 1970, 1981; Clark and McArthur, 1994; Webster, 1987, cited by Clark and McArthur, 1994). For 2-day-old Ayrshire calves, it is specified at 13°C and at 8–10°C for 1–8-week-old cross-bred calves (Radostits et al., 1999).

Considerable hypothermia in calves delays the absorption of immunoglobulin from raw milk and thus keeps low the levels of antibodies against infections in the blood during the first days of life (Olson et al., 1980, 1981).

Local hypothermia followed by tissue death (gangrene) and affected parts falling off can occur on calves' ears, tails and feet when wet parts of the body are exposed to intense cold (Ekesbo, 1963; Radostits et al., 1999).

If dairy cows in intense cold get their teats wet, for example in the milking parlour in open loose-housing plants, their teat skin might get frost injuries (Ekesbo, 1963).

Climate requirements

B. indicus is more heat tolerant than B. taurus. There are also breed differences. Thus, Holstein cattle are less heat tolerant than Jersey cattle (Kadzere et al., 2002). The danger of heat stress is highest with a combination of high temperature and humidity (Hahn, 1999). Heatwaves occurring every 2–3 years in southern and central USA have been responsible for the deaths of thousands of feedlot cattle (Mader, 2003).

Cattle orientate their bodies to minimize adverse climate effects. In cold temperatures, they place themselves at right angles to the sun's rays and they are reluctant to lie down on wet grass (Phillips, 2002). Heavy rain stops them grazing. If it starts raining when cattle are resting on the pasture, they usually remain lying. If it starts raining shortly before a lying period is due, they avoid lying down. Instead, they place themselves in rain- and wind-protected positions. In cold wind and rain, they stand with their hindquarters to the wind, the thighs compressed against the udder, the hind legs set tight together, the tail held against the body and their heads, which are more thermally sensitive, if possible protected, for example by being pushed into some sort of shelter.

In windless and dry weather, cows manage temperatures down to –25°C provided that the teat skin is not wet (Fig. 4.21). For dairy cows in open loose housing this means that the udder must be dried carefully before they leave the milking parlour (Ekesbo, 1963). Calves kept at temperatures below zero must be protected from sucking by other calves in order not to get frost injuries (see Fig 4.24).

A prerequisite is that if the animals are forced to lie on wet surfaces, the LCT increases (Webster, 1974). Unexpected cold weather, such as early snowstorms, has been responsible for the deaths of tens of thousands of feedlot cattle in the USA (Mader, 2003).

Young cattle have to learn the importance of being in an optimum temperature (Beaver and Olson, 1997). They are more likely than older cattle to use areas where the ambient temperature is below the comfort zone. In cold weather, older cattle learn to use favourable microclimates and they move to exposed areas at times when ambient temperatures are increased (Phillips, 2002).

Tied dairy cows should have a temperature of 12–15°C in the barn. The optimum temperature indoors for calves is 12–15°C, provided that they have a dry-bedded resting area. If young calves have no option but to use a slatted floor as a lying area, the barn temperature should be at least 15°C. However, keeping calves on a slatted floor or other types of perforated non-bedded lying surfaces exposes them to health risks by unbalanced cooling through draught. Keeping calves indoors in uninsulated buildings might often be a health hazard because of high air moisture and draught. An alternative then is to keep the calves outdoors in small enclosures with insulated huts provided with bedding and, during the cold season, a suitable source of heat (Fig. 4.22). The relative humidity (RH) in barns should be 70% ± 10%, which is easy to obtain in a well-ventilated barn for tied animals. In insulated loose-housing barns, it is possible to attain this RH, but only with very efficient ventilation. In uninsulated, closed loose-housing buildings, this is not possible. In open loose-housing barns, the RH is usually no problem as it follows the climatic situation outdoors.

If calves younger than 3–4 months are exposed to draught, they are readily affected by respiratory diseases. Tied cows exposed to draught are easily affected by stiffness in their muscles, which might

Fig. 4.21. Dairy cows kept in open loose housing. The temperature was –28°C when the picture was taken (image I. Ekesbo).

Fig. 4.22. Calves kept outdoors in huts (image I. Ekesbo).

influence their getting up or lying down, thus increasing the risk of traumatic udder injuries.

4.10 Vision, Behaviour in Light and Darkness

Cattle get 50% of their total sensory information from vision (Phillips, 2002) and they have optimum visual acuity at a light intensity as low as 120 lux (Dannenmann *et al.*, 1985).

The location of their eyes on the lateral part of the head gives a visual field of about 330° to compare with *c*.180° in humans. This wide field of vision is an advantage for detecting predators quickly. The construction of the cattle eye makes it easier for them to notice moving compared to static objects. They have some ability to discern colours, i.e. light with long wavelength, such as orange, yellow and red (Jacobs *et al.*, 1998; Phillips and Lomas, 2001; Phillips, 2002).

Cattle have difficulty accommodating and performing estimations of distance. This might explain why, when excited or in a panic, they will often bolt away without attempting to avoid a rope, chain or the like, but will make their way round a clearer, more visible fence (Grandin, 2000). This might also explain why they sometimes react to an accentuated shadowed area as if it were a physical obstacle (SCAHAW, 2002).

4.11 Acoustic Communication

Hearing

Although hearing might be less important to cattle than vision, it is of importance in communication with conspecifics (Phillips, 2002). Cattle are sensitive to high-frequency sounds. While a human's auditory organ works at its best within the frequency range of 300–3000 Hz, cattle are most sensitive at 8000 Hz, but can catch sounds up to 21,000–22,000 Hz. Cattle have a broad hearing range, between 23 Hz and about 40,000 Hz (Algers *et al.*, 1978a,b; Heffner and Heffner, 1983, 1992a; Heffner and Masterton, 1990; Strain and Myers, 2004). They therefore can be frightened of sounds that are not caught by humans.

Cattle have a considerable low-frequency hearing ability. Their lowest hearing threshold is 11 dB (Heffner and Heffner, 1990).

Vocalization and acoustic communication

In addition to visual signals, cattle use sounds as marks of identification, e.g. between cow and calf, especially in the first weeks after birth. Unlike solitary animals, as herd creatures they use acoustic communication.

It is possible to discern a number of different sounds used by cattle. A low soft sound can be heard when the animals detect the stockman before milking or feeding or when a cow returns to the herd. A more intensive sound occurs when cows eagerly await milking or feeding, especially if they have been waiting longer than usual, or when a cow has become separated from the herd. An even more intensive sound is uttered if a cow is separated from her calf.

Calves can recognize their mothers using vocal cues, but it is not clear whether cows recognize their offspring in this way (Watts and Stookey, 2000).

Some predators and humans are able to locate a sound accurately within an angle of 1°, whereas cattle can locate a sound only within an angle of 30° (Heffner and Heffner, 1992b). One reason might be that cattle, as prey animals, only need an approximate idea of where the danger lies before fleeing (Heffner and Heffner, 1992a; Phillips, 2002).

4.12   Senses of Smell and Taste, Olfaction

The detection of odours is of major significance to cattle for intraspecies communication, synchronization of reproduction, acting as territorial markers and for signalling the presence of predators (Phillips, 2002).

Cattle can react to different scents and pheromones, from conspecifics as well as from other species, including humans. They have a well-developed olfactory sense formed by their ability to register odours not only in the nasal cavity but also through the cells in the vomeronasal organ, located in the roof of the mouth (Reece, 2004). Their method of using the vomeronasal organ can be seen in connection with oestrus as 'flehmen'. This is when the cow or the bull in a characteristic way registers the smell of heat in a cow through stretching the head and neck forward, bending the upper part of the muzzle backward, whereby the nostrils will be partially clogged and the animal inhales via the mouth. Cows in oestrus spend much time sniffing and licking the anovaginal areas of other cows.

It is well known among experienced farmers that cattle react positively or negatively to changes in hay or straw harvested under different conditions, obtained from different locations, or of a different sort. All this suggests that the animals register taste differences.

4.13   Tactile Sense, Sense of Feeling

Tactile touch

Most, but not all, cattle show positive reactions when stroked by someone to whom they are habituated. They seem to prefer to be stroked before being patted.

Sense of pain

Cattle probably have similar mechanisms for sensing pain to those of humans (Iggo, 1984; Livingstone et al., 1992). Vocal measures are less useful in detecting pain in cattle (Rushen et al., 2008).

4.14   Perception of Electric and Magnetic Fields

The level of current that will disturb cows is much less than for humans. They readily detect that low-level electric currents, in parlours for example, make them vulnerable to stray voltage. Most cows will alter their behaviour to a 3 mA current, corresponding to 0.7 V, whereas human resistance is two to ten times greater (Phillips, 2002).

In a survey, Hultgren (1990a) concludes that the results might indicate that dairy cows actually perceive short-duration and long-duration currents at the same levels, but do not exhibit an avoidance response to the short-duration current.

Stray voltage may have direct physiological effects in cows. An altered behaviour and physiological change due to stray voltage exposure may be termed a stress response. The type of stress most likely to be encountered in such cases is chronic (Hultgren, 1990b).

4.15   Heat and Mating Behaviour, Pregnancy

Full sexual maturity for heifers, i.e. that permitting normal delivery of live young 9 months later, is normally first reached at 16–18 months of age. In order to avoid increased risks for difficult calvings caused by the heifer not being full-grown, they should not be mated before that age. There are, however, examples of unintentional matings that have occurred even before 1 year of age. Such deliveries are mostly difficult and as a rule require veterinary assistance. Therefore, heifers and bulls must not be kept in the same enclosure, not even tied side by side, after the heifers are 7–8 months old. Bulls can often mate successfully at about 1 year of age.

Heat (oestrus) lasts 1–2 days and in non-pregnant animals it recurs every 21st day. Disquiet occurs usually in a herd when a cow comes in heat and some other cows often try to mount the oestrus animal. Bulls can register heat in cows days before humans notice the signs, e.g. heat discharge. After calving, the first heat occurs usually after 3–6 weeks.

Pregnancy lasts about 284 days, with a variation between 273 and 292 days (Blood and Studdert, 2000).

4.16   Before and After Parturition

Behaviour before and during parturition

Cows on pasture leave the herd before calving and try to find a place, e.g. shrubbery, a copse or high grasses, for the delivery (Hafez and Bouissou, 1975; Lidfors, 1994). As such a place is seldom to be found nowadays, at least not in pastures where dairy cows are kept, these cows just leave the herd and are seen lying alone at a distance from the rest of the animals.

During the beginning of the calving, the cow as a rule lies on the abdomen. The first stage – the opening stage – mostly lasts a couple of hours, sometimes longer, and normally lasts longer in heifers than in cows. During this stage, the amnion membrane slowly extends the mouth of the uterus. The next phase – the expulsion stage – usually takes an hour or so. The cow then lies either on the abdomen or, if the contraction pains are considerable, on her side.

At the beginning of the expulsion stage in a normal calving, the amnion bladder is seen first and, through its wall, if not burst, the hooves of the calf's forefeet. Usually, the amnion bursts and the hooves become fully visible. Then comes the most critical moment, when the head and the shoulders have to pass the vaginal orifice. During the expulsion stage, both uterus contractions and contraction of the diaphragm occur. When the head and then the calf's shoulders have passed the vaginal orifice, the calf slides out in an arched movement, with its belly against the pelvic floor of the cow, a process that as a rule does not take long provided that the calf is normally developed. It occasionally happens that the calf's hips are so big that they get stuck in the pelvic opening of the cow. This is often

the case with breeds bred for fast and sturdy muscle development, e.g. the Belgian Blue. However, when female cattle are bred at too low an age, that is before about 16–18 months, the risk of calving difficulties generally increases because the pelvic opening of the cow is not fully developed and is therefore too tight. The expulsion stage normally lasts 1–2 h, but can take longer. It sometime happens that calves are born backwards, i.e. with the hind legs first, a posterior presentation. At twin births, this is often the case.

A common misposition that requires professional assistance is when the head is thrown sideways; others are when one of the forelegs is bent downward and only one foreleg is in the pelvis of the cow, or when the tail is presented first with the hind legs pointing forwards, i.e. breech presentation.

Number of calves born

The birth of twins is not exceptional in cattle. Breed differences occur. According to Swedish investigations (Jordbruksverket, 2007), it occurs in about 3.4% and 4.5% of all Swedish Red (SRB) and Swedish Holstein (SLB) cow calvings, respectively. The frequency increases with increasing age and milk yield. The birth of triplets does occur but is unusual.

After the birth

After calving, the cow usually rises within a few minutes and promptly starts licking the calf dry, a procedure that takes about half an hour. This is an instinctive behaviour, genetically controlled and triggered by the stimulus of the newborn calf and its smell (Fraser and Broom, 1990). At the beginning of licking, the calf is passive (Fig. 4.23), but soon tries to get on its feet. During this process, the cow utters soft, sometimes intense, sounds. By tactile, olfactory, visual and acoustic signals, the bond between cow and calf is created and strengthened. Soon after calving, the cow as a rule eats the afterbirth – a behaviour probably developed in order to prevent predators locating the calf.

The calf is usually standing within half an hour of its birth and tries immediately to get to the cow's udder to suck. The cow stays with its calf until it is licked dry. It is practically impossible to dry a calf as effectively as is done by the cow's licking. This is one of several reasons why cows always should be given the opportunity to lick their calf dry.

Fig. 4.23. Cow and calf in the calving pen in a loose-housed herd. The newborn calf is licked by the mother, which is tired after a difficult delivery (image I. Ekesbo).

A newborn calf triggers this licking behaviour in other cows in the neighbourhood. This can cause another, usually higher ranking, cow to push the mother away from its calf if the mother and her newborn calf are near to the rest of the herd.

The calf lies hidden during the day close to the spot where it was born, while the cow follows the herd and now and then returns to the calf to let it suck. Calves thus are, like pigs, 'hiders', unlike foals, lambs and kids, which follow their mother immediately after the birth, 'followers'. This behaviour occurs not only in primitive breeds, such as Maremma (Vitale *et al.*, 1986), but also in common modern dairy and beef breeds.

Newborn calves suck 5–10 times per day and each period lasts up to 10 min. At 6 months of age, calves suck 3–6 times per day (Edwards, 1982; Lidfors, 1994).

Cows, and even more so heifers, are disturbed very easily just before and during calving. If people appear whom the cow does not know or trust, calving may be interrupted and the cow may walk away and deliver its calf in a different place.

After a difficult calving, cows may be very weak, might not be able to rise, some may be lying flat on their sides. If the calf then walks or is pushed to the cow's head and its coat, with its residue of amniotic fluid, comes into direct contact with the cow's nostrils and muzzle, the cow will then often show increasing signs of vigour and will start licking the calf.

Heifers sometimes do not take care of their calves and even attack them with their heads or horns. The reasons for this abnormal behaviour have not been investigated totally.

The calf does not always seek for the udder just before the hind leg of the cow but instead searches behind the front leg at the shoulder. If the calf does not succeed in finding its way to the udder, the cow often positions herself parallel to the calf, so that the calf's head comes into the vicinity of the udder and its buttock ends up at the cow's shoulder. When sucking, the calf presses more than sucks the milk from the teat by means of its mouth and tongue. It changes teats during sucking and now and then pushes its head against the udder. As the teats often are lower than the calf's head, it must lower its head and neck and at the same time turn the head upwards. This position is considered to contribute to closure of the reticular groove, whereby the milk passes directly to the abomasum.

If the cow has a big and low-hanging udder, this can create difficulties, sometimes making it impossible for the calf both to find the teats and to suck. Therefore, such cows must be husbanded carefully after the delivery so the calf can get its colostrum as soon as possible.

In many herds, the cow and calf are kept together in the calving pen for 2–4 days. However, calves are often taken from their mothers shortly after birth and put into an individual pen. Probably, the greatest impact of individual housing on the welfare of young calves is in reduced opportunities for exercise (Rushen *et al.*, 2008). On the other hand, the individual pen, if designed correctly, might reduce the risk of transmission of infection between calves.

Play behaviour

Young calves show play activities, such as prancing, kicking, pawing, snorting and head shaking. Sometimes, even adult animals perform playful behaviours, e.g. just after being brought out to pasture the first day of spring.

Weaning

Natural weaning in feral cattle is at a higher age than in beef calves from suckler cows, which usually are force weaned at 4–9 months of age (Reinhardt and Reinhardt, 1981). When cow and calf are kept together, as in beef herds, weaning usually occurs naturally when the calf is approximately 6 months old (Lidfors, 1994).

4.17   The Calf and the Young Cattle

In feral cattle herds, the mother–calf bond persists throughout the animals' lives, even when nursing becomes irregular, and does not wane as new offspring are born (Reinhardt and Reinhardt, 1981).

Female twins seem to keep less individual distance when resting at pasture than non-twins. This behaviour seems to remain until at least shortly before their second parturition (Ekesbo, 2003).

4.18   Assessment of Cattle Health and Welfare

In order to determine whether or not an animal is healthy, special attention should be paid to bodily condition, movements and posture, rumination, condition of hair, skin, eyes, ears, tail, legs and feet.

The healthy animal

Healthy cattle have sounds, activity, movements and posture appropriate to their age, sex, breed or physiological condition. These include: clean and shiny coat, clear bright eyes, moist cool muzzle, a sweet-smelling breath; alert and keen watchfulness of the environment; posture with straight, sound feet and legs, straight spine and elevated head; normal sucking or suckling, feeding, ruminating, drinking, defecation and urination behaviours; normal getting up, lying down and resting behaviour; and otherwise normal movements and behaviour. The appearance of the faeces varies according to the nature of the feed. Heat (oestrus) appears regularly and in a normal way.

When cattle walk, graze, feed or rest, the tail as a rule hangs straight downwards. When cattle are excited, running, show exploratory behaviour in connection with heat and after covering, the tail as a rule is kept slightly elevated.

The sick animal

Symptoms

Sick cattle always show discrepancies in some part of the body posture, locomotive behaviour or other deviations from healthy animals. The sick animal eats only slightly or stops eating, or performs feed intake in an abnormal way. The hair coat quickly becomes lustreless; if the animal is feverish, the hairs stand on end, which changes the animal's appearance. The tail in sick cattle is often pulled in against the body, which is also the case when the animal is cold, or when it shows submission to another animal. Unhealthy animals often show drowsiness or apathy. The eyes become lustreless or anxious; when in pain, a cow's eyes often have an expression of suffering and do not have the vivacity characteristic of a healthy cow's eyes. Sweating can occur. The temperature can be increased, but can also be lower than normal.

With certain illnesses, the cow stands with curved spine; with others she might try, without success, to urinate. Together with these general symptoms, other local symptoms might occur, e.g. increased salivation, discharge from eyes, nostrils or vagina, swellings and perhaps soreness in the udder. Pain in the mouth can manifest itself as difficulties in chewing or ruminating; other diseases can cause rumination to cease totally. With pain in the feet, the cow tries to relieve the pressure on the sore foot.

Sick animals as a rule lose the inclination to carry out body care. Loss of grooming behaviour is thus a sign of disease. Cows stricken with certain illnesses often do not flinch from lying down on manure-dirty surfaces. This can be seen in loose-housed heifers, in bad shape after calving and hit by diseases like ketosis or endometritis.

Examples of abnormal behaviour and stereotypies

Cattle exposed to stress, kept without stimuli in barren environments for example, maybe with insufficient access to roughage during the day and night, as a rule perform different forms of stereotypies. Deprivation of resting behaviour results in an increase of licking and chewing at the pen fixtures and an increased frequency of transition between behaviours (Munksgaard and Simonsen, 1996; Munksgaard *et al.*, 1999). Examples of stereotypies are abnormal sucking in calves, abnormal licking in calves and adult animals, or biting or so-called tongue rolling in adult animals.

SUCKING   As cattle have a natural weaning age of about 6–7 months, it is not surprising that when young calves are held together in groups, mutual sucking can become a problem (Fig. 4.24). As a rule, they direct the sucking towards the ears, the navel, the prepuce and the udder, which can result in considerable irritations on the skin. By offering the calves concentrates or roughage directly after the milk or liquid milk replacement ration, this behaviour can often be reduced.

Sucking directed towards the udder in calves or young heifers might be one of several causes of the mastitis cases that are discovered at first calvings and which, in some herds, are a big problem. Sometimes, cows try to suck the teats of other cows, and occasionally also their own teats.

This abnormal sucking behaviour has been observed between cows in 29%, between heifers in 60% and between calves in 60% of 230 herds (Lidfors and Isberg, 2003).

TONGUE ROLLING   At tongue rolling, the animal lifts its head, pokes out and curves the tongue. It can appear when the animal is under stress, e.g. when understimulated (barren environments, basic needs not satisfied) or overstimulated (such as by constant noise) or when other cows take over their newborn calf (Lidfors, 1994).

REPEATED LICKING OR BITING   This behaviour (Figs 4.25 and 4.26) is also seen when cattle are exposed to under- or overstimulation, or a combination of both. Examples of understimulation are short eating periods without anything available

Fig. 4.24. During its first 8 weeks, this calf was kept in a single box in an uninsulated building during the winter. A calf in an adjacent pen sucked its right ear, which became frostbitten. Necrosis arose and one-third of the ear dropped off (image I. Ekesbo).

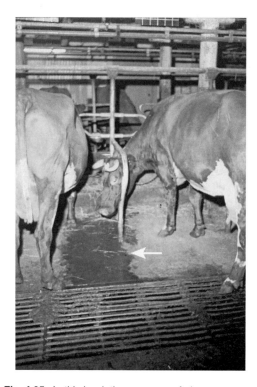

Fig. 4.25. In this herd, the cows were fed a concentrated feed with very little fibre. Total eating time was thereby very short. No straw was used for bedding. Several cows showed stereotyped behaviour. The cow to the right constantly licked in the water bowl. The arrow points out the spilt water (image I. Ekesbo).

for eating, e.g. roughage or clean straw bedding, during most of the non-resting time. An example of overstimulation is constant noise, e.g. from ventilation fans. Licking can be directed towards fittings, but also towards the cow's own body or towards other animals' tails, scrotum or prepuce. Biting is directed towards fittings, like pipes or wooden parts. Intense licking may also appear as a symptom in connection with specific diseases, e.g. ketosis. When the primary disease is remedied, this behaviour ceases.

Examples of injuries and diseases caused by factors in housing and management, including breeding

Young calves kept together are at risk of contracting contagious diseases. Single boxes with compact walls may prevent infection through direct contact with other calves. If prevented from maintaining

the body temperature because of cold or wet lying surfaces or exposure to draught (Fig. 4.27), they may contract respiratory and/or alimentary disorders. In large herds, there are generally increased risks of contagious respiratory diseases among calves.

In herds with respiratory problems, the calves sometimes are kept outdoors in insulated huts with heating devices (see Fig. 4.22). The huts are placed at some distance from each other to control infection via direct contact and are cleaned and disinfected before a new calf is brought in. After having used the huts during their 5–8 weeks of age, the calves are often kept on straw-bedded areas in open sheds. This calf husbandry seems to reduce the risk of respiratory disorders.

Dairy cows that have to lie on concrete or similar hard areas run increased risk of trampling their teats compared to cows where the lying area is supplied with enough soft bedding. If the lying area is slippery, there is increased risk of similar injuries occurring when the cow is lying down or getting up. Such injuries as a rule are followed by mastitis (Figs 4.28 and 4.29), the most costly of all dairy cow diseases, except epizootic diseases. Prevention of teat injuries is thus a very efficient mastitis prevention (Ekesbo, 1966; Bendixen *et al.*, 1986, 1988, 1989).

Unlike other animals, for example horses or sheep, cows might not always avoid consuming foreign bodies, such as nails, pieces of wire, etc., when grazing or feeding from the manger. Such objects, if eaten and swallowed, usually get stuck in the reticulum, sometimes also in the rumen. They can penetrate the wall of the rumen or reticulum, the diaphragm, enter the pleura or the pericardium and cause life-threatening injuries and secondary infections.

The increase in milk production in dairy cows has been followed by an increase of the incidence of several diseases. Thus, in Sweden, the reported incidence of parturient paresis rose from 0.5% in 1940 to 9% in 1980. Several studies indicate an association between incidences of, for example, mastitis and parturient paresis and high milk yield (e.g. Bendixen *et al.*, 1987, 1988). By using the veterinary practitioners' disease recording, compulsory in Sweden but unique internationally, geneticists and cattle breeders in Sweden have succeeded in reducing the hereditary disposition to these diseases, as well as to some others, e.g. dystocia. Thus, the reported incidence of parturient

Fig. 4.26. This young stock, kept on a slatted floor without any straw bedding, were given a limited amount of concentrates and short chopped silage, no other roughage. The total eating time was very short. Several animals performed tongue rolling and, as the near animal shows, abnormal biting (image courtesy of Bengt Vilson).

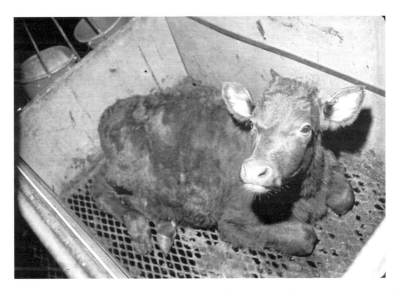

Fig. 4.27. Calves lying directly on a perforated or slatted floor without dry bedding are at very great risk of respiratory and alimentary diseases. Note the diarrhoea held in the perforated floor (image I. Ekesbo).

paresis had diminished to 7% in 1994 and to 2.9% in 2008 for the Swedish Red breed, and to 3.6% for Holsteins.

Dystocia is common in some beef cattle breeds developed genetically for extreme meatiness, e.g. the Belgian Blue. The difficult parturition in this breed has led to Caesarean sections becoming very common in order to save mother and calf. From an ethical point of view, this is a serious development. However, if genetic measures are taken, improvements are possible. Thus, in Sweden, high levels of dystocia in the Charolais breed (Lindhé, 1968)

Fig. 4.28. Cow with acute mastitis in the right hindquarters after teat trampling. Note the teat pin applied to stop the teat canal being blocked by oedema (image I. Ekesbo).

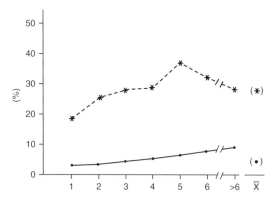

Fig. 4.29. Mastitis incidences within number of lactations in dairy cows before the mastitis was (*), was not (•), respectively, reported for teat trampling. $\overline{X}$ represents the mean value (Ekesbo, 1991).

initiated a selection for easier calvings and the incidences have decreased gradually towards the end of the 20th century, to 2.6% in 2000 and to 1.5% in 2004.

Symptoms of pain

The overt and visible responses to pain are not as great in cattle as in humans because cattle have evolved as prey animals and humans as predators. It is disadvantageous for a prey animal to indicate that it is in pain, as it may be singled out for attack (Phillips, 2002). However, in some diseases, e.g. traumatic peritonitis, cattle show very significant

signs of suffering pain. Yet there are other signs which indicate pain or disease to experienced veterinary practitioners. Some of these are: dull and depressed appearance, inappetence, groaning, bruxism, restlessness, reluctance to move, limping, stamping, kicking, rolling, licking or biting at the site of damage, intense licking of objects, tachycardia and tachypnoea.

Cattle also manifest significant avoidance behaviour responses against local intrusions, e.g. pinpricks in some parts of the skin, especially on the legs or feet, or if a teat is squeezed.

4.19　Capturing, Fixation, Handling

Capturing loose cattle can be either easy or difficult, depending on whether or not the animal is shy of people. A shy animal should be pushed carefully towards an appropriate enclosure, or, if indoors, a pen or similar, whereupon a halter can be put on it. It is important to act very carefully in order to avoid upsetting the animal. If upset, it may be seized with panic, whereupon it often runs straight towards humans or through fences.

When the head of a cow has to be fixed, the right arm should be put around the cow's head just in front of the horns, above the eyes on polled cattle, and then one should clasp the cow's head against oneself (Fig. 4.30). Then, the nasal septum should be clasped with the thumb and fingers of the left hand.

Cattle which have always been kept loose and never been tied behave totally differently from an animal that is accustomed to being tied. If a halter is put on the latter, it can, as a rule, be led easily by a halter-strap, something which might be totally impossible with an animal that never has been tied. When leading an unknown cow, and always when leading bulls, one must walk beside the animal, abreast of the shoulder, in order to avoid attack by its head or horns. Unknown bulls and also often bulls that are considered as calm must, besides having a halter and halter-strap, be led with a rod fixed to the nose ring, thus making it possible to keep the bull away from oneself should it suddenly attack.

Beef cattle or young stock kept on pasture are often difficult to separate from the group for examination or treatment. For such purposes, fenced enclosures, alleys with separation gates and treatment stalls have to be used.

Fig. 4.30. Restraint of a young bull. This man has, with his back against the shoulder of the bull, put his arms behind instead of in front of the horns and has not yet clutched hold of the nasal septum. For safety reasons, the arms should always be put in front of the horns (image I. Ekesbo).

4.20 The Importance of the Stockman

Human behaviour may have played a crucial role in the domestication process. Animal husbandry information literature from older times often emphasized the importance of behaving gently towards cattle. Thus, the Swede Hjortberg (1776) argued that the cow housemaid should present 'a cheerful temperament and sing for the animals'. Many veterinarians have experienced that women are superior to men when it comes to taking care of cattle, especially calves. Herd owners with high calf morbidity have thus, instead of medical treatment by their veterinarians, sometimes been 'prescribed' a change from a male to a female caretaker, advice which after some time has often resulted in improved health. There are also scientific studies indicating that dairy herds in which females have the primary responsibility for the care and feeding of calves have lower calf mortality (e.g. Losinger and Heinrichs, 1997).

In studies of the effect of human behaviour on animals, there are indications that cattle with little fear of people show health advantages compared to cattle that show a strong fear of humans (e.g. Hemsworth *et al.*, 1995). There are also indications that pleasant behaviour, a 'nice social intercourse', with cows increases their milk yield (Seabrook and Bartle, 1992). Fear of people can be a major source of low production and conception rates (Rushen *et al.*, 1999; Breuer *et al.*, 2000; Hemsworth *et al.*, 2000; Waiblinger *et al.*, 2002, 2006; Hemsworth, 2003; de Passillé and Rushen, 2005).

Cattle can distinguish between people by the use of vision, smell and possibly hearing. They also seem to differentiate between people, depending on how they have been treated by them (Rushen *et al.*, 2001, 2008).

Studies in ranch herds show that management at regular times and the offering of small quantities of appetizing feed increase the animals' tendency to overcome their shyness. In larger groups, it takes a longer time before the animals approach the stockman (Lundberg *et al.*, 2006).

5 Sheep (*Ovis aries*)

5.1 Domestication, Changes in the Animals, their Environment and Management

Sheep are considered to have been domesticated early, i.e. more than 8000 years ago (Zeuner, 1963; Mason, 1984), but may have been used by humans as early as 9000 or even 12,000 years ago (Lynch *et al.*, 1992; Clutton-Brock, 1999). Thus, sheep were one of the first species to be domesticated. There have been different opinions among scientists regarding when and from which wild ancestor sheep were first domesticated. However, it is now believed that all domesticated sheep were derived from the Asiatic mouflon, *Ovis orientalis* (Clutton-Brook, 1999). Their wild ancestors should have been living in dry and mountainous regions in south-west and central Asia (Rutter, 2002). Most of the features of domestication in sheep, these being alteration of horn shape (with hornlessness in ewes), fattiness and length of the tail and the woolly, white fleece, were already common in western Asia by 3000 BC – all these characters are shown in pictorial representation in Mesopotamia and are also written about in the Babylonian texts (Clutton-Brock, 1999).

In the wild, sheep are grazing animals in areas with relatively hot and dry climates and dry and firm ground, for which their feet, anatomy and physiology in general are adapted. However, because of the differences between breeds, for example fleece characteristics, domestic sheep will thrive in a wide range of climate conditions. In the wild, they do not occur in damp or marshy areas, or in rocky country, like goats, probably because usually they flee predators together in a close-knit flock, which cannot be done on rocky terrain (Lynch *et al.*, 1992).

The reason sheep were selected by humans for domestication, instead of deer for example, might be because sheep had a social system that was based on following a single dominant leader, e.g. the herdsman (Geist, 1971).

The breeds of domestic sheep are more diverse in anatomical appearance than the wild species, a result of human selection for traits of interest. However, more complex traits, such as behavioural characteristics, are, on the whole, unchanged. The management of sheep has not undergone the same dramatic changes as that of swine, poultry and cattle.

There are in excess of 1000 breeds of sheep living in the world and about 1.1 billion animals. They are kept by humans in different feed and climate environments, in the barren Orkney Islands, in the summer warmth and winter cold of British Columbia, with a great variety of feed, or in the warm climate of Arizona in the USA, or in Australia, where there is little variety to their diet. The largest sheep populations are found, in the early 2000s, in China and Australia (Lynch *et al.*, 1992; Dwyer, 2009).

The male sheep is described as a 'ram', the female as a 'ewe' and the offspring as 'lambs'. The expression 'lamb', however, is often used until the animal is about 1 year old.

5.2 Innate or Learned Behaviour

Innate behaviours are, for example, the ewe leaving the flock at the time of delivery, licking of the lamb after the birth, the lamb following the mother and suckling of the lamb.

Learning in sheep is shown in their ability to learn to recognize conspecifics. It has been shown that sheep can discriminate between 25 pairs of other sheep and can remember them for more than 18 months (Broad *et al.*, 2002). They can learn to discriminate between humans, and thus to recognize their shepherd. They can recognize the faces of other sheep and differentiate familiar from unfamiliar sheep (Houpt, 2004).

By teaching her lambs to respond to the raised head posture when young, the ewe passes on to the lamb the need to pay attention to this important signal. The signal tells the lamb that it may approach and suck, but it is also an alarm posture used by adult sheep to communicate potential danger to the rest of the flock (Dwyer, 2009).

The fact that individual sheep avoid more dominant group members is strong evidence that they can recognize other individuals within their group (Rutter, 2002).

Young lambs that see their mothers eating grain have a lifetime memory for the feed and have been shown to eat it readily some 3 years later (Lynch *et al.*, 1992). They do not accept unknown roughage until after more than a week. However, if that roughage is given an odour or flavour well known to them, they will accept it (Van Tien *et al.*, 1999).

Direct eye contact between a human or a dog and a sheep is a well-known ploy used to control the behaviour of a sheep (Phillips and Piggins, 1992).

5.3 Different Types of Social Behaviour

Social behaviour

In the wild, and except for the breeding season, sheep live in small family groups of females and up to 2-year-old males, and in separate groups of males. Each group has its own home range within the overall sheep flock home range. This is passed from generation to generation, with the lambs of a ewe occupying the same home range as their mother. Sheep have a social system that is based on a single dominant leader. They do not defend the territory that is their home range, as do some other ungulates (Geist, 1971; Lynch *et al.*, 1992; Clutton-Brock, 1999; Rutter, 2002).

A flock of sheep is held together by visual and, to some extent, acoustic contact between individuals. They show a pronounced flock and follow behaviour when moving from one place to another. Normally, sheep exhibit a slow and relaxed behaviour.

Sheep show a strong desire to remain with their group mates and become very vocal when separated from the flock. They use threat behaviour to help minimize fighting between individuals. This helps to lessen the risk of injury associated with fighting. Threat behaviour consists of a lowered head and a stretched neck. If the threat fails to deter a potential rival, the males engage in rearing and butting to establish their dominance within the group (Rutter, 2002).

Flock affiliation is so strong that an animal separated from the flock is almost impossible to control as, in these situations, it is usually panic-stricken. When a sheep at slaughter shows great anxiety, this is considered not to depend on death fear but on the fear that has arisen from being separated from the flock. Studies of factors that have a negative influence on sheep indicate that isolation from the flock gives more strong stress symptoms than, say, transport (Baldock and Sibly, 1990; Douglas-Hudson and Waran, 1993).

Flock affiliation is stronger in sheep than in cattle and manifests itself not only on pasture but also indoors. Fine-wool breeds, such as Merino, flock closer together than Northern European hill breeds, such as Scottish blackface (CoE, 1992).

The individual animal shows a highly social behaviour and often keeps itself near other sheep in the flock. An individual animal reacts if for some reason it is apparent it will be separated from the flock. If a sheep is separated from the flock, it will show all the signs of great anxiety and restlessness, and bleats uninterruptedly. If it can see the flock, it runs towards it, even if there is a threat, a dog for example, between the flock and the sheep.

A sheep is considered able to identify all the other individuals in the flock (Hall *et al.*, 1998). Studies of unknown animals confronted with each other show that this will trigger measurable stress (Hall *et al.*, 1998).

Sheep within groups maintain a certain distance from their nearest neighbour when grazing. This nearest neighbour distance tends to be a characteristic of the breed, with hill sheep breeds usually found further apart than lowland breeds. Nearest neighbour distance also decreases as vegetation quality and homogeneity increase. In Scottish hill sheep, each of the clearly defined cohesive groups will monopolize the use of an area of hill and will avoid close social contact with sheep from other groups (Rutter, 2002).

Sheep appear to pay little attention to a sick or disabled member of the group, except for butting it (Geist, 1971).

The social behaviour of sheep is well known and has been described for thousands of years, which is evident from the Bible. Shepherds in sheep husbandry have used the 'follow' behaviour from time immemorial. In the Bible, in St John 10:4, there is a good example of this: 'And when he [the shepherd] putteth forth his own sheep, he goeth before them, and the sheep follow him: for they know his voice.'

The follow behaviour is sometimes made use of in slaughterhouses, where a special ewe (the Judas sheep) is used as the leader of the slaughter sheep. In front of the leader ewe walks her well-known human handler. The sheep that are to be moved into the slaughterhouse or the stunning pen are headed by this leading ewe. Thereby, the sheep flock can be moved into an unknown environment in a calm and unexciting manner.

Communication

If a sheep is separated from the flock, it will keep bleating without interruption. If bleating is heard by the flock, the sheep in the flock usually respond with similar bleating. Such acoustic communication can last for several hours.

Lambs can find their dams when they are out of sight, by identifying their voices. A ewe calling her lamb orients towards the lamb and looks directly at it (Geist, 1971). This ability improves with age, and seems to vary between breeds. A lamb answers bleats more quickly when they are made by its own dam. When a ewe hears her lamb bleat, she may help her lamb to locate her by bleating promptly in reply. Ewes are inclined to bleat when they hear a lamb bleating, but become more specific in replying as their own lambs get older. As a result, lambs of 6 weeks can usually find their ewes very easily (Broom and Fraser, 2007).

Behavioural synchronization

When a flock of sheep moves quietly from one place to another, the flock usually walks in a row and, in conformity with deer for example, follows well-worn paths. When the flock sets itself in motion, one or a few individuals initiate the movement, whereupon all the others follow. If the leading ewe stops, the others stop. It seems, at times, as if there is not just one leading animal but also several potential leaders. Instead, some of the older ewes, sometimes one, sometimes another, function as the leading animal. It seems that, if one older ewe starts moving in a certain direction, the whole flock, including all the other older ewes, will follow.

However, a flock of sheep does not show the high behavioural synchronization at grazing as cattle (Fig. 5.1). Individual sheep graze in different directions,

Fig. 5.1. A sheep flock on pasture in early spring in northern Germany. Notice the shepherd in front of the trees (image I. Ekesbo).

not practically all in the same direction, like cattle. Also, sheep do not commence or cease grazing at the same time, like cattle. In a large flock, the time from the first to the last sheep to commence grazing can vary by up to 45 min (Lynch *et al.*, 1992).

Dominance features, agonistic behaviour

Dominance between individuals and by social status is settled by physical contact in the form of nudging and butting by the head. This is common between rams in the mating season. The ram moves back in order to take a run for a fast and hard attack with his head against the opponent.

Mutual grooming as a way to indicate domination and subdomination, as in cattle, does not occur in sheep.

During encounters, sheep adopt certain positions against conspecifics. A direct stare of one sheep at another appears to be an aggressive posture. Subordinates look away from dominants when the latter approach. Sheep tend to rest in such a manner as not to face each other directly. It appears that, at short distances, only dominant sheep are free to look in all directions.

When sheep are close together, they tend to face in the same direction, because this minimizes staring and is hence the most peaceful group structure (Geist, 1971).

A very common threat shown by sheep is by very resolute and quick lifting of one of the forelegs and immediately thereafter hitting the ground with it. Threats between individuals are shown by indicating increased size by raising the head and neck. Submission is indicated by lowering and twisting the head sideways, or by departure from the spot. Antagonists approach each other and butt head on. The clash can be preceded by the individuals, usually rams, jumping just before starting a sudden attack, and running forehead against forehead or horns against horns.

5.4 Behaviour in Case of Danger or Peril

Like other animals, sheep are disturbed easily, particularly in an unknown environment. Indoors and in limited enclosures, like yards, they can be frightened easily by sudden shadows, reflections and unknown, sudden and unexpected sounds. If a sheep raises its head and holds it up rigidly while walking with tense steps, it at once draws the attention of all the others. This is the alarm posture (Geist, 1971).

Sheep show a very ready disposition to flee from danger. When danger threatens a flock of sheep, they will usually stand still for a short moment with their heads directed towards the direction from where the threat is coming. Thereafter, each individual walks stiffly and with lifted head until the whole flock, tightly together, flee in the same direction. Like cattle and horses, they can fall into a wild flight when frightened by something new or unknown, e.g. fluttering paper or clothing.

Sheep show a substantial antipredator response to humans, dogs, foxes, etc., indicating that they regard such species as dangerous. Exceptions are people they know and have confidence in. There are several examples of this in the Bible; for example, the well-known passage in St John 10:14, 'I am the good shepherd, and know my sheep, and am known by mine.'

Comparisons of sheep behaviour towards humans, goats and dogs indicate that they usually just sniff goats but show a fear-related behaviour against humans and dogs – the strongest reaction being towards dogs (Beausoleil *et al.*, 2005).

5.5 Some Normal Physiological Frequency Values of Interest at Clinical Examination

Normal rectal temperature in adult sheep at rest is 39°C, and is up to 39.5°C in lambs. The critical point for adult animals is > 40.0°C (Robertshaw, 2004).

In adult animals at rest, pulse frequency is 70–80 beats per min (Erickson and Detweiler, 2004), and is about 100 beats per min in lambs.

At rest, the adult animal respiration frequency is 19–25 breaths per min. It varies with breed, thickness of fur and environmental temperature (Reece, 2004).

The number of rumen contractions is about 8–12 every 5 min.

5.6 Active and Resting Behaviour Patterns

Activity pattern, circadian rhythm

Like cattle, sheep divide their time on pasture largely between grazing, ruminating and resting. Sheep graze 9–11 h a day and night, in shorter periods than cattle. Grazing periods are interrupted by ruminating and rest. Rumination occupies about 8 h per day. Sheep sleep for short bouts during resting periods (Lynch *et al.*, 1992).

Play behaviour is seen in lambs from a few days old to 10 weeks of age (Lynch *et al.*, 1992).

Exploratory behaviour

Like most animals, sheep behave warily and carefully in the presence of anything strange or unknown to them. In unfamiliar environments, they stop short before changes of flooring, but walk unhampered over different flooring in an environment familiar to them. However, a piece of paper in an alley or a snake or a frog on a path can cause a sheep to refuse to walk further. In unknown drivealleys, sheep often refuse to cross areas of light illumination as well as dark, shaded passages.

Diet, food searching, eating and eating postures (feeding behaviour)

Even during their first week of life, lambs nibble on vegetation and this activity becomes more common with age (Geist, 1971).

Like cattle, sheep avoid grazing on areas contaminated by manure. They are very selective grazers, choosing not only specific plants but also preferring leaves and blades over stems (Geist, 1971). Their divided upper lip, philtrum, enables them to pick small plant parts easily.

Dietary profiles compiled from the world literature indicate that the average sheep dietary botanical composition for all seasons consists of 50% grass, 30% forb (herbs) and 20% shrub (Alcock, 1992).

Grasses are grasped between the lower teeth and the dental pad, and then torn when the head is moved posteriorly with a sudden jerking movement. The head may swing laterally and more food is seized while the fore or hind leg takes one step forward. When eating shrubs, the sheep can strip the branch of leaves, break the twig and chew it, or pick off discrete leaves (Lynch *et al.*, 1992).

Sheep, like cattle, by nature have a need to search for feed and can spend up to 12 h actively searching. However, they devote about 8 h per day in efficient grazing (Rutter, 2002). This food search behaviour is still shown even when the sheep may have satisfied their nutritional requirements within 3 h in the form of a concentrated ration of feed. In order to fulfil this behavioural need, sheep, when kept indoors, should therefore always have access to roughage, preferably straw or hay.

When grazing, an individual distance is maintained between each sheep. The average distance between neighbouring sheep when grazing varies from 4 to more than 19 m; the greatest distances are for hill breeds of sheep and the smallest for Merinos. The average distance of the nearest neighbours among sheep in all breeds is within 5 m, but breeds differ on this basis (Broom and Fraser, 2007).

Ruminating occurs for about 8 h of a day and night. Rumination occurs in both the lying and standing positions, but requires calm and peace and that the sheep is not disturbed in any way. Sheep are not able to ruminate during transport (Austin, 1996).

Water requirement varies with age and milk yield. Sheep on pasture may need to drink more than once a day, but they can still go several days without water (Lynch *et al.*, 1992). On the other hand, adult ewes given dry feed may require up to 20 l water per day, especially ewes that suckle (CoE, 1992).

The sheep's dentition is complete with permanent teeth at 3–4 years of age (Sisson, 1975). In the lamb, the set of milk teeth is complete at 5 weeks of age.

Selection of lying areas, lying down and getting up behaviour, resting postures, sleep

Sheep on pasture choose, if possible, night rest in the highest area available to them. During resting, the flock lies close together. A sheep initiates lying down behaviour by scraping the ground with the forefeet in a similar way to cattle, but often with more pronounced sniffing and examining the lying area before lying down. Lying down and getting up behaviour in sheep does not deviate from that of cattle, but the movements are much faster. They stretch after resting. Unlike cattle, sheep yawn (Geist, 1971).

A sheep's resting positions are similar to those of cattle. During part of resting, they often have one of the forelegs stretched forwards. During warm weather, sheep often rest with legs extended, but during cold weather they draw their legs against the body (Geist, 1971). Adult sheep are seldom seen lying flat on their side. When resting, social distance is reduced greatly.

Sheep are awake for about 16 h per day; SWS sleep occupies 3.5 h per day and REM sleep occurs in seven periods of an average total of 43 min (Broom and Fraser, 2007).

Locomotion (walking, running)

The gaits of sheep are walking, trotting or, occasionally, galloping, depending on the circumstances. Like cattle, but to a higher degree, they use certain paths when they move between separate grazing areas.

During normal walking, the sheep's head is held quite low, nose pointing to the ground, while the ears are held back and drop down a little (Geist, 1971).

Swimming

Sheep are natural swimmers, although they do not go into the water voluntarily, like cattle sometimes do.

5.7   Behaviour at Defecation and Urination (Eliminative Behaviour)

Unlike horses and swine, sheep defecate and urinate randomly over their enclosure. At defecation, sheep raise the tail. They defecate usually after getting up, but also while standing, walking and feeding. On urination, the ewe always stands still, raises the tail, arches the back and usually splays the hind legs before the process starts. Rams urinate from a nearly normal standing position, except that the hind legs tend to be a little further back and the back is slightly depressed.

Urination cannot be initiated or performed if the animal is disturbed or frightened. Thus, it does not occur during transport, unless the transportation is carried out in a very calm manner.

5.8   Body Care, Cleanliness

Sheep perform body care by rubbing their body against trees, bushes, etc. They scratch the brisket, the upper part of the front legs, the neck and the head with the hind leg. With the mouth, they pluck the legs, and sometimes the flanks. Unlike cattle, sheep do not groom each other.

5.9   Temperature Regulation and Climate Requirements

Temperature regulation

Sheep have less ability to regulate their body temperature via sweating than cattle. Temperature regulation in sheep occurs mainly via the respiratory organ (Andersson, 1977). Panting accounts for about 80% and sweating for about 20% of the relative proportion of total evaporative heat loss in sheep (Robertshaw, 2004).

In cold, wet weather, sheep will huddle together to afford mutual shelter and to conserve body heat (Rutter, 2002). In hot, sunny conditions, sheep will often use the shade of a tree or other shelter. There is a significant positive correlation between the proportion of the day spent in the shade and the daily maximum air temperature. Such behaviour is an important means of regulating body temperature since, for example, the heat load in newly shorn animals due to solar radiation can be similar in magnitude to the metabolic heat production of the animal. There is much individual variation in the strategies used by sheep being heated by direct sunlight (Broom and Fraser, 2007).

Climate requirements

In cold weather, sheep seek shelter (Geist, 1971). Among free-ranging sheep, sheltered areas are being identified and confirmed constantly as a result of exploration. Unshorn and recently shorn sheep choose different resting locations. During calm weather, in daylight, sheep largely remain as one flock, but, in windy weather, and at night, most shorn sheep congregate in a shelter, while unshorn sheep remain away from shelter. Unshorn sheep appear to avoid sheltered areas and prefer the colder or higher areas of grazing (Broom and Fraser, 2007).

Newborn and young lambs are affected adversely by cold, wet and windy conditions, some breeds being more affected than others. Therefore, dry lying areas protected against precipitation are necessary for lambs not to succumb. If such protected areas are available, the sheep seek protection in these against wind and rain.

The lower critical temperature (LCT) figures depend on whether or not the sheep are sheared, i.e. the thickness of their wool. The following figures show this. The LCT for sheep with 10 mm thick wool is 25°C, for sheep with 50 mm thick wool it is –5°C and for sheep with 70 mm thick wool it is –18°C. For a newborn but dry lamb it is 27°C (Broom and Fraser, 2007). However, there seem to be differences in LCT between breeds (Hammarberg, 2010).

The relative humidity in insulated sheep houses should be 70% ± 10%.

5.10   Vision, Behaviour in Light and Darkness

Sheep seem to have good sight at both near and long distance. The location of their eyes on the lateral part of the head gives them a visual field of > 300° monocular and 60° binocular. They have, depending on the position of obscuring wool, horns and ears, a field of about 70–90° to the rear where they are not able to see an object with both eyes simultaneously when the head is directed forward (Hutson, 1993).

This wide field of vision is an advantage for detecting predators quickly. The construction of the sheep eye makes it easier to discover moving rather than static objects.

Sheep have colour vision (Jacobs *et al.*, 1998), and difficulty in accommodating and estimating distances. This might explain why, when excited or in a panic, they will often bolt away without attempting to avoid a rope, chain or the like, but will make way for a clearer, more visible fence (Grandin, 2000). This might also explain why they shy away from an accentuated shadowed area as if it were a physical obstacle (SCAHAW, 2002).

5.11   Acoustic Communication

Hearing

Sheep have a broad hearing range, between 125 and 42,000 Hz, with maximum sensibility at 10,000 Hz (Heffner and Heffner, 1990, 1992b).

Vocalization and acoustic communication

Sheep have a rich sound repertoire depending on which situation the animal is in: danger, contact between lambs and ewe, isolation from the flock, warning, etc. The repertoire ranges from a 'rumbling' sound made by ewes towards newborn lambs, and also rams during courting, the 'snort' of aggression or warning, to the 'bleating' of contact and distress calls. Vocal communication may be used in a number of different settings, including mother–young contact, mating, territorial warning, alarm, aggression and distress (Lynch *et al.*, 1992).

5.12   Senses of Smell and Taste, Olfaction

Sheep have a very well developed olfactory sense that, apart from being used in the search for feed, is probably used in order to recognize conspecifics within a flock. Individual animal recognition clearly involves olfaction in mother–lamb recognition (Lynch *et al.*, 1992). They are considered as being able to scent a human being at a distance exceeding

300 m (Geist, 1971). Moreover, it is shown that they have a good olfactory memory and avoid areas and feeds that imply danger or health risks (Hutson, 1993).

Although visual cues are no doubt important in individual recognition, the fact that sheep spend a lot of time in mutual sniffing, especially with a strange animal that is introduced into the group, implies that olfaction must be important in this respect (Rutter, 2002).

Sheep have pedal scent glands on all four feet, as well as inguinal and infraorbital glands, the latter near the eye (Dwyer, 2009).

5.13   Tactile Sense, Sense of Feeling

Tactile touch

It seems as if the sense of touch around the muzzle area of the sheep's head may be important. Sheep use touch to recognize an individual plant or plant part, which is important as they have a blind area of some 3 cm directly in front of the nose (Lynch *et al.*, 1992).

Sense of pain

There is no evidence that sheep are less sensitive to pain than other animals. As relatively stoical creatures, it is possible that sheep do not display obvious signs of distress and pain, or that human observers do not have the ability or skills to identify these indicators. As prey animals, it is, in evolutionary terms, possibly advantageous for sheep not to show signs of pain, as this could make them the target of predators.

Inflammatory diseases are probably a major source of pain in ruminant species (Fitzpatrick *et al.*, 2006).

5.14   Perception of Electric and Magnetic Fields

Effects on behaviour do not seem to have been described in sheep.

5.15   Heat and Mating Behaviour, Pregnancy

Sheep are seasonally polyoestrous, with recurring oestrous periods in the autumn breeding season. Heat (oestrus) lasts 13 days and occurs 6–7 months after lambing and, if conception has not taken place, it recurs after about 17 days. Sheep in the wild have a mating season in the autumn and lambing occurs in the spring.

Libido is generally low outside the breeding season. The males become more aggressive during the breeding season.

Full sexual maturity occurs at different ages in different breeds (Papachristoforou *et al.*, 2000; Avellaneda *et al.*, 2006; Fogarty *et al.*, 2007; Kridli *et al.*, 2007). Feral rams usually mature sexually at 18–30 months of age (Geist, 1971). However, in some domestic breeds, male lambs at 4 months of age may perform mating, followed by pregnancy (Hammarberg, 2010).

When in heat, female sheep show increased motor activity and appear restless. Receptive females show an increase in the frequency of non-specific bleats. The rams use olfactory and gustatory stimuli to detect females that are in oestrus. The male smells the female's urine and then stands rigid, with his head raised and lips curled, showing 'flehmen' behaviour for about 10–30 s (Rutter, 2002). The female will also sniff the male's body and genitals at this stage of mating. This head-to-genital approach of both the male and female leads them to circle each other. The male makes courtship grunts and licks the female's genitalia during sexual approach.

Pregnancy lasts about 151 days, but a mainly breed effect variation between 145 and 155 days can occur. After lambing, the first heat usually occurs after 6–7 months – sometimes, however, after only 6–8 weeks.

5.16 Before and After Parturition

Behaviour before and during parturition

If lambing occurs on pasture, most parturient ewes usually move a short distance away from the flock. By doing so, the risk that other ewes will steal the lambs is decreased.

Before parturition, ewes sometimes show interest in other lambs. While 60% or more of most ewes lamb in isolation from the flock, the remainder lambs in the area where the flock is grazing. Sheep of different breeds often behave quite differently in seeking isolation at lambing. Most breeds seek some shelter before parturition (Broom and Fraser, 2007).

Before lambing, the ewe shows disquiet, restlessness, lying down and getting up behaviour, walking around, scraping the feet on the ground and bleating nervously.

The opening stage usually takes place over 1 h at most. During lambing, the ewe shows repeated rising and lying down; when straining, she usually stands. She mostly lies during early expulsion. The expulsion stage might take about 15 min for each lamb, and ewes often stand at the end of the expulsion stage.

Mispositions occur, although they are less common than in cows.

If lying during the expulsion, the ewe will usually stand within 1 min of giving birth. Immediately after the birth, vigorous licking by the mother has a stimulatory effect on the offspring. It is during this phase of intense licking, which lasts for about an hour, that the ewe learns to distinguish her lamb from others. Grooming by the mother helps reduce heat loss, as young lambs are particularly susceptible to chilling. Ewes will reject alien lambs, usually by headbutting them.

If the lamb is removed from its mother during this critical early phase following birth, the strong maternal bond can be broken and the lamb may be rejected if presented to her 6–12 h later.

Feral sheep eat the placenta after the birth (Geist, 1971). Whether or not the domestic goat eats the placenta depends on the breed (Rutter, 2002).

Should the lamb die, the ewe rapidly loses interest in the dead body (Rutter, 2002).

Number of lambs born

Ewes give birth to one or two, sometimes three and seldom four lambs. Differences occur between different breeds.

After the birth

When the ewe licks the lamb or lambs, which happens within some minutes after lambing, she gives characteristic low growling sounds. The ewe usually licks the lamb very intensively, beginning on the lamb's head and neck. The ewe seems to be attracted to areas of the lamb soaked with amniotic fluid.

The newborn lamb soon gets up, shaking its head during the intensive licking by the ewe. Within an hour, the lamb will stand on its own feet and search for the ewe's udder. The ewe licks the lamb's perineum in this stage, which seems to influence the lamb's intensity to seek the udder and suck. During sucking, the lamb performs intensive and characteristic sideways movements with the tail. The lamb bumps and pushes its head repeatedly against the udder. Lambs suckle from the side, sometimes while kneeling; occasionally, they suckle from the rear.

If a lamb does not get up, the ewe may kick it with a foreleg. If the lamb still does not get up, the kicks could cause serious injuries (Hammarberg, 2010).

Domestic ewes do not usually eat the fetal membranes, but lick amniotic fluids from the lamb and usually nibble pieces of fetal membranes.

Unlike cattle, where the calf stays in the calving place, the lambs follow the ewe from the very beginning (Fig. 5.2). The lamb is therefore a 'follower', unlike the calf, which is a 'hider'. There arises a stronger bond between lambs and ewe than between calf and cow.

The ewe forms a selective olfactory memory in a sensitive period of around 2–4 h after parturition and consequently will reject the approach of any strange lamb (Kendrick *et al.*, 1992; Sanchez-Andrade and Kendrick, 2009).

Other ewes without lambs show marked interest in the newborn lamb and may try to 'steal' it. Therefore, it is very important that the ewe is not disturbed during and after lambing, so that the bonds between ewe and lambs can be created and strengthened during the first day. Otherwise, the lamb may follow the 'wrong' ewe. Breeds vary in the likelihood that disturbance of the ewe–lamb bond formation will occur, fine-wool breeds being

most susceptible, and lambs, especially second-born lambs, may become separated from their mothers and die (CoE, 1992).

The ewe returns to the flock about 12 h, sometimes longer, after lambing.

As soon as she has established the bond between herself and her own lamb, a ewe will not accept another lamb other than her own. When sheep are kept indoors, the ewe should, if possible, be kept in her own lambing pen for some days in order to build up imprinting between lambs and ewe. Some parturient ewes become nervous when separated from the group. In such cases, it might be best to let them lamb in the flock pen and immediately after lambing, transfer ewe and lambs to the lambing pen. It sometimes happens that ewes do not accept their own lamb and even butt it. Such ewes must be isolated in a special pen and it may be necessary to put them in a cage in order to protect the lamb. The cage must be designed so the lamb can reach the udder easily for sucking. Most ewes accept their lambs after a few days of such fixation.

On the first day, the lamb sucks up to 15–20 times. The frequency declines successively to about 6 times per day at 2 months of age (Lynch *et al.*, 1992).

The ewe seems to recognize its lamb first by smell, then by its bleating, but also visually. The lamb obviously can recognize its mother's bleating and, at 3–4 days old, even recognizes her visually (Geist, 1971; Lynch *et al.*, 1992). At 24 h old, lambs rely more on the behaviour of the ewes to

Fig. 5.2. Sheep and lambs on pasture in southern Portugal in autumn during the lambing season (image I. Ekesbo).

select their dam than on their individual physical characteristics (Terrazas *et al.*, 2002).

The ewe protects its lamb when it is small, but this protection lessens after a couple of weeks and the ewe then ignores its lamb, even when it is in trouble (Geist, 1971).

Twin lambs develop a strong bond with one another and learn to recognize each other's voices and appearance. When given a choice, they stay nearer to each other than to other lambs in the flock (Broom and Fraser, 2007). During the first few weeks of life, lambs stay quite close to their own mothers. By 1 month of age, young lambs spend two-thirds of their time in the company of other lambs.

Play behaviour

Lambs perform play behaviour, sometimes even during their first day. At the age of 1 month, play is well developed. Play behaviour is seen in lambs up to 10 weeks of age (Lynch *et al.*, 1992).

The form of play in lambs is very typical. The play involves upward leaps, little dances and group chasing. Play is reduced as lambs grow, and becomes rare by 4 months of age (Broom and Fraser, 2007).

They show clear sex differences in the types of play exhibited, with males engaging more in play fighting and females in more rotational-locomotor play. It is suggested that the locomotor play in females ensures good physical fitness so that the animal is prepared to avoid potential predators (Rutter, 2002).

Weaning

Weaning under natural conditions is a progressive process and lambs are weaned at about 3–6 months of age as the ewe ceases to give milk. However, also after weaning, a stronger bond exists between the ewe and her lambs than between these and other sheep (Broom and Fraser, 2007). Domestic sheep are usually weaned at about 8 weeks of age. In the wild, natural weaning first occurs when the lambs are 6 months old when their mothers enter the rut (Dwyer, 2009).

The formation of 'weaner-only' flocks breaks this bond and a new social organization has to be developed with the formation of small subgroups, in which inter-animal distances are low. Gradually, these groups become larger, until eventually a flock is formed. The size of subgroups increases with age from weaning up to 4 months old; this is unrelated to the size of the paddock or space available to the animals. Even as late as 11 months of age, subgroups may be formed. Normal adult flocking behaviour appears to be established by 15 months of age. The flock identity is strong for adult sheep, and members within contact distance immediately run together when disturbed (Broom and Fraser, 2007).

5.17   The Lamb and the Young Sheep

In lambs and young sheep, there is little aggressive behaviour. Such behaviours increase with age (Lynch *et al.*, 1992).

5.18   Assessment of Sheep Health and Welfare

Special attention should be paid to bodily condition, including behaviour, movements and posture, rumination, condition of the fleece, ears, eyes, tail, legs and feet.

The healthy sheep

Healthy sheep make sounds, show activity, movements and posture appropriate to their age, sex, breed and physiological condition. These include: general alertness, good uniform fleece, clear bright eyes, good teeth, free movement, absence of lameness, good appetite; normal feeding, drinking, nursing and suckling behaviour; rumination, normal getting up, lying down, defecation and urination behaviour; as well as freedom from external parasites and with no visible wounds, abscesses or other injuries. The faeces have – taking feeding into consideration – normal amount, consistency, colour and smell.

The sick sheep

Symptoms

A sick sheep falls behind when the flock starts to move; sometimes, it even withdraws itself from the flock. The sick sheep always shows divergences in some part of its body posture, locomotive pattern, or other differences from healthy sheep. It ceases to eat, or eats only slightly, and cessation of rumination may occur. The animal can show listlessness or apathy. The glance becomes weak or nervous – when in pain, sheep often have a suffering glance and the eyes lack the mobility that characterizes a healthy sheep. The temperature can be increased, but also can be lower than normal. Together with these general symptoms, local signs may occur; for example, increased salivation, discharge from eyes, nostrils or vagina, swellings and perhaps soreness in the udder. Pain in the mouth can manifest itself as difficulties in chewing or ruminating; other diseases can cause rumination to cease totally. At pain in the feet, the sheep tries to relieve the pressure on the sore foot. Frequent scratching or rubbing is not a normal condition.

Examples of abnormal behaviour and stereotypies

Stereotypies have not been described formally in detail for the sheep. However, sheep in confinement may exhibit repeated actions, like wool or pen chewing (Lynch *et al.*, 1992).

Examples of injuries and diseases caused by factors in housing and management

Sheep kept on moist pastures which irritate the sensitive skin in the interdigital cleft may be affected by foot rot. This infectious disease arises when the skin is infected by bacterial agents. Foot rot is one of the most common health and welfare problems in sheep husbandry (Egerton, 2000).

The highest lamb mortality occurs during the first 3 days. For sheep kept indoors, lamb mortality is low, 2–3%. If kept outdoors without a dry lying area with sufficient protection against wind and precipitation, bad weather is an important cause of high lamb mortality, 15–20% (Broom and Fraser, 2007).

Symptoms of pain

Sheep do not show obvious signs of pain – this is not to imply that they do not feel pain like humans or other animals. They show no or little reaction even if exposed to seemingly significant pain, e.g. tissue damage like wounds or cuts in connection with shearing. However, a sheep with a sore foot avoids treading down with that foot, indicating that it hurts. Sheep produce high levels of cortisol and β-endorphin after surgical mutilation, which indicates that they feel pain (Shutt *et al.*, 1987).

As a prey animal, it might try, as far as possible, to avoid showing any sign of illness or pain. However, sheep suffering from foot rot indicate pain by showing lameness (Egerton, 2000).

5.19   Capturing, Fixation, Handling

Sheep must be handled calmly, as they are more likely to be willing to be led or driven when treated in this way. When sheep are moved, their gregarious tendencies should be exploited. Activities which may frighten, injure or cause agitation to animals must be avoided. Sheep should not be lifted by the head, horns, legs, tail or fleece.

When a sheep must be caught and held for medical examination, etc., there are various methods of doing this. One example is the following: the stockman who has to hold the sheep sets the knee against the standing sheep's chest, places one hand under the sheep's jaw in order to hold its head upwards and to prevent it from rushing ahead. At the same time, he should grasp with the other hand the wool and skin on the flank on the other side of the sheep and lift the flank upwards so that the hind leg of that side is lifted from the ground. In the same moment, he should with his knee push the chest of the sheep and momentarily take the sheep up on his lap. By that, he prevents the sheep from bracing its other hind leg against the floor. Thereafter, the sheep should be set down very carefully, with the hind part on the floor and the withers of the sheep placed between the knees of the stockman. The sheep thus is put into a sitting position. Through easy pressure against the breastbone, one can usually get the sheep into a more relaxed posture.

When a sheep finally has been caught, it usually shows very little reaction in the form of release attempts. If captured, and once held, they show very little behavioural response (Hutson, 2007).

5.20   The Importance of the Stockman

Careful handling of the animal and other contact from an early age contribute to developing a positive relationship between humans and sheep.

6 Goats (*Capra hircus*)

6.1 Domestication, Changes in the Animals, their Environment and Management

Goats were domesticated more than 8000 years ago (Zeuner, 1963) but may have been used by humans as early as 9000 or even 12,000 years ago (Clutton-Brock, 1999).

The first domestication of sheep and goats began in western Asia and spread rather rapidly westwards into Europe and probably north and east into Asia and the Far East. Radiocarbon dating indicates that the domestication of sheep preceded that of goats. Archaeological findings from several places indicated that it was more common to keep goats rather than sheep as suppliers of meat during this early period. The domestic goat is considered to be descended from the bezoar, *Capra aegagrus*, and not from the ibex, *C. ibex*. The earliest domestic goats had straight scimitar horn cores, but twisted horns appeared and, by the Bronze Age, twisted-horned goats predominated over those with straight horns in western Asia, as they do at present (Clutton-Brock, 1999).

The reason why humans selected goats for domestication, instead of animals like deer, might have been because goats, like sheep, had a social system that was based on following a single dominant leader, e.g. the herdsman (Geist, 1971).

There has been a lot more research into sheep than into goats, though we know that goats are adapted best to dry firm ground. Their ability to climb (Fig. 6.1) is considerable. Unlike sheep, they may force fences by climbing.

Because of their adaptation to a particularly harsh environment, goats are perhaps the most versatile of all ruminants in their feeding habits, a factor that has affected greatly their success as a domestic animal. They are also extremely hardy and will thrive and breed on the minimum of food.

Hornlessness in goats is not nearly as common as in sheep, and it is not known to have occurred until the third millennium BC in ancient Egypt. Hornless goats were fairly common in Roman times (Clutton-Brock, 1999).

Whereas sheep are grazing ungulates that inhabit hilly regions and the foothills of mountains, goats, as good climbers, are browsers whose natural habitat is on the high, bleak mountain ranges of Europe, Asia and Ethiopia. It is sometimes argued that it has been the browsing of goats over the past 5000 years that has, to a great degree, caused the expansion of the desert areas of the Sahara and the Middle East. They have never penetrated as far north as mountain sheep, which explains why goats did not cross into North America over the land bridge of the Bering Straits during the Pleistocene (Geist, 1971; Clutton-Brock, 1999).

Goats, and sheep, are mentioned about 2100 BC in the Epic of Gilgamesh, the poem from Ancient Mesopotamia. The goat has played an important role as a mythological character in many cultures, often as a symbol for the demoniacal, the licentious or as a fertility symbol. The cart of Thor in Nordic mythology was pulled by male goats. In Greek mythology, the god Pan had the hindquarters of a goat. In the Bible, the male goat served as a propitiatory sacrifice for the sins of the people. The devil is often depicted in popular art with goat horns and goat feet. The steinbock, ibex, a wild goat, held a prominent position in the art of the nomadic peoples of south Russia and central Asia in the third millennium, thus seeming to have played a part in their cultural life.

Goats are regarded as showing more intelligence and having a greater tendency to stress than sheep.

The breeds of domestic goats are more diverse in anatomical appearance than the wild species, a result of human selection for traits of interest. However, more complex traits such as behavioural characteristics are, on the whole, unchanged. The management of goats has not undergone the same dramatic changes as that of swine, poultry and cattle.

Fig. 6.1. Young goats and kids, 2–4 months old. Goats like pine bark and twigs and are good climbers (image courtesy of Kalle Hammarberg).

There are nearly 600 breeds of goat living in the world and nearly 800 million animals. About 75% of the world's goats are found in the developing countries (Rutter, 2002). About 65% are found in Asia and about 30% in Africa, but only 2% in Europe. Meat is the most important goat product in the developing countries. World goat meat production more than doubled between 1980 and 2000 (Dwyer, 2009; Smith and Sherman, 2009; Hammarberg, 2010).

The term 'doe' or 'nanny' is used for the female goat, 'buck', 'billy goat' or 'he-goat' for the male, and 'kids' for the offspring.

6.2 Innate or Learned Behaviour

Innate behaviours are, for example, the female goat leaving the flock at the time of delivery, licking of the kid after delivery, the kid hiding itself near the place of kidding, the mother keeping herself in the neighbourhood and the sucking behaviour of the kid.

Gentle treatment of goats indicates that they learn from experience (e.g. Lyons and Price, 1987).

It is common experience among goat farmers that goats can learn to manipulate things in order to gain benefits, for example how to open latches on farm gates, and that they learn quickly to move through passages and areas with which they are initially unfamiliar.

Daily learning in goats decreases as a result of the stress caused by the mixing of unfamiliar animals, i.e. the provoking of agonistic behaviour (Baymann *et al.*, 2007).

6.3 Different Types of Social Behaviour

Social behaviour

In the wild, goats live in social groups usually composed of near-related females, kids and young animals, and often also males younger than 2 years. The adult male goats form separate subgroups, except during the mating season. A flock of goats is held together with visual, olfactory and acoustic contact between individuals. They show pronounced flock behaviour and a strong desire to remain with their group mates. They become very vocal when separated from the flock (Rutter, 2002).

Forced isolation of a goat from the flock may cause behaviour disturbances and serious, even fatal, consequences such as failure to feed (Boivin and Braastad, 1996). Kids show more intense vocalization and excitement than lambs if isolated from the group (Price and Thos, 1980). However, goats show less anxiety than sheep when separated from the flock (Hammarberg, 2010)).

Goats have a social system that is based on a single dominant leader. They do not defend the territory that is their home range (Dwyer, 2009), like, for example, some sheep breeds (Rutter, 2002).

Goats, especially bucks, often show an aggressive behaviour against conspecifics that come too close or when dominance relations are not settled. If dominant animals feel themselves threatened, they attack by butting. When new animals are introduced into a flock, this must be considered.

Goats can identify conspecifics visually, whereby aggression can be avoided.

Communication

Goats are far more mobile and far more observant of control signals from other goats within the flock than sheep. Goats separated from the flock will bleat. If the flock hears the bleating, the goats in the flock usually respond with similar bleating.

During the first days after parturition, a doe communicates with her kid(s) by vocal signals, either for warning or for the kid to join her (Lickliter, 1984).

Behavioural synchronization

Goats exhibit 'flocking' and 'following' behaviour, but not the same behavioural synchronization as cattle and sheep. Goats, unlike cattle, do not commence or cease grazing at the same time. They express individuality to a greater extent.

The social ranking and group behaviour in goats also differ from those in sheep. Thus, when a flock of goats has to be driven, they generally move in family groups, with the oldest females first.

When a flock of goats is moving from one place to another, they do not walk in a line to the same extent as sheep.

Gaze following is a benefit to animals that live in social groups. Goats, like some primate species, can follow the gaze direction of conspecifics, a behaviour seemingly not studied as yet in other non-primate mammals (Kaminski et al., 2005b).

Dominance features, agonistic behaviour

Goats have a strong hierarchical structure in the flock or herd. Both males and females will establish social dominance in their respective groups through head-to-head fighting. They use their horns to advantage when fighting to establish their social dominance (Smith and Sherman, 2009).

Of the agonistic behaviours observed in a study of 72 individuals, a third of the occurrences were shown to be biting and the rest butting. Biting was more frequent in animals 3 years or older than in younger animals, and more frequent in hornless than in horned animals. Butting behaviour was more frequent in individuals higher up the social hierarchy, whereas social hierarchy did not influence the frequency of biting behaviour (Tolu and Savas, 2007).

Antagonistic goats stand about 1–2 m apart, then rear up with their body at right angles to their opponent. They then pivot, lunge forward and come together with a loud crack. Unlike sheep, which move back in order to take a run for a fast and hard attack, goats directly rear on their hind legs and bump forward and downwards.

Although the majority of the animals in a group will use aggression at some time, most individual relationships between goats in a group can be peaceful, i.e. do not involve any aggression. They use threat behaviour to minimize fighting between individuals. This helps to reduce the risk of injury associated with fighting. Threat behaviour consists of a lowered head and a stretched neck. If the threat fails to deter a potential rival, the males engage in rearing and butting to establish their dominance within the group (Rutter, 2002).

6.4 Behaviour in Case of Danger or Peril

When alarmed, goats will stamp one foot and produce a high-pitched noise that sounds like a sneeze. These signals alert the other members of the flock to possible threat. When approached by a potential predator, for example a dog, goats tend to break away from the flock. This makes it relatively difficult to 'herd' goats, as they do not group together (Rutter, 2002).

When an unfamiliar person approaches a sheep flock, the sheep usually flee immediately, while goats might stop or come to meet the person. Goats warn of perceived sources of danger, which may include humans, by sounds such as snorts. Sneezing in humans is a sound that goats seem to perceive as a warning signal and can cause panic in a goat group. At danger, goats either flee, but choose sometimes to turn and line themselves up for defence, or go to attack (Hammarberg, 2010).

Indoors and in paddocks they will get frightened easily by, for example, sudden shadows, sun reflections and high, unexpected sounds.

At loud rumblings, as from thunder or the sudden appearance of low-flying aeroplanes, feral goats rush to find protection by pressing themselves against stone or even house walls. This behaviour has been explained as an indication that mountain goats head for cover against an impending avalanche (Geist, 1971).

6.5 Some Normal Physiological Frequency Values of Interest at Clinical Examination

The normal rectal temperature of adult goats at rest is 39°C, with a range of 38.5–39.7°C; the normal temperature of kids can be up to 39°C. The critical point for adult animals is >39.7°C (Robertshaw, 2004).

In adult animals at rest, the pulse has a frequency of 70–80 beats per min (Erickson and Detweiler, 2004); in kids it can be about 100 beats per min.

In adult animals, the respiration frequency is 12–25 breaths per min; in kids it is 20–40 breaths per min. It varies with breed, thickness of hair coat and environmental temperature.

The number of rumen contractions is about 8–15 every 5 min.

6.6 Active and Resting Behaviour Patterns

Activity pattern, circadian rhythm

Like cattle, goats divide their time on pasture between mainly grazing, ruminating and resting. However, as goats devote more time to browsing than sheep, their total feeding time occupies up to 11 h of a 24-h day (Rutter, 2002).

Exploratory behaviour

When left undisturbed in a new environment, goats spend much time in exploratory behaviour, maybe more than some other ungulates.

Diet, food searching, eating and eating postures (feeding behaviour)

Goats prefer a mixed diet and they can digest coarse roughage very efficiently, such as the leaves of many trees and shrubs. They select their diets from a greater range of forage sources than sheep. They obtain their food by browsing more than by grazing (Fig. 6.2). This encourages them to wander over longer distances in the search for food, which leads to gaps in their eating activity as they move from one tree or shrub patch to another. As agile climbers, goats will even climb into trees to browse (Dwyer, 2009). They try all the time to mix their diet with grass, herbs, buds and bark from trees and bushes (Fig. 6.3).

According to dietary profiles compiled from the world literature, the average goat dietary botanical composition for all seasons consists of 29% grass, 12% forb (herbs) and 59% shrub (Alcock, 1992). By studying aggressions when feeding on hay and silage, Jørgensen et al. (2007) have concluded that goats prefer hay to silage. Like cattle and sheep, goats avoid grass contaminated by conspecifics' manure. On the other hand, they have naturally inquisitive feeding habits and are more inclined than sheep to examine the taste and consistency of novel things, e.g. discarded human clothing, whenever the opportunity occurs.

The upper lip of the goat is more mobile than that of the sheep, allowing the goat to be more selective and to be able to 'pick' preferred plant parts with relative ease.

Goats need to drink almost daily, although, due to their browse diet, they may be adapted to coping with periods without water better than other farmed animals (Dwyer, 2009).

The dentition is complete with permanent teeth at 4 years of age. In the kid, the set of milk teeth is complete at about 35–40 days of age.

Selection of lying areas, lying down and getting up behaviour, resting postures, sleep

Goats often choose to rest on a higher part of their territory, e.g. an elevated stone. Individuals standing highest in the group's rank scale usually select the most elevated resting places.

The goat initiates lying down behaviour by scraping the ground with the forefeet and, in conformity with cattle (but often more pronounced), sniffing and examining the lying area before lying down. The lying down and getting up behaviour of goats does not deviate from that of cattle, but the movements are much faster. Goats seem to prefer hard lying to straw-bedded areas (Bøe, 2007).

Fig. 6.2. Goats offered bark from felled pine tree logs. In the background, other goats browsing on meagre grass vegetation (image courtesy of Kalle Hammarberg).

Fig. 6.3. Goat does, one with and one without horns, browsing shoots and buds from osier (willows). The goat in the foreground is emaciated because of chronic disease (image courtesy of Kalle Hammarberg).

Locomotion (walking, running)

The gaits of goats are walking, trotting or gal-loping, depending on the circumstances. They are very active animals with lively movements and are skilful when it comes to jumping and climbing.

In Sweden, in earlier times, a goat was often tethered to the chimney of the house in order to graze the vegetation growing on the turf roof.

Swimming

Goats are natural swimmers but do not like to go into the water.

6.7 Behaviour at Defecation and Urination (Eliminative Behaviour)

Goats wag their tails back and forth while defecat-ing. When urinating, the doe goes into a squat position, similar to the one assumed by a female dog. Even male kids will arch their back and bend their legs while urinating. This behaviour is not displayed in healthy adult bucks but may occur if they have difficulties in urinating, e.g. because of suffering from a urinary stone.

Goats defecate and urinate randomly over their enclosures.

6.8 Body Care, Cleanliness

Goats usually try to keep their hairy coats clean and dry, which helps protect them against cold and wet weather.

In goats, oral grooming takes the form of scrap-ing the lower incisors through the pelage in bouts of upward motions directed to a single area. Adult goats use their horns to scratch their back. Newborn and young goats groom more frequently than similarly maintained adults. It has been shown, for example, that goat kids from 2 weeks of age orally groom and scratch-groom signifi-cantly more frequently than adult females (Hart and Pryor, 2004).

6.9 Temperature Regulation and Climate Requirements

Temperature regulation

Like sheep, goats have less ability than cattle to regulate their body temperature via sweating.

Panting accounts for about 80% and sweating for about 20% of the relative proportion of total evaporative heat loss in goats (Robertshaw, 2004).

Caprine horns function as thermoregulatory organs. Both the surface of the core and the sur-face of the sinus of goat horns are covered with a rich vascular plexus. The horns vasodilate in response to heat stress and exercise. They vasocon-strict when the resting goat is in a cold environ-ment. At an ambient temperature of 30°C, a resting goat normally loses 3–4% of its heat pro-duction through the horns, while a goat heated up from running could lose as much as 12% of its heat production after stopping, until the horns vasoconstrict (Taylor, 1966).

Climate requirements

Newborn and young kids are affected adversely by cold, wet and windy conditions. Otherwise, goats generally are quite hardy animals, being able to weather heat and cold comparatively well, as long as they are provided with a well-constructed shed.

However, goats dislike rain. Farmers report that goats will run to the nearest available shelter on the approach of a storm or rain, often arriving before the first drops of rain have fallen. They also have an intense dislike of puddles and mud.

While hot weather poses no great problem to most goats, a high level of humidity does cause them stress.

The combination of cold and rain implies appar-ent health risks for goats. However, dry cold is not a problem for goats if their hair and coat have enough density and thickness. Unlike sheep, goats do not have a thick layer of subcutaneous fat. Their hair is their protection against cold. As in cattle, their greatest problem is not dry cold but draught and humidity indoors and wind and rain outdoors.

Goats respond to low temperatures by reducing their lying time while increasing the time spent active and eating (Bøe, 2007).

Goats living in hot, dry and treeless areas will congregate and huddle during the midday heat. Although this latter behaviour seems to go against common sense, it occurs when the heat that is being taken in exceeds the ability of the goat's body to dissipate it. By huddling together, goats can reduce the intake of direct and reflected solar energy, and so are able to maintain their body tem-perature better than if they stood alone in the sun (Rutter, 2002).

The relative humidity in isolated goat houses should be 70% ± 10%.

6.10 Vision, Behaviour in Light and Darkness

The prominent eyes of goats give them a panoramic field of 320–340° and a binocular vision of 20–60°. This wide field of vision is an advantage for locating predators, and the construction of a goat's eye makes it easy to detect moving objects quickly.

Tests have been carried out on male goats to determine their capacity for colour vision and they have been found to be able to distinguish yellow, orange, blue, violet and green from grey shades of similar brightness (Buchenauer and Fritsch, 1980; Jacobs et al., 1998).

6.11 Acoustic Communication

Hearing

Goats have a broad hearing range, between about 60 and 40,000 Hz, with maximum sensibility at 2000 Hz (Heffner and Heffner, 1990, 1992a), and react suddenly to unexpected or high sounds.

Some predators and humans are able to locate a sound accurately within an angle of 1°, whereas goats can locate a sound only within an angle of 18°. They are, however, a more accurate sound localizer than cattle, where the corresponding figure is 30° (Heffner and Heffner, 1992b).

Vocalization and acoustic communication

Goats are often vocal in their responses to one another and to humans. They communicate with each other largely with different sounds and snorts, the latter often used as a sound warning of danger.

6.12 Senses of Smell and Taste, Olfaction

Goats seem to have a well-developed olfactory sense, which is used to discriminate between different sorts of feed and to identify conspecifics. They spend a lot of time in mutual sniffing, especially with a strange animal that is introduced into the group. This implies that olfaction must be important in this respect (Rutter, 2002). Goats do not have infraorbital, interdigital or inguinal glands like sheep. They have pedal scent glands on two feet and have a gland under the tail, which may explain why usually most goats' tails are raised constantly (Dwyer, 2009; Hammarberg, 2010).

Olfaction is used by the doe to recognize the kid.

6.13 Tactile Sense, Sense of Feeling

Tactile touch

Most, but not all, goats show positive reactions when stroked by someone to whom they are habituated. They seem to prefer to be stroked before being patted. Goats are less hesitant towards humans than are sheep.

Sense of pain

Like cattle, goats probably have mechanisms for sensing pain similar to those of humans (Iggo, 1984; Livingstone et al., 1992).

Goats seem to have a relatively low pain threshold when compared with some other farm animals, such as cattle, and they do not tolerate ill health very well (Harwood, 2006).

6.14 Perception of Electric and Magnetic Fields

Effects on behaviour do not seem to have been described in goats.

6.15 Heat and Mating Behaviour, Pregnancy

The length of the breeding season is influenced by such factors as day length, temperature and geographic origin. Those breeds that originate from high, mountainous areas have an abbreviated breeding season. All goats have a peak oestrus cycle in the autumn, thus allowing most kids to be born during the favourable springtime.

Goats are seasonally polyoestrous, with recurring oestrous periods in the autumn breeding season. Heat lasts 18–24 h and, if no pregnancy occurs, it recurs after about 21 days. Goats in the wild have a mating season in the autumn and kidding occurs in the spring. Pregnancy lasts about 150 days, but may vary between 140 and 160 days (Rutter, 2002). The sexual activity of the female is influenced by day length in temperate animals (shortening days) and by rainfall in tropical goat breeds (Dwyer, 2009).

Libido is generally low outside the breeding season. The males become more aggressive during the breeding season.

In the wild, the bucks approach the female group and compete with each other in order to win mating possibilities. The oldest and strongest may be able to gather a harem of females. When in heat, female goats show increased motor activity and appear restless. Receptive females show an increase in the frequency of non-specific bleats. As in cattle but unlike in sheep, nanny goats tend to mount and be mounted by other females. The bucks use olfactory and gustatory stimuli to detect females that are in oestrus. The male smells the female's urine and then stands rigid, with his head raised and lips curled showing 'flehmen' behaviour for about 10–30 s (Rutter, 2002). The female will also sniff the male's body and genitals at this stage of mating. This head-to-genital approach of both the male and female leads them to circle each other. The male makes courtship grunts and licks the female's genitalia during sexual approach. Male goats will urinate frequently on to their forelegs during sexual excitement. The male will nudge the female and she will turn her head back towards the male. Courtship terminates when the female becomes immobile and adopts a posture that allows the male to mount. If the female is receptive, copulation can occur rapidly and is very brief. The male's head moves backward rapidly at ejaculation. The male then dismounts.

Full sexual maturity occurs at different ages in different breeds (Papachristoforou *et al.*, 2000; Freitas *et al.*, 2004; Delgadillo *et al.*, 2007). Domestic goats mature much earlier than feral goats and may be sexually active at about 9 months of age.

6.16   Before and After Parturition

Behaviour before and during parturition

Before the birth, goats show behavioural indications of impending parturition, increased restlessness, pacing, pawing the ground, frequently standing up or lying down, frequent vocalizations and increased intolerance of conspecifics in about 50% of the does. Like most cattle and sheep, they often leave the flock before parturition. According to Lickliter (1984), over 75% of does leave the flock more than 4 h prior to giving birth. The majority of multiparous does seclude themselves in a vacant or unfrequented part of the barn or enclosure, often adjacent to vertical surfaces, a behaviour seen in both feral and domestic goats. Vocalizations are of low amplitude and frequency and are characterized by short grunts, low-pitched bleats and 'mmm' sounds similar to those made by cattle and sheep. Intolerance of conspecifics is characterized by lunging, butting or chasing any goat approaching within 3–4 m of a parturient doe's seclusion site. Defence of a birth site and enforcement of isolation from conspecifics occurs more frequently in multiparous females than in primiparous females. The birth may take 2–5 h and during the expulsion stage the doe usually lies with her neck raised but may stand during the end of the expulsion stage (Lickliter, 1984).

Difficulties at birth are uncommon and mispositions are relatively exceptional.

Goats differ from other feral or domestic ungulates in that females do not display maternal interest in the young kids of other females (Lickliter, 1984).

Number of kids born

The goat gives birth to one or two kids, seldom more.

After the birth

Immediately following the birth, the doe gets up, if not standing already, turns around – this breaks the umbilical cord – and begins licking and nudging the kid.

Newborns begin snorting, sneezing and shaking their heads within seconds following their expulsion from the birth canal, thereby expelling lung fluids and cleansing their throats and nasal passages.

The vigorous licking by the mothers has a stimulatory effect on their offspring. Goats are particularly vigorous in grooming and orienting to their firstborn. This means that the secondborn, which is usually the weaker of the two, has a greater opportunity to suck. This process of maternal bonding is much quicker in the goat than in the sheep, only requiring a few brief moments after birth. Grooming by the mother helps reduce heat loss, as young kids are particularly susceptible to chilling. During these initial movements, mothers and kids perform vocal interchanges, most intensive for the first half an hour but intermittently during the next 4–6 h.

Most kids stand within 15 min of birth and begin seeking the udder. Most teat seeking starts on the body of the mother on the upper portions of her legs, the angle between the legs and the body surface, and finally the kid will find the udder region of the dam. As soon as the kid has oral contact with the mother's teat, successful suckling starts, which is accompanied by vigorous tail wagging by the kid. Goats invariably nose or sniff their kids' rumps, tails or anogenital regions during suckling bouts. There seems to be little grooming of kids after the first 2–4 h post-partum.

The placenta is expelled within 3 h following birth and is eaten by about 40% of the does. More than 50% of does actively defend the area close to their newborns during the post-partum period. This active defence appears very similar to that observed in does prior to parturition, and is more evident in older, multiparous does than in younger, primiparous females (Lickliter, 1984).

In the wild, the kids hide after the kidding, but the female goat is never far away. Domestic goats together with their kids also remain isolated from conspecifics at this time. The kid hides for between 1 and several days following parturition. During this time, the does leave the birth site and their kids, although never far away, for shorter or longer periods. When returning to the precise location where the kid is lying, she makes a muted vibrato call, which signals the kid to rejoin its mother. Disturbance does not seem to frighten the kid while it is lying out, although it may freeze if the mother makes an alarm call. This freezing response makes it more difficult for a predator to locate the kid. The 'lying-out' phase of the kid's development has been reported to last between 3 days and several weeks. After the lying-out period is over, the nanny goat and kid use bleating to keep in contact when moving or grazing (Geist, 1971; Lickliter, 1984; Rutter, 2002).

The end of the seclusion period is marked by kids leaving their dams and birth sites to lie down at the perimeter of the enclosure near a vertical surface, e.g. a barn stall wall, a fence or gate, or a feeder trough. Does do not leave the birth site and rejoin the herd until after their young have left the area and consistently lie down at a distance from them (Lickliter, 1984).

Play behaviour

The kids show play activity from a young age (Rutter, 2002). However, in a study of mountain goats, very little play was observed among young goats (Dane, 1967).

Weaning

Domestic goats are usually weaned at about 8 weeks of age. In the wild, natural weaning occurs first when they are 6 months old, when their mothers enter the rut (Dwyer, 2009).

6.17 The Kid and the Young Goat

Young goats stay with their mothers, forming family groups in the flock. The males stay often with their mothers during their second year (Rutter, 2002).

6.18 Assessment of Goat Health and Welfare

Special attention should be paid to bodily condition, including behaviour, movements and posture, rumination, condition of the fleece, ears, eyes, tail, legs and feet.

The healthy goat

Healthy goats make sounds, show activity, movements and posture appropriate to their age, sex, breed and physiological condition. These include: general alertness, good uniform coat condition, clear bright eyes, good teeth, free movement, absence of lameness; good appetite, normal drinking and suckling behaviour, rumination; freedom from external parasites, with no visible wounds, abscesses or other injuries. The excreta have, taking feeding into consideration, a normal amount, consistency, colour and smell.

In the healthy goat, the head is held up, with a bright alert attitude.

The sick goat

Symptoms

A sick goat may withdraw itself from the flock. When a goat is sick or in pain, it will often begin to vocalize; the volume and pitch will change depending on the intensity of pain or discomfort felt (Harwood, 2006). The sick goat shows divergences from the healthy animal in some part of its body posture and locomotive pattern. It ceases to eat or eats only slightly, and cessation of rumination may occur. The animal can show listlessness or apathy. The glance becomes weak or nervous; when in pain, goats often have a look of suffering and not the usual look of alertness that characterizes the eyes of a healthy goat. The temperature can be increased, but can also be lower than normal. Together with these general symptoms, local symptoms may occur, for example, increased salivation, discharge from eyes, nostrils or vagina, swellings and perhaps soreness in the udder. Pain in the mouth can manifest itself as difficulty in chewing or ruminating; other diseases can, at times, result in total cessation of rumination. Local pain in the feet results in the goat trying to relieve the pressure on the sole of the foot. Like other animals adapted to dry and hard surfaces, they develop hoof disorders easily if kept on mudddy or otherwise wet surfaces. Frequent body scratching or rubbing is not a normal condition in goats.

However, if kept indoors or otherwise where the area is limited, even small behavioural changes in a sick animal, often difficult to observe by humans, will be detected by the other animals, resulting in attacks on the sick animal (Hammarberg, 2010).

Examples of abnormal behaviour and stereotypies

Stereotypies do not seem to have been described in detail for goats. However, social isolation provokes abnormal behaviour.

Examples of injuries and diseases caused by factors in housing and management

The curiosity and climbing instincts of goats ensure that they will find a way to become entangled whenever possible. Therefore, fractures are common in goats. Environmental hazards are often implicated in the occurrence of fractures (Smith and Sherman, 2009).

Like dairy cows, goats exposed to extreme cold and wind may get frostbitten teats if the teats are wet from inadequate drying after udder preparation for milking.

Symptoms of pain

Goats, like cattle and sheep, do not usually show obvious signs of pain – this does not imply that they do not feel pain in the same way as humans or other animals. As prey animals, they may try to avoid, as far as possible, showing any sign of illness or pain.

However, abdominal pain may be manifested by depression, restlessness, bleating, teeth grinding, reluctance to move, increased shallow respiration, increased heart rate or an abnormal posture with an arched back and tucked-up abdomen.

6.19   Capturing, Fixation, Handling

Goats should always be treated as individuals, even in large herds. When goats have to be kept singly, they require more frequent contact with, and supervision by, the stockman. Unlike sheep, which usually have to be forced, goats respond positively when shown the way. Given space, they will stay in their group when driven. Goats do not respond well to forcing or crowding in confined areas. Commonly, an animal that breaks from the group under pressure will return to the group if given time. Each breed of goat has its own unique characteristics and the stockman should be aware of the particular requirements of the animals in his care.

Goats should not be driven or forced to walk long distances over long periods of time, as they crowd and 'pack up' and can be easily smothered. Unlike sheep, dog use with goats should be kept to a minimum.

When forming new groups, care must be taken to avoid fighting and stress.

Heavily pregnant goats should be handled with special care to avoid distress and injury, which may result in premature kidding. Goats should not be lifted by the head, horns, legs, tail or coat.

Goats should be kept loose in groups, where possible. They should not be restrained permanently, but can be tethered temporarily.

6.20   The Importance of the Stockman

In order to develop a positive relationship between humans and animals, there should be appropriate careful handling and other contact from an early age of the animal. Goats that have been treated gently by humans show fewer stress symptoms when exposed to humans than goats that have been treated unkindly (Jackson and Hackett, 2007).

Kids kept socially isolated from the goat group show more readiness to seek contact with humans than lambs exposed to the same type of isolation from the sheep group. This might indicate a higher degree of social behaviour in goats than in sheep (Price and Thos, 1980).

PART II
Domesticated Birds

7 Domestic Fowl (*Gallus gallus domesticus*)

7.1 Domestication, Changes in the Animals, their Environment and Management

The ancestor of the domestic fowl is the Red Jungle Fowl (*Gallus gallus*). Modern forms of these jungle fowl are still found today in South-east Asia and the domesticated bird can be regarded as a subspecies (*G. gallus domesticus*). Estimates vary but domestication is thought to have occurred about 8000 years ago, first in India and China, then spreading along trade routes. Initially, birds were probably used for sacrifice for ceremonial purposes, because of their beautiful plumage, and for cockfighting. Although there existed wood-heated incubators in old Egypt, it was not until the Roman times that the bird's potential as an agricultural animal was developed and laying breeds selected for high egg production, and even a poultry industry, were established. This industry collapsed when the Roman Empire collapsed and large-scale selection of birds for commercial use did not resume again until the 19th century (Wood-Gush, 1971; West and Zhou, 1989; Crawford, 1990; Siegel *et al.*, 1992).

The species hen is promiscuous, with males mating with several females. The male may defend his harem of females from other males.

Jungle fowl are omnivores, spending a large portion of their day pecking and scratching in the ground for seeds, worms and insects. Studies of the way in which semi-wild, free-ranging jungle fowl in zoos allocated their time between different activities showed that in 60% of all minutes during the active part of the day, hens were seen to be ground pecking and in 34% they were ground scratching (Dawkins, 1989).

Domestic hens also spend a large proportion of their day pecking and scratching.

The beak is the main exploratory organ of the bird. It is well innervated at its tip, with collections of touch receptors which allow the birds to peck accurately (Desserich *et al.*, 1983).

Several characteristics predisposed jungle fowl to domestication. They are social and gregarious, living in groups of 1–2 males and 2–5 females plus young, which has allowed them to be managed in groups. They have a hierarchical structure, probably based on individual recognition, which reduces the risk of injury caused by fighting. They show promiscuous sexual behaviour, which allows any male to be mated with any female and so facilitates artificial selection. Fowl have flexible dietary requirements and are adaptable to a wide range of environments. All these traits have been used to advantage in commercial poultry production.

Domesticated hens have retained the jungle fowl's strong motivation to perform nesting, perching and dust bathing behaviour (Keeling, 2002).

The domestic fowl (*G. gallus domesticus*) is, like its ancestors, precocial, that is, shortly after hatching the chicks are able to move, feed, explore, etc., but they are incapable of regulating their own temperature fully until they are past the first 2 weeks of life (Riber *et al.*, 2007).

Until the mid-1900s, the small, mixed-sex and -age group consisting of maybe up to 30 individuals was the size and composition of many poultry flocks in most farms in Europe (Fig. 7.1), as is still the case in many parts of the world. This group composition and size were comparable to those occurring naturally in wild jungle fowl. Since World War II, however, the number of such backyard flocks has diminished radically in industrialized agriculture and has been succeeded by a limited number of large flocks where a great number of individuals of the same age and, in the case of laying hens, sex are kept segregated in large groups. Thus, within a very few years after the 1950s, hen husbandry was transformed into large, caged herds kept indoors year-round in buildings without windows (Figs 7.2 and 7.3). The cage system has been criticized from an animal health and welfare point of view. As a result, alternative systems where hens are kept loose in larger groups, aviaries, have come into use, with bedding, perches and egg-laying nests available. Such aviaries were introduced during the 1980s in, for example, Switzerland and Sweden (Fig. 7.4), and also later in other countries. However, in most countries, the majority of hens are still kept in cages. Cages with perches, egg-laying nests and sometimes also sand baths, have

Fig. 7.1. Until the mid-1900s, most farms in Europe had small, mixed-sex and -age chicken flocks. This practice still occurs in many parts of the world. Observe the cock (image I. Ekesbo).

been introduced in some countries. Domestication has changed the appearance, egg production and rapidity of growth dramatically. During the 19th century, selection for breeds to specialize in either egg production or meat production began. Selection for production has been more intense in the domestic fowl than in any other domestic farm animal species regarding both egg production and growth, and rapid growth for broiler strains. From the 1950s to the beginning of the 21st century, a broiler chicken reaches full adult body size in 6 weeks, as compared with 17 weeks for egg-laying chickens. The increase of ovulation intensity in domestic fowl since World War II is without parallel in the domestication of any other animal. Even if the laying hen is not selected specifically for growth, brown hybrids grow to about double the size of jungle fowl.

There were about 56,000 million chickens raised in the world for eggs or meat at the end of the first decade of the 21st century (Spinka, 2009).

The jungle fowl laid one clutch of 8–10 eggs and then brooded them – in total, not more than 60 eggs per year. The commercial laying hen laid, without pause, about 130 eggs per year in the 1950s, and, at the beginning of the 21st century, she could lay up to 350 eggs in 1 year. Each egg is more than double

Fig. 7.2. Within a very few years of the mid-1900s, hen husbandry was transformed into industrialized egg production with numbers of birds being kept all year round in cages in windowless buildings (image courtesy of Jan Svedberg).

Fig. 7.3. The hen's ability to move is restricted in the cages (image courtesy of Stefan Gunnarsson).

Fig. 7.4. Aviaries for hens were introduced in the 1980s in, for example, Switzerland and Sweden (image I. Ekesbo).

the size of those of jungle fowls (Schütz, 2002). The commercial broiler chicken at 42–45 days of age weighs 2.5 kg, which is more than three times even the adult jungle fowl (Keeling, 2002).

Domestication has altered some behaviours, such as brooding behaviour and feeding behaviour (appetite). But the behaviour of the domestic birds is more or less unchanged compared to the jungle fowl.

The term 'hen' is used for the adult female domestic fowl, 'cock' or 'rooster' for the male and 'chick' for young birds up to about 3–4 months of age. The word 'chicken' is sometimes used as a generic term for the species, *G. gallus*.

7.2 Innate or Learned Behaviour

An example of an innate behaviour is chicks' responses to an object passing over them. At first, they show an antipredator response: they stop what they are doing, crouch and freeze. If the action is repeated a number of times without any unpleasant consequences for the chicks, the duration of freezing becomes shorter and eventually they ignore the stimulus altogether. This is an innate response modified by the bird's experience, habituation (Appleby *et al.*, 2004).

Feeding is an innate behaviour and chicks will feed readily in the absence of the hen, but the nature of their feed intake and their long-term feed preferences are influenced strongly by the hen's behaviour (Nicol, 2006). Chicks that have an opportunity to observe a trained hen pecking a key to obtain food learn this task much more quickly than chicks that do not have this experience (Johnson *et al.*, 1986). If kept with the hen, they learn quickly to identify food. During feeding, the brooding hen will express feeding displays (Fig. 7.5), such as picking up and dropping food items or pecking exaggeration, while she vocalizes by food or titbit calls. The food call of the brooding hen also indicates food quality. Chicks have to learn which substances are actually food items (Wood-Gush, 1971; Keeling, 2002; Riber *et al.*, 2007).

While the tendency to peck is innate, chicks have to learn which substances are actually edible. They can do this by trial and error, or the mother hen can help by giving the titbit call and attracting the chicks over to where she is feeding. Maternal display not only attracts chicks to profitable food items, but also redirects their attention away from harmful or non-profitable items. It is an intriguing finding, with implications for animal sentience, that brooding hens are sensitive to the skill level of their chicks during the time that chicks are learning what to eat and what to avoid. Hens adjust their maternal display behaviour in a flexible manner that suggests they may be 'teaching' their chicks (Nicol, 2006).

The mother hen helps the chicks to learn about food, but she also helps them to learn to roost up on branches at night. Chicks reared without perches for the first 4 or 8 weeks of life have greater

Fig. 7.5. During feeding, the brooding hen will express feeding displays, such as picking up and dropping food items (image courtesy of photo archive HMH Department).

difficulty learning this behaviour later, and this can affect other aspects of their behaviour in commercial housing systems, such as finding the nest boxes, which are often raised up off the ground. No access or late access to perches during rearing has been shown to increase the later prevalence of floor eggs and cloacal cannibalism in loose-housed laying hens. When birds are reared without perches, they can later have difficulty using perches due to low muscle strength, a lack of motor skills and an inability to maintain their balance, or they can have impaired spatial skills, which are necessary for moving around in a three-dimensional space (Gunnarsson et al., 1999, 2000).

Previous experience of perching influences the usage of perches (Appleby et al., 1983). This could indicate that perching is a learned behaviour. However, birds that are able to perch, which is most birds when they have access to perches that are easy to reach, seem to prefer roosting at night to sitting on the ground. Birds that are thwarted from access to perches show signs of frustration (Olsson and Keeling, 2000). Access to a perch is apparently a behavioural need (Gunnarsson, 2000). Perching therefore must be regarded as an innate behaviour.

Dust bathing and the recognition of dust as a suitable dust-bathing material also have to be learned in the first days of life. Chicks learn to recognize suitable dust-bathing substrates, and the different elements that make up a dust-bathing bout begin to appear in their behaviour repertoire, with full dust baths finally being performed when they are several weeks of age (Spinka, 2009).

Rearing conditions – and management during rearing – are shown to affect the behaviour of laying hens and their capacity to adapt to specific conditions in the laying period (Keeling, 2002).

Social learning plays an important role in helping chickens decide what to eat and what to avoid. The importance of social learning varies with age. In early life, the behaviour of the hen is important in encouraging chicks to peck at edible items. Older chicks can enhance their foraging success by observing the behaviour of conspecifics within their own social group. Day-old chicks avoid pecking at an aversive stimulus after observing the disgust responses of another chick. The young chick differs from the adult bird because the reward of food ingestion is not involved strongly in pecking motivation. Because of this, young chicks may be more attentive to the behaviours of conspecifics,

and less attentive to their own experiences of ingestion, than older birds (Nicol, 2006).

Social learning can contribute to the spread of cannibalistic behaviour in domestic fowl (Cloutier et al., 2002).

Individual recognition occurs mainly by visual cues and it has been shown, in studies where the physical characteristics of the birds have been manipulated, that it is the features of the head, such as comb shape, colour and size, that are most important (Keeling, 2002). Hens are unable to discriminate between familiar and unfamiliar birds, except when they are very close to them (Dawkins, 1995).

7.3 Different Types of Social Behaviour

Social behaviour

The small, mixed-sex and -age group of jungle fowl consisting of perhaps 5–30 individuals has a well-developed social system (Keeling, 2002). It may be difficult for the birds to set up and maintain stable social systems in the commercial flocks of today where large numbers of individuals of the same age and sex are kept together. These differences from the natural group composition have consequences for the social organization of the flock.

The pecking order is a social control system. Moving, pecking or mating of each bird is regulated by its dominant neighbours. In undomesticated fowl, males prohibit aggression between females within 3 m of them (McBride, 1970). Keeping large groups of hens without a cock seems to increase the risk of pecking behaviour. It has been shown that the presence of males reduces aggression between females (Odén et al., 1999, 2000).

Communication

As in all social species, domestic fowl have a well-developed system of communication. Visual and acoustic signals are the most important. The main means of communication in fowl are the calls. Calling begins while the chick is still in the shell and communication between the chick and the mother starts before the chick hatches (Wood-Gush, 1971; Evans et al., 1993; Keeling, 2002).

Brooding hens utter food calls while pecking, especially when their chicks are not feeding and/or have been at some distance for several seconds. A chick's response to its mother's feeding activities is more pronounced in the presence than in the absence of a food call. A chick responds to this call by approaching its mother and increasing its pecking; its response will become more efficient as it grows older. Food calls can be regarded as an arousal vocalization that directs the chick's attention to a food item chosen by a hen. A hen's food call serves as a form of cultural transmission in which information regarding available food is disseminated from hens to chicks (Wauters and Richard-Yris, 2002; Allen and Clarke, 2005).

In male chickens, alarm and food calls – vocalizations addressed to conspecifics – have been found to depend on social context, i.e. the presence of a conspecific audience increases the probability of calling (Evans and Evans, 1999).

Behavioural synchronization

Although hens are flock animals, their behaviour synchronization is not general. They have pronounced behaviour synchronization when seeking night rest. If a chicken in a flock finds something edible and then another chicken also discovers it, all the others will rush over. However, this behaviour cannot be described as behaviour synchronization.

During daytime, chicks kept with their mothers perform frequent alternation between active and inactive phases. Riber *et al.* (2007) have investigated this behaviour and found that the higher degree of social synchronization in the active phases observed in brooded chicks is likely to be due to social facilitation by the brooding hen. So, if the brooding hen is carrying out a particular behaviour, this motivates the chicks to join her in the activity. Thus, the activity of chicks is associated with the activity of the hen. More chicks perform the same activity as the brooding hen as compared to a different activity. The brooding hen has a greater synchronizing effect than any of the chicks in the group because she modifies her behaviour in the presence of chicks and thereby actively encourages the chicks to join her in many of her activities. The absence of the brooding hen in today's commercial broiler and laying hen production would therefore be expected to lead to increased social desynchronization of activity in groups of chicks. However, under commercial conditions, chicks seem to have preserved this pattern of successive active and inactive phases. Brooded chicks seem to have a higher social synchronization in activity in adulthood than non-brooded chickens (Riber *et al.*, 2007).

Dominance features, agonistic behaviour

The natural group of jungle fowl consists of perhaps 5–30 individuals; a dominant male, females and their young. Juvenile males are subordinate to the dominant male and may even be expelled from the group when they mature. A flock moves within its home range, which may overlap with the home ranges of other flocks, but each flock returns to a specific roosting site as dusk falls (Dawkins, 1989; Keeling, 2002).

These differences from the natural group composition can have consequences for the social organization of the domesticated flock. Thus, the presence of males reduces aggression between females, and this is the case even in large flocks with only a few males (Odén *et al.*, 1999, 2000, 2004).

Increased aggression can be a problem under commercial conditions, resulting in a high prevalence of pecking wounds (Fig. 7.6). In caged chick-

Fig. 7.6. Emaciated hen from a cage with a pecking injury on the back of the head. Note the blood coagulation on the outer edge of the injury and the thin pectoral muscles. The beak, held in the forceps, is hidden by the comb (image courtesy of Jan Svedberg).

ens, agonistic behaviour increases with increased number of birds per cage (Al-Rawi and Craig, 1975).

The broody hen behaves in an aggressive way towards other hens (Kent, 1992) and, even though aggression is low in loose housing, it is high in front of the nest boxes (Odén *et al.*, 2002).

Originally, it was thought that there would be problems with aggression in floor-housed and aviary flocks, because it would not be possible for a bird to recognize all the other individuals. But aggression in large groups is not especially high, and it has been proposed that there is no hierarchy in larger groups of birds because the costs outweigh the benefits in terms of priority of access to resources. Instead, it is proposed that birds use direct assessment of status based on cues such as comb and body size when they meet (Mench and Keeling, 2001).

Even in cages where the birds are kept in small groups, a pecking order exists and the absence of males increases aggressive behaviour. Aggressive behaviour increases with the number of animals per unit of cage floor area (Vestergaard, 1981).

Aggression in poultry can take the form of subtle threats and avoidances, pecks and even fights and chases. Although the more severe forms of aggression are rare in stable groups of birds, they are more common when males or unfamiliar birds meet. During a fight, a bird leaps at the other bird, holding forward the spurs on the back of its legs. The beak is also a dangerous weapon and pecks from one bird to the comb of another often leave small wounds or scratches. Fights consist of repeated pecks, but usually pecks occur singularly and even one peck can be sufficient to establish dominance. Pecks are nearly always directed at the head of the other bird and are made up of a hard downward stabbing movement. Since threats vary in severity from intention movements to higher aggression, a bird can give a threat when it just raises its head as though to peck another bird. Submissive gestures are when a bird lowers its head or turns away. Sometimes, these gestures are so subtle that it is difficult for the observer to see that there has been a threat or a submissive signal given. Males and females usually have separate hierarchies and young birds are almost always subordinate to adults (Keeling, 2002).

In mixed-sex flocks, males and females generally develop separate dominance hierarchies and rarely show aggression towards one another (Spinka, 2009).

7.4 Behaviour in Case of Danger or Peril

Domestic fowl have three types of alarm calls at impending danger. One is emitted as a warning when hawks and other aerial predators are discovered; one is a warning for approaching ground predators; and finally, one is given by a hen that is being held by a predator. The alarm calls contain sufficient information for the other birds to act appropriately (Wood-Gush, 1971; Evans *et al.*, 1993).

When attacked directly, the individuals in a flock of hens flee in panic, squawking, in all directions by feet or on the wing.

In some birds, instead of creating flight behaviour, the escape behaviour gives way to an alert posture with the bird standing motionless and fixating visually on some aspect of the environment. The posture is not rigid, in fact the bird looks relaxed but, once adopted, the posture tends to be maintained. Studies of heart rate show that 'flighty' birds have a huge increase in heart rate – up to 450 beats per min, which might be the maximum possible physiologically – and recover quite quickly, whereas 'placid' birds show slightly less panic and less increase in heart rate, but the high heart rate remains as long as the motionless posture is maintained. Thus, docile birds may be as frightened as flighty birds in physiological terms (Duncan and Filshie, 1980).

Tonic immobility is an unlearned state of profound but reversible motor inhibition and reduced responsiveness which is induced by physical restraint. It can be induced in domestic fowl and also in other species. It is elicited, but not maintained, by a relatively brief period of physical restraint. Typically, the fowl struggles and attempts to escape when first held but soon adopts an immobilized posture which may persist from a few seconds to even hours after termination of restraint. It is thought to represent the terminal reaction in a chain of antipredator responses and is considered to be related positively to fear, because procedures intended to elicit fear prolong the response whereas fear reducers attenuate it. Heart rate is generally increased on exposure to a frightening situation. But the relationship between heart rate and tonic immobility in the fowl is not clear. Respiration rate increase initially and is associated with the intense struggling common to many birds before tonic immobility ensues. The birds are still capable of processing information during tonic immobility (Jones, 1986; Cashman *et al.*, 1988).

7.5   Some Normal Physiological Frequency Values of Interest at Clinical Examination

The body temperature of chickens varies between 40.6°C and 43°C, with an average of 41.7°C (Robertshaw, 2004).

The heart rate in undisturbed but active hens varies between 280 and 320 beats per min. It may decrease to 230 beats per min at night (Duncan and Filshie, 1980).

The respiratory rate in resting birds with a body weight of 5.2 kg is stated, according to Duke (1977), to be 13 breaths per min, and for 'adult birds' is stated, according to Kassim and Sykes (1982), to be 23 ± 9 breaths per min. Fedde (1993) states 17 breaths per min for males with a body weight of 4.2 kg, 27 breaths per min for females of 3.4 kg and 23 breaths per min for females of 1.6 kg.

7.6   Active and Resting Behaviour Patterns

Activity pattern, circadian rhythm

Chickens are day-active animals that seek night rest on perches when darkness is falling. The birds show a relatively standardized diurnal pattern of behaviour. Hens leave the perches just before sunrise, even though there might be a light intensity less than 2 lux, start preening and then enter a longer feeding period. If the bird is an adult laying hen, it usually lays an egg in the morning. Egg laying lasts between 30 and 90 min. In the middle of the day, there is a peak in dust bathing and, in the afternoon, preening behaviour. Birds may rest in the afternoon if food is freely and easily available; otherwise, there is more foraging behaviour. Mating behaviour in flocks usually occurs in the late afternoon. Finally, there is a second peak in feeding behaviour before the birds go to roost in the evening. Some birds are ground roosting, but the majority of the domesticated species roost up off the ground in bushes and trees whenever possible. Birds usually start to move towards the roost and jump up to it shortly before darkness. Free-living hens roost on branches in trees at night, and laying hens in aviary systems or in modified cages provided with perches also make extensive use of these for night-time roosting. There is some evidence that birds have preferred places to roost and operant studies have shown that birds are motivated to find access to a perch for night-time roosting. So roosting on perches seems to be important to hens. Under natural conditions, roosting up off the ground probably has survival value by reducing predation from night-hunting ground predators (Huber, 1987; Keeling, 2002; Olsson and Keeling, 2005; Riber et al., 2007).

Under natural conditions, chickens can spend over 90% of their time during the day in feed searching and feeding (Keeling, 2002).

Preening behaviour occurs primarily in the morning and late afternoon. However, dust-bathing behaviour takes place early in the afternoon. On average, fowl dust bathe for about 30 min every 2–3 days (Vestergaard, 1982).

Exploratory behaviour

On the first day of its life, the young chicken has already begun to explore its environment by pecking at potential food objects.

Chickens perform exploratory behaviour by pecking with the beak and scraping with their feet (Gunnarsson, 2000).

The domestic fowl shows examining behaviour by watching the object that has drawn its attention, sometimes by turning its head and looking at the object with one eye (Dawkins, 1995, 2002).

Diet, food searching, eating and eating postures (feeding behaviour)

Hens have retained the food-searching behaviour that characterizes the jungle chicken. This behaviour consists of pecking with the beak and scraping with the feet and claws after feed (Simonsen, 1983). Hens drink often and in small quantities and therefore a permanent water resource is a necessity. An adult chicken requires 150–200 ml water each day and night at normal outdoor temperature (Appleby et al., 1992). Poultry drink by putting their beak end just under the water surface, then scooping up water in their beak and raising their head so that the water runs down the oesophagus.

In most commercial herds, hens are given nipple drinkers instead of troughs or cups. Some nipple drinkers require the birds to drink in an unnatural way, or do not give enough water flow (Cornelison et al., 2005).

As omnivores, domestic fowl eat seeds, grubs, worms and insects. They search and obtain their food by scratching and pecking the ground and alternate feeding and drinking over the day. During eating behaviour, the bird scratches the ground, pecks and swallows the food.

Selection of resting areas, resting postures, sleep

In groups of jungle fowl and medium-weight domestic hybrids, all the birds perch during most of the night, except for some anticipation of sunrise. Jungle fowl only rarely (1% of the birds) perch during the day, whereas 10% of perching is recorded during the day for layer hybrids (Blokhuis, 1984).

Perching behaviour starts as early as 5 days of age in chickens and is well developed at 4 weeks of age. Some strains can start using a perch very quickly as adults, but some other strains apparently never do so (Hughes and Elson, 1977).

Birds having the opportunity to perch at 4 weeks of age show a higher proportion of perchers than those not given the opportunity, for up to 34 weeks of age. Chickens reared without perches begin to perch only slowly as adults (Appleby and Duncan, 1989). Hens can lie down on the ground but resting, in the proper sense, is always carried out on a perch if possible. Hens reared without perches often have problems in perching, and this can have detrimental effects on the birds. Access to perches by 4 weeks decreases the prevalence for cloaca cannibalism in commercial flocks. Rearing without early access to perches impairs the cognitive spatial skills of the domestic hen and the effect is both pronounced and long lasting (Gunnarsson, 2000). Roosting is a highly motivated behaviour. Thus, hens will push through a weighted door to gain access to a perch at night (Olsson *et al.*, 2002).

In the sleeping posture, when perching, the head is tucked into the feathers above the wing base or behind the wing and feathers are slightly fluffed; sometimes, the wings droop. If standing, a slight crouching posture is shown and the tail is down (Broom and Fraser, 2007).

Locomotion (walking, running)

Chickens move by walking slowly and by running or flying short distances. Generally, they are poor flyers.

Swimming

Domestic fowl do not swim.

7.7 Behaviour at Defecation and Urination (Eliminative Behaviour)

Like all birds, chickens defecate randomly over their enclosure.

7.8 Body Care (Preening), Cleanliness

Birds' body care consists mainly of taking care of their plumage and of stretching movements. The care of the plumage consists of dust bathing followed by shaking the plumage. They use their beaks to perform control and care of the feathers and plumage.

Stretching is mainly by wing flapping, which consists of rapid, simultaneous movements of both wings, during which the wings are stretched and flapped together above the back of the bird (Simonsen, 1983).

Dust bathing is performed once every 2–3 days and consists of the bird lying and rubbing litter material through its feathers (Vestergaard, 1982). The dust-bathing sequence starts with the bird scratching and bill-raking in the substrate (sand, dry earth, or the litter material). Gradually, it erects its feathers and squats down in the substrate. The part of the sequence that occurs when the bird is lying down contains four main elements: vertical wing shaking, head rubbing, bill-raking and scratching with one leg. This vigorous phase of dust bathing is followed by a phase during which the feathers are flattened against the body and the bird spends most time sidelying or side-rubbing, although interrupted with some of the elements from the first phase (van Liere, 1992). Then, the bird stands up and shakes the loose litter material from its feathers. In this form of 'dry shampoo', any excess lipids, as well as ectoparasites, are removed from the feathers (Keeling, 2002).

Laying hens, deprived of dust for 33 days, showed an average increase in the amount of lipids on the back feathers from 10.3 to 14.5 mg lipids per g feathers at the end of the dust-deprivation period. After the hens could dust bathe again, the original level was restored within 2 days. Also, the downy parts of these feathers appeared to be fluffier. Dust bathing presumably regulates the amount of feather lipids and maintains down structure in good condition (van Liere and Bokma, 1987).

Dust bathing shows a clear diurnal rhythm and, under unrestricted conditions, hens dust bathe about every 2 days. With respect to external factors, it has long been believed that dust bathing is socially facilitated, but this has been questioned in recent studies. The presence of a suitable substrate is an important stimulus for eliciting dust bathing, and hens seem to prefer substrates with a fine structure such as sand and peat (Olsson and Keeling, 2005). There is a strong demand for peat moss for dust bathing (de Jong *et al.*, 2007).

Birds usually dust bathe in litter, but in the absence of this they sham dust bathe as a vacuum activity (Vestergaard, 1982). Sham dust bathing is not satisfying or perceived as normal dust bathing, even for birds that develop dust-bathing behaviour in the absence of litter. This is because birds without previous experience of peat as litter have been shown in experimental studies to be as motivated to work to gain access to this substrate as birds used to dust bathing in peat (Wichman and Keeling, 2008).

7.9 Temperature Regulation and Climate Requirements

Temperature regulation

Birds lack sweat glands and, under heat stress, they rely on increased evaporation from the respiratory system for heat transfer. Domestic fowl regulate the temperature mainly by emitting evaporated water via respiratory organs, but a certain amount of warmth loss also occurs via the skin. Day-old chickens, as well as adult animals with undamaged plumage, have a greater capacity to cope with decreased rather than increased temperature. However, day-old chicks are quite susceptible to temperatures below 30°C. Hens that have lost parts of their plumage are extremely susceptible to cold. The upper lethal body temperature of the fowl is about 45°C, and this is little influenced by the age of the bird (Freeman, 1984).

The feather-pecking recipient birds with large areas of exposed skin have difficulty regulating their body temperature (Spinka, 2009).

Climate requirements

The optimal environmental temperatures are between 20 and 21°C for fully feathered domestic fowl. However, they can stand lower temperatures,

5–10°C, for even longer periods if given enough feed and provided that the climate is dry.

Hens are susceptible to rain and seek protection immediately. Dry chill is, on the other hand, a smaller problem if the birds have been able to adapt to cold and provided that they have a dry protected area free from draught and precipitation. Hens avoid damp surfaces.

During daytime, the frequent alternation between active and inactive phases seen in chicks has been interpreted as an adaptation to lack of thermoregulation. Under natural conditions, the chick will stay under the brooding hen to keep warm during the inactive phases. The domestic fowl is incapable of regulating its own temperature fully until it has passed the first 2 weeks of life. Chicks raised under commercial conditions have preserved the pattern of successive active and inactive phases despite the fact that the temperature is kept at a constant high level, i.e. the need for the chicks to seek a heat source to keep warm has been removed (Riber *et al.*, 2007).

The relative humidity in hen houses should be 70% ± 10%. The air usually contains traces of ammonia. In several countries, animal welfare legislation unfortunately accepts up to 25 ppm ammonia in the indoor air for hens and broilers. However, broiler fowl avoid ammonia at concentrations commonly found in poultry units, regardless of previous experience, causing it to show aversive behaviour at concentrations above approximately 10 ppm (Jones *et al.*, 2005).

7.10 Vision, Behaviour in Light and Darkness

Jungle fowl and domestic fowl are prey species and as such are well designed to detect and avoid predators. Vision is important. They have a well-developed colour vision and a visual field of about 330° (Wood-Gush, 1971). The bird's field of view only has a 30° overlap in which binocular vision could occur. The viewing distance is short, maybe 50–60 mm. They have a visual system that is well adapted to collecting spectral information. However, the penalty may be that a high illuminance is required for the system to work to its full potential. Current light environments in hen husbandry may impose some sensory deprivation on the hen, rendering aspects of her vision redundant. If hens could be reared under natural light, perhaps with some supplementation for those times of the year when the photoperiod is not conducive to development or laying, then most lighting concerns could be alleviated (Prescott *et al.*, 2004).

Hens use either a lateral visual field to explore an object by monocular fixation at an angle of either 30–39° or 60–69° from the median line. If they have to use binocular vision, they are limited to an angle of 18° at each side of the median line, and this is used only for very close objects, less than 20 cm from their head. Hens are unable to discriminate between familiar and unfamiliar birds, except when they are very close to them (Dawkins, 1995).

Change in object distance is associated with a change from lateral to binocular viewing. Hens tend to view distant objects laterally, while they preferentially view objects less than 20–30 cm away frontally, whether looking at another bird or at an inanimate object. As well as switching between lateral and frontal viewing, the hens also swing their heads from side to side with movements so large that the same object appears to be viewed with completely different parts of the retina, and even with different eyes, in rapid succession. When confronted with a novel object, the hens walk more slowly but continue to show large head movements. This may suggest that, unlike mammals, which gaze fixedly at novel objects, hens investigate them by moving the head and looking at them with different, specialized parts of their eyes (Dawkins, 1995, 2002).

There is a serious discrepancy between chicken and human spectral sensitivity. Thus, the 'man-calibrated' lux-meter is no longer the most appropriate instrument for most purposes in poultry research. Nevertheless, in studies on the effects of illumination levels, illumination is usually measured in lux, which may explain in part the inconclusive effects reported in the literature (Nuboer, 1993).

Chickens reared under permanent light develop 40% larger relative asymmetry than those subject to changing light and dark conditions (Möller *et al.*, 1999).

In a review of the scientific knowledge, Perry (2004) concludes that domestic fowl prefer light to dark, brighter rather than dimmer intensities and that they are sensitive to UV light. He also remarks that the hen's eye is structurally different from that of humans; hence, issues such as brightness and spectral intensity should not be extrapolated from human responses.

There is some evidence that low light intensities can have deleterious effects on the behaviour of birds. Pullets housed in a light intensity of 17–22 lux were found to be more fearful, showing marked avoidance activity in response to novel objects, compared with birds housed in a light intensity of 55–80 lux (Hughes and Black, 1974). Manser (1996) has made a comparison between light intensities measured in lux in buildings used by humans and in poultry houses (Table 7.1). Although vision is important to domestic poultry, and birds have better visual acuity in bright rather than dim light, they are often kept in light intensities of 5–10 lux or lower. Some impairment in the ability of the birds to engage in exploratory behaviour and social interaction is likely at these levels. Furthermore, intensities of 6 lux or below can lead to increased mortality in brooded chicks. These light levels are also associated with leg problems, eye abnormalities and breast blisters in growing birds. Hens have been shown to be more fearful when housed at light intensities of up to 22 lux, as compared with higher levels (Manser, 1996).

For layer and broiler chicks, the initial preference for bright light weakens with age, and this preference is attributable only to a change in partiality for resting in bright light when young and in dim light when older. This implies the need for spatially or temporally variable illuminance in poultry houses. For laying hens, fluorescent light is preferred to incandescent light (notwithstanding differences in perceived illuminance). There are indications that recognition of familiar from unfamiliar laying hens on the basis of feeding preferences and aggressive interactions is possible under bright white light (77 lux) but not dim white light (5.5 lux), and recognition

Table 7.1. Recommended light intensities (lux) in buildings used by humans and in poultry houses (after Manser, 1996).

Recommended light intensities in buildings used by humans	Lux	Commonly used light intensities in poultry houses	Lux
Office	500–750	Laying poultry	5–10
Living room (general)	100	Broilers	5
Hospital corridor in day	300	Turkeys	0.5–5
Hospital corridor at night	5–10		

is not possible under red or blue light at either illuminance (Prescott *et al.*, 2003).

7.11 Acoustic Communication

Hearing

For jungle fowl and domestic fowl, hearing is important for detecting dangers. They are sensitive to frequencies in the range of 15–10,000 Hz (Wood-Gush, 1971; Appleby *et al.*, 1992). Their sounds usually are emitted within the frequency of 400–6000 Hz. Their most efficient hearing is between 3000 and 5000 Hz (Temple *et al.*, 1984).

Vocalization and acoustic communication

Poultry usually have a big repertoire of utterances. A great number, roughly 30, of different vocalizations have been described for adult and juvenile chickens. Calling begins while the chick is still in the shell. In artificial incubators, chicks give peeping calls. Embryos respond to the calls of the parents with calls that further influence the parent behaviour. The embryos also vocalize to one another, which causes the development of the less advanced embryos to accelerate and thus ensure that all of the eggs hatch at around the same time (Spinka, 2009). The calls of adult birds are grouped into crowing, warning, food, alerting, laying and several other types of calls. The crowing of the cockerel, usually given in the early morning, is associated with territorial defence (Wood-Gush, 1971; Evans *et al.*, 1993; Keeling, 2002; Manteuffel *et al.*, 2004).

The laying call in hens, crowing in males and predator alarm calls are the vocalizations most easily recognizable in fowl. The laying call is also called the 'gakeln' call and consists of an elongated note that increases slightly in pitch, followed by a succession of shorter notes. It functions possibly to attract the male to escort the hen to and from the nest, although it is also thought to be an indicator of frustration. The frequency of crowing is related to comb size, with larger, more dominant males crowing at a higher rate. Crowing rate is probably used by females and rival males as an indicator of fitness, since it is only larger, healthy males that can perform crowing at a high rate. Ground predator alarm calls are harsh, with a quick start to the call, and elicit standing erect with a vigilant posture, whereas aerial predator calls gradually increase in intensity and elicit crouching or running for cover. Studies investigating the audience effects of alarm calling have shown that males give most alarm calls when there is a female of the same species nearby (Leonard and Horn, 1995; Keeling, 2002).

During movement, the broody hen will utter clucks that stimulate the chicks to follow her, and at night she will emit roosting calls to encourage her chicks to follow her up on to perches (Riber *et al.*, 2007).

7.12 Senses of Smell and Taste, Olfaction

It has long been thought that the sense of taste in hens is not very well developed. However, hens might have a better-developed sense of taste than previously believed (Wood-Gush, 1971). They avoid sour, acrid and salt feedstuffs (Appleby *et al.*, 1992).

Fowl are very sensitive to water temperature, are reluctant to drink warm water above ambient temperature and can discriminate differences as small as 3°C (Appleby *et al.*, 2004).

The nasal cavity of many bird species has an olfactory epithelium that is structurally similar to that found in mammals. Experiments and observations of behavioural responses to olfactory cues provide evidence of a well-developed sense of smell in birds (Appleby *et al.*, 2004).

Olfaction is an important factor in the chicken's perception and regulation of its world. It has been shown that chickens prefer the familiar soiled substrate from their home nest or cage rather than clean wood shavings or those soiled by a strange conspecific. Their preference for familiarity has been shown to embrace artificial smells (pure odorants) as well as the complex mixture of odours associated with naturally occurring stimuli such as soiled substrates (Jones and Roper, 1997).

Chicks can discriminate between the odour of their own soiled bedding material and the home cage of an unfamiliar individual. Chickens can detect and respond to a wide range of olfactory stimulants in diverse contexts. Olfactory familiarization does not seem to be utilized for the rapid development of individual discrimination by chicks. However, this does not necessarily imply that olfaction and audition play no role in social discrimination by chicks. Olfaction has been implicated in the formation of attachments, the elicitation of fear responses by alarm- and predator-related odours, the control of feeding and drinking and the avoidance of noxious substances. Chickens can taste but they may not perceive similar sensations to those of humans (Jones and Gentle, 1985; McKeegan, 2004; Porter *et al.*, 2005).

7.13 Tactile Sense, Sense of Feeling

Tactile touch

Domestic fowl do not groom each other. However, they perform allopreening, which may be related to feather pecking (Wood-Gush and Rowland, 1973; Vestergaard, 1994; Leonard *et al.*, 1995).

Sense of pain

Birds possess pain receptors and show aversion to certain stimuli, which can be interpreted as experiencing pain. They show fearful behaviour and avoid frightening situations, implying that they experience fear. They show behaviour indicative of frustration. If suffering is defined as 'a wide range of unpleasant emotional states', then it is clear that domestic fowl are capable of both suffering and indicating fearfulness (Duncan and Woodgush, 1971; Jones and Faure, 1982; Jones, 1986; Gentle and Hunter, 1991; Gentle and Wilson, 2004).

By gait scoring, it is shown that broiler chickens feel pain at gastrocnemius tendon slippage, tendosynovitis and severe contact dermatitis (Berg and Sanotra, 2001; Bradshaw *et al.*, 2002). Studies show clearly that hens feel pain when affected by joint disorders, e.g. arthritis (Gentle and Wilson, 2004).

Skeletal fractures are common in hens when they are taken from their cages, but are also reported from aviary systems. As pain following skeletal fractures in humans is common, it might be that such injuries also cause pain in hens. However, no studies seem to have been performed to see if this is true.

Osteoporosis is the major factor predisposing laying hens to bone fractures. If hens are given the opportunity to exercise, this can decrease the severity of osteoporosis markedly (Whitehead, 2004).

The beak is abundantly innervated. There is clear behavioural, psychological and anatomical evidence that hens feel both acute and chronic pain after de-beaking (Nyström and Hagbarth, 1981; Desserich *et al.*, 1983, 1984; Gentle and Wilson, 2004).

7.14 Perception of Electric and Magnetic Fields

A study performed by six different laboratories in four different countries referred to by Fernie and Reynolds (2005) concluded that exposure of chicken embryos to a 10-mG pulsed magnetic field increased the incidence of abnormal embryogenesis. Chickens have the ability to orient using magnetic cues (Freire *et al.*, 2005).

7.15 Heat and Mating Behaviour, Nesting, Egg Laying, Brooding

Domesticated breeds do not perform all the courtship displays of some wild species, but visual displays are nevertheless important. The cock takes the initiative in sexual behaviour, moving around the hens (Fischer, 1975).

Domestic hens are sexually mature at 16–18 weeks of age and they start egg laying at 18–20 weeks. Mating in domestic fowl is performed as in other birds.

In the times when on each farm there would be a flock of hens and a cock mostly loose among the buildings, it would often happen that a hen would suddenly stop delivering her eggs in one of the nest boxes in the hen house and would instead disappear regularly during parts of the day. After a while, she would be totally absent for some weeks. Although the farmer's wife often tried to find where the hen had laid her eggs, she often did not do so — the nest was usually in a place difficult to find and reach. Generally, the hen reappeared after some weeks with a brood of newly hatched chicks.

Several studies from the 1970s (e.g. Duncan *et al.*, 1978; Huber *et al.*, 1985; Kent, 1992; Keeling, 2002; Weeks and Nicol, 2006) have demonstrated many details in pre-laying, nest-building and egg-laying behaviour, the hormonal processes behind them and the triggering factors. Pre-laying behaviour has not changed in any noticeable way, even as a consequence of selection, and hens still show the same phases of nesting behaviour as their ancestors and as birds kept in the wild. The first phase of nesting is nest-site selection. The bird may walk considerable distances during this phase and explore many potential sites before selecting one where she starts to build a nest. Laying hens appear to have an instinctive need to perform pre-laying (nest-building) behaviour and have a strong preference for a discrete, enclosed nest site, for which they will work hard to gain access as oviposition approaches. Access to a nest site is thus a high-ranking priority for laying hens, preferred over food at this time. During nest building, the bird scrapes out a hollow with her feet and rotates in it, trying to deepen the nest. She also rakes loose material towards the edge of the nest to build up a raised edge. In commercial keeping of egg-laying hens, different sorts of surfaces are used in the nest boxes. Preference tests show that hens avoid artificial surfaces, like wire mesh, perforated plastic or synthetic grass for egg laying. A hen usually lays eggs in a clutch, at which point she stops laying and starts incubating them. However, nowadays, modern strains of egg-laying hens do not usually show brooding behaviour, a result of the selection for intense egg production. While brooding, hens rarely eat and leave the nest mainly to drink and defecate. The broody hen behaves in an aggressive way towards other hens. Pairs of broody hens keep at a greater distance from each other than pairs of non-broody hens.

The daily cycle of mating in breeding birds is related to the egg-laying cycle, because the hen's fertility is lower around the time of ovipositon. Mating is therefore most frequent in fowl in the afternoon because their eggs are laid in the morning (Spinka, 2009).

7.16 During and After Hatching

Behaviour during and after hatching

In chicks, part of the yolk sac remains in their abdomen, which usually provides sufficient nourishment for the first 2–3 days of life. However, young chickens are precocial and will, on the first day of life, start pecking at many potential food items and readily explore their environment. During the first 10–12 days, the chicks are in close contact with the hen and are brooded regularly, whereas later they feed independently from her but still sleep under her. This stage lasts until they are 6–8 weeks of age (Wood-Gush, 1971; Keeling, 2002; Riber et al., 2007).

Litter size

The jungle fowl lay about 10 eggs per clutch.

Play behaviour

Young chicks frolic and spar with one another, a play behaviour that resembles adult chasing and fighting (Spinka, 2009).

7.17 The Chick

When the hen ceases to brood, the chicks will be driven away, which might occur when they are somewhere between 6 and 12 weeks old (Wood-Gush, 1971). Older chicks can enhance their foraging success by observing the behaviour of conspecifics within their own social group (Nicol, 2006).

7.18 Assessment of Domestic Fowl Health and Welfare

For thorough overall inspection of the flock or group of birds, special attention should be paid to bodily condition, movements, respiration, condition of plumage, eyes, skin, beak, legs, feet and claws and, where appropriate, combs and wattles; attention should also be paid to the presence of external parasites, to the condition of droppings, to feed and water consumption, to growth and, during the egg-laying period, to egg production levels.

The healthy bird

The healthy bird has sounds and activity appropriate to its age, breed or type, clear bright eyes, good posture, vigorous movements if unduly disturbed, clean healthy skin, good plumage, well-formed shanks and feet, effective walking and active feeding and drinking behaviour.

The sick bird

Symptoms

The intense selection for growth, combined with selection for feed efficiency, has resulted in broilers being less active and having difficulty walking. In a study of 39–49-day-old broiler chickens, the birds spent 76–86% of their time lying down (Weeks et al., 2000). The study suggests that lameness, with its associated disability and probable pain, increases the time spent lying down significantly. Selection for high breast-meat yield has moved broilers' centre of gravity forward, which may unbalance them and predispose them to sitting behaviours. It is likely that sternal recumbency (here termed 'lying') is the most comfortable, and least physically demanding, posture for a bird with leg problems. Inhibition of activity, including reduced grooming and exploratory behaviours, is often noted in animals experiencing chronic pain. Although deemed an important activity for both layers and jungle fowl, in the study dust bathing occupied less than 0.5% of broilers' time. Reduced dust-bathing activity in lame broilers is a function of pain. The time the broilers spent ground pecking was under 3%, thus dramatically less than the 60% seen in their genetic ancestors, Red Jungle Fowl (Weeks *et al.*, 2000).

Nervousness ending in panic may occur in laying hens if frightened by sudden noise or movements, e.g. people opening the door to their room. Hens in cages, as well as hens kept in loose systems, can be affected and the risk increases with flock density (Hansen, 1970; Broom and Fraser, 2007). In large flocks of laying hens in loose-housing systems, such panic can result in birds rushing to one end of the house, thus forming heaps, such that those underneath are in danger of suffocation.

Some years after the introduction of cages for egg-laying hens, reports came of spontaneous fractures in the extremities or the breastbone caused by osteoporosis resulting from inactivity, as well as feather wear caused by contact with the cage walls (Obel, 1971). Birds in cages without the space to flap their wings have wing bones half as strong as those of birds in percheries where wing flapping can be performed (Knowles and Broom, 1990).

Examples of abnormal behaviour and stereotypies

Feather pecking is a potentially harmful behaviour which may be found in all housing systems. It is a pecking activity involving the feathers only and is often thought to be a form of redirected behaviour (Blokhuis, 1986; Vestergaard and Lisborg, 1993). Moreover, the feather pecking activity eventually may degrade feather cover, which may interfere with the bird's body heat regulation and, in crowded systems, may give rise to skin damage caused by wear and tear from the environment and flock mates. Measures such as the additional provision of grain during rearing have been shown to reduce feather pecking damage during laying, even when no additional grain was provided during the actual laying itself (Blokhuis and van der Haar, 1992). Access to litter during the rearing period has an important effect in reducing the amount of feather pecking in

adults. This effect can be through either redirected ground pecking or abnormal dust bathing (Blokhuis, 1986; Vestergaard and Lisborg, 1993; Vestergaard *et al.*, 1993) in birds deprived of litter during rearing. Feather pecking can be prevented by offering adequate substrate, which should also include the rearing period (Blokhuis and van der Haar, 1992).

However, there are also genetic factors involved and an important preventive measure is identification of the genetic characteristics of feather peckers and non-feather peckers in order to breed birds that have a reduced tendency to develop feather pecking (Rodenburg and Koene, 2004).

Hens exposed to frustration show a considerable increase in stereotyped behaviour (Duncan and Woodgush, 1971).

Examples of injuries and diseases caused by factors in housing and management, including breeding

Ammonia concentrations at 25 and 50 ppm induce eye lesions in young broiler chickens after day 7 of initial exposure. Light intensity of 0.2 and 20 lux for 14 days further exacerbates eye lesions. However, light intensity alone yields no significant eye lesions (Olanrewaju *et al.*, 2007).

Selection for fast early growth rate and feeding and management procedures which support growth have led to various welfare problems in modern broiler strains. Problems which are linked directly to growth rate are metabolic disorders causing mortality by the sudden death syndrome and ascites. Fast growth rate generally is accompanied by decreased locomotor activity and extended time spent sitting or lying. The lack of exercise is considered a main cause of leg weakness, and extreme durations of sitting on poor-quality litter produces skin lesions at the breast and legs (Bessei, 2006). The increase in egg production has influenced the disease panorama. There is a positive association between increased egg production and increased incidence of salpingitis (Lindgren, 1964).

The prevalence of footpad dermatitis at the time of slaughter is estimated to be 5–10% for severe lesions and 10–35% for mild lesions in Swedish broiler chickens (Berg, 1998). There is a positive association between bad litter quality and insignificant litter depth and the prevalence of footpad dermatitis in broilers.

Cannibalism sometimes follows from feather pecking, but can also arise independently (Spinka, 2009). As caged hens have no possibility of escaping, they cannot avoid being pecked if attacked. This can result in wounds that usually lead to more pecking. The hen subjected to pecking may find it difficult to compete for feed and may change her behaviour, which in turn often leads to intensified attacks from the other hens, finally causing its death (Fig. 7.6). Hens in cages may develop foot injuries (excessive calluses) from standing on the grid floor (Fig. 7.7).

Symptoms of pain

The beak is abundantly innervated (Desserich *et al.*, 1984). Beak trimming involves both cutting and

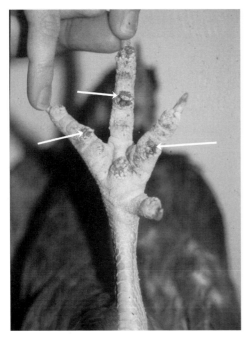

Fig. 7.7. A caged hen showing foot injuries (excessive calluses) from standing on the grid floor (image courtesy of Stefan Gunnarsson).

cauterizing the beak, and a significant amount of the remaining beak, including some of its nerves, is damaged by the cautery. After about 15 days, neuromas are present at the end of the nerves. It is clear that the neuromas are formed as a result of the amputation and that they probably give rise to abnormal spontaneous nervous activity. The activation of specific nociceptors in humans and spontaneous discharges originating from stump neuromas are implicated in acute and chronic pain syndromes (Breward and Gentle, 1985; Gentle, 1986). Clinical studies of pain in the amputated extremities in humans where neuromas have arisen show that tapping the neuromas accentuates not only local pain but also phantom pain from the parts originally amputated (Nyström and Hagbarth, 1981). There are reasons to believe that the reactions in the amputated beak of chickens will be the same during a part or the whole of the life of these animals (Gentle, 1986; Gentle and Wilson, 2004).

Lameness and other locomotion behaviour is the main symptom of pain caused by joint disorders.

Feather pecking may predispose to cannibalistic pecking activities. For the recipient, having feathers pulled out is painful (Spinka, 2009).

7.19 Capturing, Fixation, Handling

When hens are to be captured, the operation must be performed with great caution and prudence in order to avoid panic reactions. Hens should be carried with one hand under the breastbone to take the weight of the bird and simultaneously to help restrain the bird's legs – at the same time, the other hand should be kept above the back of the hen.

7.20 The Importance of the Stockman

Chickens or hens that receive frequent human contact of an apparent positive nature such as gentle touching, talking and offering of food on the hand from an early age have improved growth rates and better resistance to infection than birds that either receive minimal human contact or have been scared deliberately (Hemsworth, 2004).

8 Turkeys (*Meleagris gallopavo*)

8.1 Domestication, Changes in the Animals, their Environment and Management

The modern domestic turkey is probably descended from the south Mexican wild turkey (*Meleagris gallopavo gallopavo*) and domestication was started approximately 2500 years ago by the North American native peoples (CoE, 2001).

Although turkeys were probably already domesticated in Mexico in prehistoric times, they are a relatively recent addition to modern agriculture, having been kept for only a few centuries. The most common commercial breeds are derived from crossings between the wild turkey of North America and turkeys that were originally taken to Europe (Keeling, 2002). It might have been either Columbus who took turkeys to Spain in 1498 or Cortez in the early 1500s. However, turkeys spread rapidly throughout Europe during the 1500s (Brant, 1998). The first occasion when turkeys were on an official menu in Europe may have been at the wedding dinner for Charles XI of France and Elisabeth of Austria in June 1570 (FAWC, 1995).

Wild turkeys live in North America, from Pennsylvania to south Mexico. There exist seven subspecies in a variety of habitats ranging from forests to plains, and they require cover for nesting and trees for roosting and protection from predators. They spend the majority of the daytime on the ground searching for food, fly only in short bursts, and are non-migratory, but some subspecies move locally between breeding and winter areas (Brant, 1998).

The body weight of the wild turkey varies markedly from one subspecies to another. There is marked sexual dimorphism in both wild and domestic turkeys. For instance, in the subspecies *M. gallopavo gallopavo*, the average weight of a male is 7.5 kg and a female is 4 kg. Turkeys are diurnal. Where trees are available, most wild turkeys rest at night towards the top of special roost trees (CoE, 2001).

Wild turkeys, *M. gallopavo*, exhibit a particularly striking array of ornamental characteristics, including a bare, brightly coloured head covered with bumpy caruncles and a distensible snood, a black hair-like beard, tarsal spurs and a large fanned tail (Buchholz, 1995).

The turkey has undergone intensive selection since the 1960s to produce a fast-growing meat bird. The majority of commercial turkeys are white-feathered (Fig. 8.1), though there are still some bronze- and black-feathered breeds. Some heavy domestic turkey strains have difficulty in performing certain aspects of the behavioural repertoire, for example flying, locomotion, preening, perching and mating, because of their increase in weight and change in body form. Domestic turkeys are less active than their ancestors (CoE, 2001). Broad-breasted strains of turkeys weigh more than 30 kg, often have pectoral hypertrophy and are frequently lame. Traditional turkeys, closer to the wild birds, weigh 9 kg on average (Abourachid, 1991). There were about 650 million turkeys raised in the world at the end of the first decade of the 21st century (Spinka, 2009).

Genetic selection of fast-growing strains has increased locomotor problems and made natural mating impossible. Since World War II, the turkey-housing environment and management have changed, mainly through crowding more birds per area of floor and by not having perches or windows in the poultry houses. These changes have led to impaired air quality through dust, a higher per cent of manure gases, mainly ammonia, and impaired bedding quality through high moisture and dirtiness. A result of this has been an increase in diseases and injuries, either directly through environmental factors or through changed behaviour patterns in the animals. In order to counteract the changed behaviour, especially pecking, interferences are made with both the birds and the environment. The birds' beaks are partially amputated, so-called beak

Fig. 8.1. Young turkeys in a commercial farm in the USA (image I. Ekesbo).

trimming, and the light level in poultry houses is reduced to a minimum.

In a survey of animal welfare in intensive production of turkey broilers, Martrenchar (1999) mentions how a range of different photoperiods is used in practice, for example constant day length, increased day length with age, intermittent lighting and others, in order to increase production efficiency or prevent behavioural disorders. The survey underlines how high stocking densities have caused several drawbacks such as poor air quality, cannibalism and impairment of the birds' gait.

As artificial insemination is applied, males and females are usually kept in separate compartments on farms.

The term 'hen' is used for the adult female turkey, 'cock' or 'stag' for the male, and 'chicken' or 'poult' for young birds up to about 3–4 months of age. A hatched group of chickens may be called a 'brood'.

8.2 Innate or Learned Behaviour

Turkeys show the same innate behaviour as other domestic fowl. Thus, turkey pullets will feed readily even in the absence of the hen. However, some artificially incubated poults are unable to initiate feeding or drinking and may die as a consequence of the lack of maternal association and the learning facility this provides (Broom and Fraser, 2007). Social facilitation of feeding behaviour is provided by other poults or by the hen as she pecks the ground and calls (Hale *et al.*, 1969).

Turkeys can learn to distinguish between conspecifics of their own group and conspecifics of another group (Buchwalder and Huber-Eicher, 2003, 2005).

8.3 Different Types of Social Behaviour

Social behaviour

The social organization of wild turkeys is complex. They have a cohesive social structure. A linear social hierarchy is established within the different groups. According to the season, males and females form groups and subgroups of different size and function. Male and female turkeys live in large separate flocks out of the breeding season – winter flocks. In spring, these flocks split into small bands of adult males (male bands) and larger groups of females (display flocks). In autumn, adult males and females move separately to the winter areas and form new 'winter flocks'. Juvenile males leave the 'brood flocks' and establish their own 'winter flock' (CoE, 2001).

Great social cohesion within poult groups is notable from the first day after hatching, but attachment to the hen turkey is also evident (Broom and Fraser, 2007).

Communication

Turkeys communicate by calls, tactile means and visual displays (CoE, 2001). Vocal and visual signals are used in the maintenance of close contact (Broom and Fraser, 2007).

Behavioural synchronization

Turkeys have pronounced behaviour synchronization when seeking night rest (Brant, 1998).

Turkeys behave like domestic fowl in that, if an individual in a flock finds something edible and then another turkey also discovers it, all the others will rush over. However, this behaviour cannot be described as true behaviour synchronization.

Dominance features, agonistic behaviour

Groups of young turkeys living together show only sporadic sparring until about 3 months old. Over the next 2 months, however, there is a gradual increase in threatening and fighting, until a peak is reached at about 5 months of age. By that time, social hierarchies are well established and fighting is reduced. Males fight more vigorously than females, but the pattern of fighting is the same for the two sexes, except that the associated morphological structures are reduced in females (Hale *et al.*, 1969).

When a turkey identifies a potential competitor, the most common social interaction is a simple threat. One bird may submit to the other, otherwise both birds circle each other warily with wing feathers spread, tails fanned and each emits a high-pitched trill. One or both turkeys may then leap into the air and attempt to claw the other. The one that can push, pull or press down the head of the other will usually win the encounter. Such bouts normally last a few minutes. Haemorrhages may occur during a tugging battle, since richly vascularized skin areas may be torn, but actual physical damage is usually slight and birds do not fight to the death (Broom and Fraser, 2007).

Threat displays in turkeys are highly ritualized and provide signals eliciting attack. Fighting typically occurs when two strange birds meet for the first time, and the winner of the fight subsequently dominates the loser. Turkeys show a variety of threat displays, ranging from 'mild threat', head raised, looking towards the opponent, to strutting against other males and postures which diverge from strutting in various respects. The only common feature of these latter postures is the extended snood and the enhanced coloration of the carunculated areas of the head and neck, with a brilliant red predominating during threat display. Body posture during threat tends to be lateral to the opponent, with the head held high and facing the other bird. The beak is downward, snood pendant, wings drooped and held slightly away from the body, feathers over the entire body are sleeked, and the tail feathers may be only partially fanned and held horizontally or at an angle between a perpendicular and horizontal plane. Vocalization during the threat display is a distinctive trill of relatively high pitch and is emitted repeatedly as the birds face each other. The birds may circle slowly while maintaining a wide stance and leaning slightly away from each other. Thus, they maintain a posture from which they may jump quickly towards or away from the other bird (Hale *et al.*, 1969).

There are significantly more fights towards non-group members than towards group members. Not only wild turkeys but also domesticated broad-breasted fattening turkeys distinguish between conspecifics of their own group and conspecifics of another group, and they antagonize non-group members markedly more than group members (Buchwalder and Huber-Eicher, 2003, 2004, 2005).

Actual fighting begins as both birds jump simultaneously, or one just slightly before the other. As the birds jump, their feet are extended forwards, with the toes spread, and moved in a raking fashion while directed at the opponent's body. During the jumping phase of the fight, domestic turkeys rarely injure one another since their spurs are poorly developed and the pattern of movement does not usually bring the spurs into contact with the opponent's body. If one bird lands a stroke on the other one's back, the latter gives up. However, if neither bird submits after a few jumps, there is a shift to a tugging battle, but as few as one or more than 20 jumps may occur first. The head is darted forward to grasp the caruncles, snood, wattle or upper or lower beak of the opponent. Necks may become entwined as the birds grasp each other simultaneously and tug and push in an attempt to force the opponent's head downwards. During this phase of the fight, the skin about the head may be injured and bleed. In prolonged fights, one bird may interrupt tugging by pressing its head over the opponent's back or under the opponent's wing. After a brief rest, the fight is resumed. If the tugging is interrupted for longer than a few seconds, because the birds stumbled over an obstacle for example, the jumping starts over again before the tugging is continued. A fight is terminated by the sudden submission of one contestant, with no prior indication of the impending event. After a bird submits, it retracts its snood, lowers the head as the wings and tail assume normal positions, attempts to hide its head under the opponent's breast, and then flees; occasionally hiding is omitted. The winner may follow closely with snood extended, head high and may threaten or peck the defeated bird. Usually, the threat display is terminated quickly (Hale *et al.*, 1969).

Extreme aggression can be a problem under commercial conditions. It usually occurs between males, often leading to wounds on the head and neck. This makes it difficult to keep these species commercially unless under very low light intensities, which in itself can be a problem. Extreme aggression can also be directed towards a few extremely low-ranking individuals within a flock, with the result that they have difficulty in feeding undisturbed and lose body condition (Keeling, 2002). An increase in floor space reduces the number of aggressive pecks and threats aimed at an introduced unfamiliar conspecific (Buchwalder and Huber-Eicher, 2003).

8.4   Behaviour in Case of Danger or Peril

The female defends the brood against intruders and may attack by hissing, running in a low crouch and jumping at the intruder with extended claws and beating wings. As the intruder approaches, the hen may hover over the poults and remain motionless. Like some other birds, the hen, if disturbed, may then fly off in a wobbly flight, feigning injury and directing the intruder's attention away from the poults (Hale *et al.*, 1969).

For several days after hatching, young turkeys do not respond directly to predators but react to various alarm calls sounded by the mother hen. She sounds various alarm calls to her poults as strange objects are sighted on the ground or in the air. The response of poults to these calls is to seek protection. On sighting an aerial object, the hen gives one of two related alarm calls, depending on the nearness of the object. When high-flying predators are detected, the hen gives a 'singing' note, which elicits an attentive vigilance response in the young poults. A lower-pitched call which resembles a sequence of clucks is sounded by the hen as a predator comes close enough to indicate impending danger. Poults respond to this call by dashing violently to cover or to the corners of a pen, pile in clumps and remain motionless for several minutes. In response to alarm calls, poults cease all other activities and become quiet, in addition to giving distinctive escape reactions (Hale *et al.*, 1969).

In older poults, the escape responses to alarm calls disappear and direct responses to predators become predominant. Strange objects or animals on the ground elicit a highly segmented alert call or cluck. Young turkeys respond to this call by dispersing slowly over a wide area in a creeping posture. At the time when young poults cease to give escape responses to alarm calls, they begin to respond directly to various objects instead of reacting indirectly through communication from the hen (Hale *et al.*, 1969).

Domestic turkeys retain many antipredator responses, such as freezing, alarm calling, running rapidly away from danger, flying away, or at least attempting to take off, and vigorous struggling if caught. Even sudden events, in particular noises, may elicit such responses (CoE, 2001).

8.5   Some Normal Physiological Frequency Values of Interest at Clinical Examination

The body temperature of adult turkey hens in complete rest is about 40°C (Yang *et al.*, 2000). In adult turkeys, the heart rate is 165–200 beats per min and the respiratory rate about 20–25 breaths per min (Schumacher *et al.*, 1997).

8.6   Active and Resting Behaviour Patterns

Activity pattern, circadian rhythm

Following initial activity and feeding soon after daybreak, turkeys taper off to a period of relative inactivity near midday and engage in considerable feather preening and dusting (Hale *et al.*, 1969).

Resting behaviour consists of numerous phases of short duration. In the meat-type turkey, it takes the major share of their time budget. There is little difference, however, in the time budget between extensive and intensive husbandry conditions (Bessei, 1999).

Exploratory behaviour

The turkey's beak is richly innervated and has a complete array of sensory corpuscles in the area just behind the beak tip. These sense organs are used in investigatory pecking.

Enriched environments facilitate exploratory behaviour. Turkeys kept in conditions approximating commercial rearing show high incidences of pecking injuries, whereas turkeys kept in environmental enrichment may, by the much varied pecking of substrates, reduce significantly injuries due to wing and tail pecking. Injuries are thus more common without enrichment. Turkeys in enriched environments also increase the latency to sit after the minimum of forced standing – indicating improved musculoskeletal function (Sherwin et al., 1999; Michel and Boilletot, 2003).

Domestic turkeys have retained the typical feeding pattern of their ancestors, which consists of investigation of the surroundings and pecking, followed by ingestion, and covers up to 50% of the whole activity time. Turkeys are often very willing to approach and investigate humans who enter turkey houses (CoE, 2001).

Diet, food searching, eating and eating postures (feeding behaviour)

Newly hatched poults peck indiscriminately at bright spots and small objects which contrast with the background. These responses soon lead the poults to peck at feed and at images in water (Hale et al., 1969).

Wild turkeys are omnivorous, feeding on plants, seeds, insects and worms. Shrubs are the most important food throughout the year (42%), followed by grasses (31%) and tree and forb species (together 28%). Manzanita, a shrub, is the principal plant food in Mexican wild turkeys. Analyses indicate low utilization of insects, mainly grasshoppers (Morales et al., 1997).

Scratching behaviour is an important component of foraging for food and gallinaceous birds scratch with one foot while standing on the other. Wild turkeys extend the foot forward and scratch backward and outward in a sweeping motion to form an inverted 'V' pattern on the ground. When feeding on acorns covered with leaves, wild turkeys may dig up vast areas. In domestic turkeys, scratching behaviour is observed rarely on ranges or in litter provided in pens and seems to depend on the rearing conditions. Scattering grain in the litter readily induces vigorous scratching behaviour in domestic fowl, but is generally ineffective with domestic turkeys. The apparent reduction of ground scratching in domestic turkeys either suggests that a highly specific stimulus is required to elicit scratching, or that there has been a marked reduction of this behaviour in turkeys under domestication (Hale et al., 1969).

Turkeys, like chickens, possess the basic avian pattern of feeding without specialized anatomical and/or behavioural adaptations. Swallowing is accomplished without the necessity to raise the head, although some adults may lift the head frequently while gulping large quantities of mash. Drinking is accomplished differently from feeding, with the turkey dipping its beak into the water to the level of the nares and making several rapid partial closures of the beak. It then raises its head, extends the beak upward and repeats several rapid closures of the beak. The latter pattern suggests some degree of active swallowing in addition to flow by gravity to the oesophagus (Hale et al., 1969).

Selection of resting areas, resting postures, sleep

Wild turkey broods spend the first days of life on the ground until poult flight capabilities are attained. Hens select ground-roost locations which have visual obstruction from multiple observation sites. Plots surrounding ground roosts have greater visual obstruction, increased tree decay, higher per cent grass, shrub, litter and forb cover and lower per cent bare ground cover than random sites. Grass, shrubs and fallen trees appear to provide the desired cover for ground-roosting broods. Plots used by broods less than 10 days old and with average survival contained more visual obstruction and shrubs than plots used by broods 10–16 days old with lower than average survival, signifying a shift in habitat use by successful broods as poults attain flight abilities (Spears et al., 2007).

Wild female turkeys do not appear to increase movements prior to roosting, suggesting that roosting needs may have a superior influence on female movements throughout the day – allowing females to be at their preferred roosting sites by dusk. They seem to favour pine trees. Alternatively, females simply may roost in the trees in the nearest suitable habitat at the end of the day (Chamberlain *et al.*, 2000).

An investigation of the roosting sites of wild turkeys in Mexico showed that a large percentage (87.5%) was located in oak–pine forest. Of the roost trees, 81.2% were pines, 17.6% were oaks and 1.2% were madrones (the tree *Arbutus menziesii*). The average roost tree height was 16.6 ± 4.4 m. The selection of pines as roost trees by wild turkey was thought to be due to tree structure. However, in places without pines, wild turkey used oaks, madrones and sycamore with similar characteristics to pine for roosting (Marquez-Olivas *et al.*, 2007).

In the meat-type turkey, resting behaviour takes the major share of their time budget, as much as 60% in adult animals. Resting behaviour consists of numerous phases of short duration (Bessei, 1999).

As dusk approaches, turkeys become inactive and rest on the ground in compact groups if roosts are not available. During the night, the inactive turkeys remain motionless if approached or exposed to a beam of light and can be caught or picked up with ease. Sleeping turkeys may place their heads under the wing but usually simply relax the neck and rest their head on the body, with the snood flaccid and extended (Hale *et al.*, 1969).

Studies by Ayala-Guerrero *et al.* (2003) reveal that sleep periods in turkeys show a polyphasic distribution but with a tendency to concentrate between 2100 and 0900 h, in spite of the fact that the recordings were carried out under constant illumination. The sleep period occupied 45.71% of the nycthemeral (day and night) cycle, 43.33% corresponded to SWS sleep, while 2.38% corresponded to REM sleep.

Locomotion (walking, running)

Turkeys move by walking slowly, running or flying short distances. The gait of healthy turkeys is a walk with perfectly symmetrical and repeatable hind limb movements. All lame turkeys' movements show abnormalities in space–time coordination. These abnormalities are bilateral but non-symmetrical, intermittent and non-systematic. Their intensity increases with increase in the degree of clinical handicap (Reschmagras *et al.*, 1993).

The high sternum and poorly developed pectoral muscles of even the wild turkey is indicative of inferior capabilities on the wing relative to most gallinaceous birds (Hale *et al.*, 1969). Many heavy domestic turkey strains have difficulty in flying.

Only light strains respond to free-range systems by increased locomotion activities. The walking ability of turkeys declines with increasing body weight. There are, however, strain differences in walking ability that cannot be explained by body weight. Light turkeys make extensive use of perches for resting. Heavy breeds perch at a lower level, because of problems in reaching higher levels (Bessei, 1999).

No significant differences are evident between the gaits of traditional and broad-breasted strains of turkeys. However, the walk of both strains shows lateral oscillations, which are very slight in the traditional turkey but very marked in the broad-breasted turkey. In the latter, modification of the centre of gravity may be associated with problems of lameness (Abourachid, 1991).

Swimming

Turkeys do not swim.

8.7 Behaviour at Defecation and Urination (Eliminative Behaviour)

Like all birds, turkeys defecate randomly over their enclosure.

8.8 Body Care (Preening), Cleanliness

Preening in the form of dusting is carried out in a stereotyped fashion, with the sitting bird extending its head and making raking movements in towards the body with its beak. The raking is continued until a fan-like semicircle has been raked and dirt piled against the body. Dust is then worked into the feathers by flapping the wings and kicking. The complete pattern may be carried out by birds on concrete floors in the absence of litter, and by young birds which have never had any experience with litter (Hale *et al.*, 1969).

Domestic turkeys, if given the opportunity, will exhibit the same wide range of comfort and grooming activities as their ancestors, including preening, which involves the arrangement, cleaning and general maintenance of the structure of the feathers by the beak or feet; raising and ruffling the feathers; stretching the wings; and dust bathing (CoE, 2001).

8.9 Temperature Regulation and Climate Requirements

Temperature regulation

Birds lack sweat glands and under heat stress they rely on increased evaporation from the respiratory system for heat transfer. Turkeys, like domestic fowl, thus regulate their temperature mainly by evaporating moisture via their respiratory organs, but some heat loss also occurs via the skin.

Climate requirements

As turkeys are susceptible to the effects of rain, a freely accessible shelter must be provided under free-range conditions to protect them from adverse weather conditions.

8.10 Vision, Behaviour in Light and Darkness

As diurnal gallinaceous birds, turkeys are presumed to have visual characteristics similar to chickens, namely colour vision, rapid accommodation, high visual acuity, a visual field of about 300° and a limited binocular field. As the eyes fit tightly in the orbits, there is minimum eye movement and moving objects are followed by moving the head and neck (Hale *et al.*, 1969).

The naked carunculated areas of the head and neck of turkeys are richly pigmented and chromatophore changes during display provide a rapidly altering spectrum of colours from brilliant red to blue, purple and white. When the bird is disturbed, all colours tend to pale; however, during threat or the early stages of a fight, the naked areas take on a brilliant red hue, and during prolonged courtship, white becomes predominant. The red colour of the neck is produced by dilated blood vessels. Carotenoid pigments contribute to the red colour of the caruncles covering the head. Blue areas have heavy deposits of melanin in the deeper epidermal layers (Hale *et al.*, 1969). That turkeys have great capacity to register colours seems logical in view of these changes in colour according to the emotional status of the bird.

Although vision is important to domestic poultry, and birds in general have better visual acuity in bright rather than dim light, turkeys are often kept in light intensities of 5–10 lux or lower. Manser (1996) has compared the light intensities in poultry houses with those recommended in buildings used

by humans (Table 8.1). Some impairment in the ability of the birds to engage in exploratory behaviour and social interaction is likely at these levels. Furthermore, intensities of 6 lux or below can lead to increased mortality in turkey poults. These light levels are also associated with leg problems, eye abnormalities and breast blisters in growing birds. Leg abnormalities have been seen more commonly in turkeys at 19 lux than at higher light intensities (Manser, 1996). The rearing of turkey broilers is possible at 15–20 lux in enriched pens, without any reduction of normal behaviour or growth, while the light has to be decreased to 5–10 lux in non-enriched pens in order to avoid pecking injuries (Michel and Boilletot, 2003). In commercial UK farms, turkeys are usually reared at a light intensity of 1–4 lux (FAWC, 1995).

For turkey poults, Barber *et al.* (2004) recorded 12 defined behavioural categories in environments with four different light intensities, i.e. <1, 6, 20 and 200 lux. When given a choice, the poults spent most time in the 200 lux environment at 2 weeks of age, but in the 20 and 200 lux environments at 6 weeks of age. This change in overall preference was reflected in the partition of different behaviours between the light environments. At 2 weeks of age, all behaviours were observed to occur most often in 200 lux. At 6 weeks, resting and perching were observed least often in <1 lux, whereas all other activities were observed more in the two brightest light environments. These results show that turkey poults have significant preferences for illuminance.

By using twilight periods in turkey houses, dimming and raising the light could reduce the risk of pecking behaviour and injuries to the birds (CoE, 2001).

Table 8.1. Recommended light intensities (lux) in buildings used by humans and in poultry houses (after Manser, 1996).

Recommended light intensities in buildings used by humans	Lux	Commonly used light intensities in poultry houses	Lux
Office	500–750	Laying poultry	5–10
Living room (general)	100	Broilers	5
Hospital corridor in day	300	Turkeys	0.5–5
Hospital corridor at night	5–10		

In commercial turkey flocks, differing lighting programmes are used – even constant light. Classen *et al.* (1994) found that turkeys up to the age of 188 days kept in houses with constant (24 lux) light day and night showed more difficulties in walking and an increased overall mortality, primarily because of an increase in the incidence of skeletal disease and spontaneous cardiomyopathy, in comparison to birds kept in houses where the light was increased gradually from 6 to 20 lux. In the former regime, the birds were less active and often sat down and then gradually decreased in condition each day.

Under commercial farming conditions, turkeys are usually reared under incandescent or fluorescent lamps, both of which produce minimal UV radiation. It is possible that this artificial visual environment, with its reduced spectral range, hinders foraging or signalling information and so might be responsible for abnormal or redirected pecking activity such as the high intensities of injurious pecking often observed when using conventional light sources. When presented with the choice of two chambers illuminated by fluorescent (an optical phenomenon in cold bodies, in which the molecular absorption of a photon triggers the emission of a photon with a longer (less energetic) wavelength) light, turkeys preferred to enter a chamber that was also illuminated by a supplementary UV light source (Moinard and Sherwin, 1999). For turkeys, ultraviolet light with a longer wavelength (UV_A-supplemented light) may be beneficial; it has a role in mating, it is preferred by turkeys and reduces pecking damage. Turkeys seem to prefer fluorescent light to incandescent light, notwithstanding differences in perceived illuminance (Prescott *et al.*, 2003).

8.11 Acoustic Communication

Hearing

Hearing is well developed, with frequency discrimination and the ability to determine the direction of sound comparable to those of humans. However, the band of frequencies to which the ear is sensitive is generally more restricted in birds than in mammals (Hale *et al.*, 1969).

The lowest frequency (number of waves that pass a fixed point in one unit of time) detected is 200 Hz, the greatest sensitivity is 2 kHz and the highest frequency detected is 6 kHz (Heffner and Heffner, 1992b).

Vocalization and acoustic communication

Vocalizations include the trill emitted during threat behaviour and the vocal component of the male strut. Hens give various yelps attracting poults, and young turkeys give a variety of vocalizations, including a distress peep, a trilling contentment call and a high-pitched, scream-like call when pecked. Another variation is heard as poults quieten down before going to sleep. In older birds, the peep is replaced by the yelp – a lower-pitched call given frequently by adult females and less frequently by males. In addition to the rolling trill given during threat display, turkeys emit a somewhat similar call, at a lower intensity and in a shorter phrase, as another bird comes near. A soft trilling call is given when relaxed or browsing; if exceptionally desirable food is present, this may be combined with the cluck commonly given to unusual or strange objects. The gobbling call of the male turkey very likely serves inter-individual recognition within a flock, especially over wider distances; in the wild, it may also play a part in the spacing of male subflocks and their roosts (Hale *et al.*, 1969).

8.12 Senses of Smell and Taste, Olfaction

Fowls are very sensitive to water temperature, are reluctant to drink warm water above ambient temperature and can discriminate differences as small as 3°C (Appleby *et al.*, 2004).

The nasal cavity of many bird species has an olfactory epithelium that is structurally similar to that found in mammals. Experiments and observations of behavioural responses to olfactory cues provide evidence of a well-developed sense of smell in birds (Appleby *et al.*, 2004).

Inferences from studies of taste in chickens suggest a well-developed gustatory sensitivity in turkeys (Hale *et al.*, 1969).

8.13 Tactile Sense, Sense of Feeling

Tactile touch

The skin of birds is rich in sensory receptors but the functional characteristics have not been clarified.

Tactile receptors may be presumed as being highly developed in turkeys since sexual responses in the female are elicited readily by tactile stimulation over extensive areas of the body. The well-authenticated ability of birds to detect distant explosions has been demonstrated as being based on the excitation of sensory organs in the legs receiving vibrations through the ground. These structures are particularly sensitive when the bird is in a resting posture on branches and may serve to detect the approach of predators (Hale *et al.*, 1969).

Although male courtship behaviour provides visual stimuli which arouse sexual receptivity in the female, during the copulatory sequence, tactile stimuli play the predominant role in eliciting female responses up to and including the termination of the mating (Hale *et al.*, 1969).

Sense of pain

The turkey's beak is richly innervated and has a complete array of sensory corpuscles in the area just behind the beak tip. These sense organs are used in investigatory pecking and are very sensitive to injury.

Beak trimming results in the loss of significant amounts of beak tissue. By 42 days after trimming, the beak is healed with extensive regrowth, including bone and cartilage formation. In the normal bird, the dermis at the tip of the upper beak contains large numbers of nerve fibres and sensory receptors. However, in beak-trimmed birds, the dermal tissue, although well supplied with blood vessels, is devoid of afferent nerve fibres and sensory nerve endings (Gentle *et al.*, 1995).

8.14 Perception of Electric and Magnetic Fields

Domestic chicks have the ability to orient using magnetic cues (Freire *et al.*, 2005). Although this does not seem to be described in turkeys, they may have the same ability.

8.15 Heat and Mating Behaviour, Nesting, Egg Laying, Brooding

Both females and males attain potential sexual maturity at approximately 7 months of age. Females do not show complete ovarian development unless they attain the proper age at a time of the year when sufficient daylight is present to stimulate reproductive maturation (Hale *et al.*, 1969), or if stimulated by artificial light programmes.

As the breeding season approaches, the males establish territories or strutting areas. During the breeding season, males compete by giving displays for females at communal display grounds called 'leks' (Keeling, 2002). Females become solitary and move freely from territory to territory. Territorial males attract females by vocalizing – gobbling and pulmonary puffs – and carrying out elaborate movements and tail fanning. Where groups of young males associate, they will display in synchrony, but only the most dominant will copulate. Additionally, single, highly aggressive males may dominate entire local populations. Hens move between territories until they approach a male and invite copulation. They then establish nests, but may return to copulate daily until ready to incubate. If mating is not successful, females form 'broodless flocks'. Domestic turkeys have retained most aspects of courtship behaviour (CoE, 2001). Wild turkey females prefer males with longer snoods and wider skullcaps (Buchholz, 1995).

Mating behaviour in both wild and domestic turkeys is characterized by promiscuous mating without the formation of pair bonds before or after copulation. The females are either presented with the male or are attracted to special mating areas. Males may alternate between more than one breeding area, with dominant males chasing subordinate males away, but with several dominant males remaining in the same area. Non-receptive females avoid the males, while receptive ones approach and respond with a sexual crouch (Hale *et al.*, 1969).

In domestic fowl, movements during courtship and mating are more rapid and the feather display of the male is less elaborate than in turkeys. Although male domestic fowl typically force matings with hens, this behaviour is not seen in turkeys and the turkey male at most presses his breast gently against a female or extends his wing over her back. Turkey females sometimes approach or follow the male before crouching and, if he moves away, they get up and crouch near him again. Female chickens show this behaviour less and frequently do not approach the cock before crouching (Hale *et al.*, 1969).

Sexual receptivity of turkey hens is indicated by squatting, sexual crouch (Fig. 8.2). The presence of the opposite sex during rearing influences the expression of a crouch. During tests for sexual responsiveness to humans, over 80% of females in

Fig. 8.2. Young turkey females exhibit squatting, sexual crouch, when the stockman approaches (image courtesy of Lotta Berg).

separate-sex and adjacent to mixed-sex groups assumed a sexual crouch, whereas less than 20% of females in mixed-sex groups crouched. In particular, young turkey females exhibited a sexual crouch at the beginning of egg laying (Siopes, 1995; Widowski *et al.*, 1998).

Mating behaviour in turkeys tends to follow a complicated chain reaction pattern. The behaviour of one sex partner elicits a specific response from the other, and that response in turn elicits a further response from the first partner. Courtship display in the male turkey is characterized by slow and restricted movements, 'strutting', and elaborate feather display. The tail is fanned vertically; feathers on the back, breast and flank are erected; the wings are lowered, with the primary feathers spread; the crop is inflated. All combine to create an impression of apparent larger size by enhancement of all dimensions. The neck forms a tight 'S' shape, bringing the head in close to the body. The snood becomes elongated and turgid, and the naked carunculated areas of the head and neck change from bright red to a predominantly white colour, with blue contrasting cheeks after prolonged courtship. While courting, the male slowly paces back and forth and, at intervals, takes a series of three to five steps, drags the primaries on the ground and accompanies the

sequence with a low vocalization described as a pulmonic puff. Sexually receptive females enter the sexual crouch by holding the head high, the tail flexed ventrally and the legs gradually flexed as the wings are drooped and fluttered. In the complete crouch, the head is usually held close to the body. As a courting male approaches, the female raises her head and then extends it to the maximum height as the male mounts (Hale *et al.*, 1969).

Females often form 'nesting groups' of 2–5 hens and lay their eggs in the same nest (CoE, 2001). Nesting sites are often selected under low, concealing vegetation at the base of a large tree, or even at times in more open areas. Good fertility of eggs laid over a period of 15–20 days is ensured since the average duration of retained fertility on the part of the female following a single mating is approximately 5–6 weeks. If the hen leaves the nest, she covers the eggs with leaves by moving the head alternately to each side, picking up leaves in her beak and depositing them over the eggs (Hale *et al.*, 1969).

In commercial turkey production, sperm is collected from turkey stags and females are artificially inseminated, since natural mating is no longer possible in broad-breasted breeds (Keeling, 2002).

8.16   During and After Hatching

Behaviour during and after hatching

Parental behaviour is typically absent in both wild and domestic male turkeys and all aspects of parental care are assumed by the female. Turkeys hatch approximately 27 days after the eggs are set, with pipping of the shell the first outward sign of activity. The turkey poult, like the chick, is active within the shell before hatching. Pipping is followed by a passive period of 12–24 h, with a renewed attack on the shell and hatching taking place on average about 28 h after the initial pip. After hatching, the poult is very mobile. Within about 8 h after hatching, the young poults are dry and start to move. Newly hatched poults peck indiscriminately at bright spots and small objects which contrast with the background. These responses soon lead the poults to peck at feed and at images in water (Hale *et al.*, 1969; Broom and Fraser, 2007).

Wild turkey broods spend the first few days of life on the ground until poult flight capabilities are attained. This is a critical period in the life of a wild turkey, with poult survival ranging from 12 to 52% (Spears *et al.*, 2007).

Young turkeys imprint on the mother and are soon able to recognize her individual calls. At first, poult movement is slow, but by 48 h the pace approximates to that of adults. The poults crouch, rest or sleep under the mother's wings or body as she hovers over them. Young wild turkeys spend each night on the ground under the mother until about 4–5 weeks of age (Hale *et al.*, 1969).

Litter size

Wild turkey hens lay 8–15 eggs (CoE, 2001).

Play behaviour

Play behaviour as it occurs in mammals does not seem to be described in turkeys.

8.17   The Turkey Chicken

After about 5 weeks of poult age, the mother roosts on the lower branches of trees and the young fly up to follow her (Hale *et al.*, 1969).

After 3 months of age, social hierarchies are formed in turkey poult groups (Broom and Fraser, 2007). When the poults grow older, the hens and poults join together to form larger flocks – brood flocks. The poults stay together with the hens until they are about 6–7 months old (CoE, 2001).

8.18   Assessment of Turkey Health and Welfare

For assessment of turkey health and welfare, special attention should be paid to vocalization, movements, respiration, bodily condition such as the condition of plumage, eyes, skin, beak, legs, feet and claws; attention should also be paid to the existence of any injuries, the presence of external parasites, to the condition of droppings, to feed and water consumption, to growth and to egg production. Furthermore, in order to recognize leg problems, the birds should be encouraged to walk.

The healthy turkey

The healthy turkey has sounds and activity appropriate to its age, breed or type, clear bright eyes, good posture, clean healthy skin, good plumage, well-formed snood, shanks and feet, vigorous movements if unduly disturbed, normal walking and active feeding and drinking behaviour.

The sick turkey

Symptoms

Common disease symptoms in turkeys are difficulty in walking, injuries through pecking, abnormal behaviour such as feather pecking, excessive aggression or cannibalism (CoE, 2001).

Examples of abnormal behaviour and stereotypies

The most common type of abnormal behaviour that develops in turkeys is some pattern of feather pecking. Under confinement, some varieties show no evidence of pecking, while others may show characteristic patterns of pecking restricted to the tail feathers, to a small area at the base of the wings, or a fraying of the feathers at the base of the neck. In some highly inbred lines or certain crosses, feather pecking may be so extreme that many members of a flock may be almost completely denuded (Hale *et al.*, 1969).

As turkeys can be particularly aggressive towards one another in the way commercial flocks are kept, they are usually kept in dim lighting to reduce pecking behaviour.

The fact that birds of breeding stock of modern turkey breeds, because of their heaviness and big breast muscles, are not able to mate naturally must be regarded as a serious inability to perform normal behaviour.

Examples of injuries and diseases caused by factors in housing and management, including breeding

Besides some specific virus and bacterial infections, early mortality, feet disorders, mild cannibalism and aggressive behaviour are the main disease problems in commercial turkey production (Ferket, 2003).

Impaired walking ability is often a consequence of the intense selection for growth combined with selection for feed efficiency, which has resulted in these birds being less active (Weeks *et al.*, 2000). Domesticated turkeys have a reduced tendency to show ground pecking and scratching and spend a large proportion of the day sitting and resting. The opposite problem to that of inactivity is that of nervousness, which in large flocks can result in birds rushing to one end of the house, where many can be suffocated (Keeling, 2002). Walking ability deteriorates as stocking density is increased per square surface of floor space. Hip and foot lesions are shown to be more frequent at high bird density (Martrenchar *et al.*, 1999).

High litter moisture alone is sufficient to cause footpad dermatitis (Fig. 8.3) in young turkeys and it has been shown that footpad dermatitis can be minimized by the maintenance of dry litter (Mayne *et al.*, 2007). The prevalence of footpad dermatitis in Swedish turkey herds in the 1990s was 20% for severe lesions, ulcers, and 78% for mild lesions. Only 2% of feet were classified as being without lesions (Berg, 1998). It is often necessary to add fresh litter during the rearing period to keep litter quality dry enough (Ekstrand and Algers, 1997), especially if turkey poults are grown until 18–20 weeks of age or longer. It has been indicated that the use of underfloor heating can improve litter quality substantially, and thereby turkey foot health. Monitoring programmes for turkey foot health – similar to those for broiler foot health – have been developed as an indicator of litter quality and are currently practised in Sweden.

Breaking the skin during feather pecking may produce bleeding wounds (Fig. 8.4) and by and by favour the development of cannibalism (Hale *et al.*, 1969).

Females of certain strains have a tendency to peck at and eventually injure the pendant snood of strutting males. If a bird is weak or cannot stand, the other birds peck it so actively on the head and back that severe damage and death are likely to follow unless the bird is removed (Hale *et al.*, 1969).

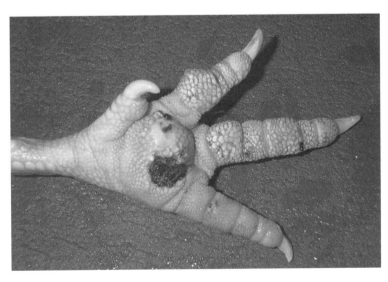

Fig. 8.3. Turkey foot injuries. In the foreground, a deep lesion covered by a crust, behind it a scar formation from an earlier wound (image courtesy of Lotta Berg).

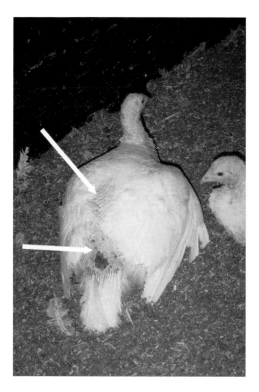

Fig. 8.4. As a result of feather pecking, this female turkey breeder hen displays a completely naked back, partly covered by the wings. On her lower back, injurious pecking has caused a wound (image courtesy of Lotta Berg).

Symptoms of pain

There can be no doubt that lame turkeys can experience pain during locomotion (Hocking, 1999). Duncan *et al*. (1991) performed a study of the extent to which degenerative hip disorders in adult male turkeys resulted in pain. It was concluded from the post-mortem examination that degenerative hip disorders occurred to such an extent in each animal that the pathological processes must have resulted in a state of chronic pain which inhibited spontaneous activity and sexual activity. For animal welfare reasons, birds displaying obvious gait abnormalities should be euthanized.

Gross pathological and histological examination shows that beak trimming damages both the upper beak and the lower beak tip, including the bill tip organ – causing the birds extensive damage and suffering from considerable long-term pain (Fiedler and König, 2006).

In some countries, the snood of males is removed surgically to prevent pecking wounds, a procedure which causes pain, even though not as lingering as beak trimming.

Birds suffering from severe cannibalism, displaying large and deep wounds occurring mainly on the back, are apparently suffering from these lesions. However, such birds often do not appear to try to hide or escape from the pecking conspecifics.

Usually, for artificial insemination, semen is collected from the males twice a week and the hens inseminated once a week. The semen-collecting procedure can be stressful to large males, as they have to be caught and placed partly on their backs, although the procedure is rather rapid. Also, the females have to be caught and turned around for insemination, which involves a stress situation. As this process is rapid, it is considered to be less stressful compared to the process for the male birds.

8.19   Capturing, Fixation, Handling

Turkeys should not be lifted by a single leg only. When turkeys are carried, they should be carried individually, using techniques appropriate to the size and weight of the birds. Small birds should either be held by both legs or be supported between the catcher's arm and body. Larger birds should be carried by one leg and the diagonally opposite wing. They should be carried with their heads upwards, except for the short period when they are first picked up (CoE, 2001).

8.20   The Importance of the Stockman

Turkeys should be caught and handled in a careful manner and, in order to develop a positive relationship between humans and birds, there should be frequent quiet but close approaches from the first few days after hatching so that the birds are not unduly frightened (CoE, 2001).

9 Geese (*Anser anser* f. *domesticus*, *Anser cygnoides* f. *domesticus*)

9.1 Domestication, Changes in the Animals, their Environment and Management

The greylag goose (*Anser anser*) is considered to be the ancestor of domestic goose breeds of European origin, while breeds of Asiatic origin, for example the Chinese goose, descend from the swan goose (*A. cygnoides*). Geese are probably the first poultry to be domesticated; originally, they were kept mainly on marginal grazing land. Geese are kept primarily for the production of meat, down and feathers, but also for ornamental purposes and, at least historically, also to keep watch.

They may have undergone multiple domestication but they may have been domesticated first in China or South-east Asia, probably earlier than the duck. Both the greylag goose and the Chinese goose are Palaearctic, the *A. anser* being distributed from Europe to Manchuria, while *A. cygnoides* is found in Siberia. Pottery models of geese in China dating to about 2500 BC suggest an early domestication. Geese are well documented from the Old Kingdom of Egypt (2686–1991 BC), but there is no good evidence that the birds were domesticated there at that time. Domestic geese were well known in Europe by 700 BC, as they were mentioned by the Greek poet, Homer. The crested goose was valued by the Romans for guarding duties and reputedly saved the capital from the Gauls in 390 BC by raising the alarm. The Romans had a well-developed system for goose husbandry, with the birds being kept for meat, fat and feathers. However, it came to an end with the fall of the Roman Empire, with the birds reverting to farmyard scavenging (Zeuner, 1963; Crawford, 1984; CoE, 1999c; Appleby *et al.*, 2004; Cherry and Morris, 2008).

The goose is a semi-aquatic bird. However, when domesticated, geese are quite capable of surviving and reproducing without access to swimming water. If water is available for domestic geese, they will spend part of the day swimming and resting on it. But most of their feeding activity takes place on land. They have a long neck, a relatively small head, a beak with horn plates and a tight plumage that entraps air and makes it possible for them to float high on water. Geese have strong beaks and legs, well adapted for seeking feed on land. In most breeds, wild as well as domestic, differences between winter and summer plumage and between male and female are minor. Greylag geese are partially or wholly migratory and wild geese migrate over long distances. They are strong flyers and gather in large flocks when migrating to winter areas. During migration, the older birds lead and the younger birds learn the migration routes from them. When moulting, geese lose all their wing quills at the same time and during this period, about 2 weeks, they are unable to fly. Throughout the ages, humans have taken advantage of this by hunting geese intensively during this phase. Compared to other poultry, geese have a long lifespan. There are anecdotes of individual domestic birds living for more than 100 years. In commercial breeding, however, they are kept until at least 5 years of age (Crawford, 1984; CoE, 1999c; Appleby *et al.*, 2004).

The body weight of the wild species is around 3.5 kg; most domestic breeds are considerably larger (Crawford, 1984; Appleby *et al.*, 1992, 2004; CoE, 1999c). Although being a semi-aquatic bird, geese search feed on land. They have strong legs and a short bill (Fig. 9.1).

Geese form less stable social groups than, for example, the galliforms. They frequently occur in large aggregations, so in this sense they are well adapted to domestication. However, geese show individual recognition, at least to the extent of long-term sexual pairings, so this is one aspect of behaviour which is disrupted in domestic conditions. Geese are monogamous in the wild but mate promiscuously under domestication. Greylag geese form less stable groups than jungle fowl and turkeys (Appleby *et al.*, 1992, 2004).

Eleven breeds are listed by the American Poultry Association. The most common breeds are Emden, Toulouse and tufted Roman geese (Crawford, 1984; Appleby *et al.*, 2004). However, in Europe there are several native breeds (Fig. 9.2).

Geese are kept mainly to produce meat. In Europe, especially in France and Hungary, geese are force-fed to yield fatty livers for use in pâté de foie gras. Enterprises that traditionally use

Fig. 9.1. The greylag goose is considered to be the ancestor of domestic goose breeds of European origin (image courtesy of Kent Ove Hwass).

Fig. 9.2. Scanian (Skåne) geese, one of the European native breeds (image courtesy of Eva Ekesbo).

geese for foie gras production are nowadays using ducks for that purpose. Feathers are harvested for clothing insulation; both down feathers and outer feathers are used and are collected at the time of slaughter. Plucking of live geese still happens in some countries. Intensive geese rearing indoors on bedding without any water source other than water via nipples occurs in several countries. Goose production has never reached the large-scale proportions of duck production. They are kept until about 4 months of age, after which they are slaughtered. Small-scale operations in which geese wander freely during the day are widespread on farms and rural properties. This is the only truly free-range system in that the birds are completely unrestricted in their movements, except that they are usually shut up at night for protection from predators (Appleby et al., 1992; Keeling, 2002).

The general term for the bird is 'goose' and the plural is 'geese'. The adult male is called a 'gander', the adult female a 'goose' and the young ones 'goslings'.

9.2 Innate or Learned Behaviour

Geese learn quickly and have a good memory. Semiferal goslings that were imprinted on humans learned to open a box to get a food reward after observing a human 'tutor' (CoE, 1999c; Appleby et al., 2004).

9.3 Different Types of Social Behaviour

Social behaviour

Geese are gregarious birds and in the wild they congregate in large flocks that stay together except during the breeding season, when they develop a monogamous bond and disperse into pairs. The pairs are usually developed during the second winter migration (CoE, 1999c).

Domesticated geese have decreased social behaviour because domestication has a reducing effect on aggression (Molnar et al., 2002).

Communication

Communication by a variety of vocalizations is an important part of their behaviour (CoE, 1999c).

Behavioural synchronization

Egg laying of domestic geese, housed separately but sharing a common grazing area, shows a synchronized laying pattern. In a study, housed egg production of each out of eight flocks was recorded over 83 consecutive days of the laying season. This pattern of synchronized laying was found in each of the eight flocks. The synchrony was not observed when the egg production of all eight flocks was combined, indicating a within-flock mechanism (Kent and Murphy, 2003).

Dominance features, agonistic behaviour

Agonistic behaviour makes up less than 1% of the time budget throughout the year, as measured by focal animal sampling. Most encounters (84%) were won by the initiator (Randler, 2004).

Males are more vigilant at the edge of the group than in the centre (Frafjord, 2004).

Interactions of various intensities are common in a goose flock. Serial aggressive interactions between greylag goose families and other flock members occur. This may cause social stress, modulating heart rate, which may serve as a measure of energetic investment and also of individual emotional involvement. Repeated attacks towards the same individual are common and up to five serial attacks by family members might follow an initial attack. Juvenile geese evidently benefit most from active social support during serial attacks. Serial attacks may signal the agonistic potential of a family to other flock members. In 1602 social interactions of various intensities in which the focal individual either attacked another member of the flock or was attacked itself, individuals showed higher heart rates when attacking than when being attacked (Scheiber et al., 2009; Wascher et al., 2009).

9.4 Behaviour in Case of Danger or Peril

Domestic geese have retained many antipredator responses such as freezing, alarm calling, threatening, attacking or attempts to run away from danger, and vigorous struggling if caught. Such behavioural responses may be associated with emergency physiological responses (CoE, 1999c).

9.5 Some Normal Physiological Frequency Values of Interest at Clinical Examination

The body temperature of geese is about the same as in other fowl. The pulse rate in geese at rest is measured to vary between about 100 and 200 beats per min (Nolet *et al.*, 1992). In ducks and geese, the respiratory rate has been registered as being about 13–16 breaths per min, according to Funk *et al.* (1992), and 14 breaths per min in a goose weighing 5 kg, according to Fedde (1993).

9.6 Active and Resting Behaviour Patterns

Activity pattern, circadian rhythm

Geese spend more time grazing than swimming (CoE, 1999c).

Wild geese are more active in the rearing period than domesticated geese. Wild geese show a higher ratio of preening, playing, feeding and drinking activities. This phenomenon in the case of feeding behaviour can be explained by the stronger need of wild geese to search for food, while the differences in preening can be rooted in the inability of domesticated geese to fly (Molnar *et al.*, 2002).

Exploratory behaviour

When a flock of geese is grazing, there are always one or more birds that do not graze but have guard duties, scanning the surroundings for predators. The same can be seen if just two geese are on a field: one is grazing but the other is exploring the surroundings and gives a warning immediately if danger threatens.

Diet, food searching, eating and eating postures (feeding behaviour)

The goose is a grazing bird, able to utilize large quantities of forage in its diet. They are specialized in utilizing a diet containing plenty of dietary fibres in their digestive system with its two caeca. Geese are regarded as more efficient than most herbivores in transforming grass to meat. Goslings eat a variety of food items, including different small invertebrates. Adult birds prefer to forage on open land, where they seek short-growing grass or plants of a tender quality. The goose characteristically feeds for prolonged periods, even in darkness. It has no crop, but there is an enlargement at the end of the gullet proximal to the gizzard, which serves as a temporary food-storage organ (Crawford, 1984; CoE, 1999c).

Selection of resting areas, resting postures, sleep

Geese seek rest on water as protection against predators, e.g. foxes.

Locomotion (walking, running)

Wild geese walk and run readily. The ability to fly is reduced in many domestic breeds, especially the heavy ones, and very fat individuals may even have difficulty in walking (CoE, 1999c).

Swimming

Geese are good swimmers.

9.7 Behaviour at Defecation and Urination (Eliminative Behaviour)

Like all birds, geese defecate randomly over their enclosure.

9.8 Body Care (Preening), Cleanliness

Geese spend considerable time performing complex preening behaviour. Even though they spend more time grazing than swimming, water is an important factor in their grooming behaviour. Important elements of bathing are the emersion of the head and shaking water from the head over the body. After bathing, geese carry out a variety of shaking, cleaning and snapping movements to remove water and foreign bodies and to arrange feathers. An elaborate sequence of movements is then carried out to distribute oil on feathers from the uropygial gland above the tail (Getty, 1975). Domestic geese have retained a variety of behavioural patterns from their wild ancestors, of which social activities and the possibility to use water in their grooming ritual seem especially important (CoE, 1999c).

Access to an outside run and water for bathing is necessary for domestic geese – as water birds – to fulfil their biological requirements (CoE, 1999c).

9.9 Temperature Regulation and Climate Requirements

Temperature regulation

Like other waterfowl, geese use the oil from their uropygial gland for waterproofing, and thereby heat regulation.

Climate requirements

Breeds of Asiatic origin, which are characterized by a knob at the base of the upper beak, are believed to be more heat resistant than those of European origin. In general, geese are frequent only in northern latitudes. Few are kept in hot climates and only the Chinese breed adapts well (Crawford, 1984; CoE, 1999c).

Goslings under the age of 3 weeks and reared artificially must be protected from getting soaked and must be kept away from bathing water because their down feathers are insufficiently oiled during this period (CoE, 1999c).

9.10 Vision, Behaviour in Light and Darkness

Geese have well-developed sight (CoE, 1999c).

9.11 Acoustic Communication

Hearing

Geese have a well-developed sense of hearing (CoE, 1999c).

Vocalization and acoustic communication

Ganders do not crow. Male and female geese vocalize with similar frequency and mostly when they gather in groups (Appleby et al., 2004). Their warning sound is a loud cackle.

9.12 Senses of Smell and Taste, Olfaction

Geese have a well-developed sense of smell (CoE, 1999c).

9.13 Tactile Sense, Sense of Feeling

Tactile touch

Geese do not groom each other.

Sense of pain

Birds possess pain receptors and show aversion to certain stimuli, which can be interpreted as experiencing pain. They show fearful behaviour and avoid frightening situations, implying that they experience fear. They show behaviour indicative of

frustration. For example, Duncan and Filshie (1980) point out that, if suffering is defined as 'a wide range of unpleasant emotional states', then it is clear that domestic fowl are capable of both suffering and indicating fearfulness (Duncan and Filshie, 1980; Jones and Faure, 1982; Jones, 1986; Gentle and Hunter, 1991; Gentle and Wilson, 2004). There is no reason to assume that the same is not valid for waterfowl. Thus, the practice of plucking feathers from live geese must cause considerable pain.

9.14 Perception of Electric and Magnetic Fields

This does not seem to be described in geese.

9.15 Heat and Mating Behaviour, Nesting, Egg Laying, Brooding

Pairing, usually for the entire reproductive life, is characteristic of wild geese but domestic birds can be mated successfully in large flocks, where they form polygamous groupings (Crawford, 1984).

Wild geese mate almost exclusively in open water. Geese are variable in their mating behaviour. Copulation is preceded by a 'dance' where the goose and gander swim side by side and perform characteristic diving movements with their heads and necks (CoE, 1999c; Appleby et al., 2004).

Domestic geese may be mated with four to six females to one gander, and it is important that the flocks are established as soon as possible. Although swimming water seems to encourage domestic geese to mate, they can mate satisfactorily without it (CoE, 1999c).

The nests may be established in small groups; the wild greylag goose lays on average 4–6 eggs, which are incubated for 27–28 days by the female alone. During the incubation period, the gander stays close to the nest (CoE, 1999c).

9.16 During and After Hatching

Behaviour during and after hatching

Males are more vigilant and females feed more during brood rearing. Vigilance decreases in both adults when the goslings become older (Weisser and Randler, 2005).

When the goslings are a few days old and leave the nest, they are cared for by both parents – at this time, small family groups may be formed. The gander will protect goslings and nest aggressively, and an intruder will be met by hissing and threatening attitudes, or may be attacked (CoE, 1999c).

Litter size

In the wild, the litter size is usually 4–6 goslings.

Play behaviour

Play behaviours do not seem to be described in geese.

9.17 The Gosling

Adoption of unrelated offspring by successful breeders is one form of brood mixing and alloparental care that is widespread among geese. Most hypotheses assume that the separation of the gosling from its original family is accidental. However, lone goslings seem to choose the dominant family. Parents show very little aggression towards lone goslings at 3 days after hatch, but aggression increases until 9 days and remains high thereafter. At the same time as aggression increases, the chance of successful adoption decreases. In the first 5 weeks of life, adopted goslings are no further away from parents than original goslings during grazing. Findings indicate that goslings might choose foster families according to dominance (Kalmbach et al., 2005; Kalmbach, 2006).

9.18 Assessment of Goose Health and Welfare

The healthy goose

The healthy bird has sounds and activity appropriate to its age, sex, breed or type, clear bright eyes, good posture, vigorous movements and vocalizations if unduly disturbed, clean healthy skin, good plumage, well-formed shanks and feet, effective walking, bathing and preening, and active feeding and drinking behaviour.

The sick goose

Symptoms

Heavy domesticated geese may show signs of difficulties in walking.

Examples of abnormal behaviour and stereotypies

Pecking towards other birds is seen in domestic birds kept confined without the possibility of performing natural behaviour (CoE, 1999c).

Examples of injuries and diseases caused by factors in housing and management

In very barren environments which leave the geese little possibility of expressing their food-seeking and otherwise investigative pecking, there is a risk that they may direct their pecking towards other birds in the flock (CoE, 1999c).

Symptoms of pain

Lameness and other locomotion behaviour are the main symptoms are of pain caused by joint disorders in heavy birds (CoE, 1999c).

9.19 Capturing, Fixation, Handling

Birds should not be carried with the head hanging downwards or by the legs alone. Their weight should be supported by a hand placed under their body and an arm around the body to keep the wings in the closed position. Heavy birds should be carried individually and put into containers/crates one by one (CoE, 1999c).

9.20 The Importance of the Stockman

As the social structure of the family group is missing in intensive rearing systems, and as goslings are imprinted easily on any item or person, correct handling, especially during the first days of their lives, is important in order to increase or decrease the possibility of imprinting (CoE, 1999c).

10 Ducks: Domestic Ducks (*Anas platyrhynchos*) and Muscovy Ducks (*Cairina moschata*)

10.1 Domestication, Changes in the Animals, their Environment and Management

Domestic duck

All breeds of domestic duck are descended from the wild duck or mallard, *Anas platyrhynchos*. Under wild conditions, the mallard is largely aquatic (Fig. 10.1). Wild ducks winter as far north as open water and food can be found (Fig. 10.2). In most parts of their range, however, there is a southward migration in autumn and a return north in early spring (McKinney, 1969).

Ducks have been domesticated for 4000 years, probably earlier in China. Originally perhaps employed for use as live decoys, they are now kept primarily for meat and egg production, but also for ornamental purposes. However, long before domestication occurred, the aboriginal people in Australia, probably like hunter-gatherers in other parts of the world, collected duck eggs and trapped adult birds when they moulted and were temporarily flightless. Wild ducks seem to have been trapped and kept in large aviaries in the swamplands of the Nile more than 4000 years ago. In the century before Christ, Roman authors described collections of duck eggs which were set under farmyard hens and how the ducks from these eggs were later kept in enclosures provided with swimming water, a grass range area and supplied with nest boxes (Appleby *et al.*, 2004; Cherry and Morris, 2008).

The duck is easy to tame and has given rise to a large number of different breeds. There are, for example, 18 breeds recognized in the UK. Intensive selection over the past 100 years has led to differentiation between egg-laying breeds, such as the Khaki Campbell, and meat breeds, such as the Pekin and Rouen. In recent decades, commercial Pekin ducks have been selected for rapid growth and good feed conversion. The Pekin breed is known from the Ming dynasty (1368–1644) in China. The use of ducks as egg layers is declining, possibly because of the strong flavour of the duck egg. All breeds retain many biological characteristics of their wild ancestors (CoE, 1999a; Appleby *et al.*, 2004; Cherry and Morris, 2008). There were about 27,000 million ducks raised for eggs or meat in the world at the end of the first decade of the 21st century (Spinka, 2009).

Domestic ducks can mate with the wild duck (mallard) and the offspring is fertile.

Wild bird weights range from about 1 kg, whereas domesticated forms may weigh up to 4.5 kg. Some commercial hybrids might even weigh up to 8 kg. Their webbed feet cover a considerable surface. Their uropygial gland permits them to protect their plumage against being wet. The males of the wild duck have a green head and a richly coloured plumage, particularly apparent during the mating season, when the colours are very intensive. In domestic breeds, the colours differ and are less intense. The female plumage remains brown year-round. Males and females are the same size (McKinney, 1969; Getty, 1975; Sainsbury, 1992; Appleby *et al.*, 2004; Cherry and Morris, 2008).

Intense indoor rearing of ducks on bedding without access to water other than via nipples occurs in several countries (Fig. 10.3). This makes normal preening behaviour impossible (CoE, 1999a).

The term 'duck' is used for the adult female duck, 'drake' for the male and 'duckling' for young birds. To distinguish clearly between sexes, the terms 'female duck' and 'male duck' may be used, as the term 'duck' is used in a general sense to cover both males and females.

Muscovy duck

The Muscovy duck (*Cairina moschata*), or musk or mute duck, originated in South America. It was domesticated by the Colombian and Peruvian Indians before the European invasion and then

Fig. 10.1. All breeds of domestic duck are descended from the wild duck or mallard, *Anas platyrhynchos*. Under wild conditions, the mallard is largely aquatic (image I. Ekesbo).

introduced to the Old World by the Spaniards and the Portuguese in the 16th century. The Muscovy duck has been domesticated in many parts of the world. Female Muscovy ducks were used in the past as natural brooders for hatching common duck eggs (McKinney, 1969; CoE, 1999b; Appleby *et al.*, 2004).

Among the domesticated animals, the Muscovy must be one of the last developed (Mason, 1984).

The Muscovy has had many names, both scientific and popular. Its name is a mystery and has no link to Moscow, nor has *Cairina* anything to do with Cairo (Appleby *et al.*, 2004).

The bird has no odour that might suggest musk or musky. Popular names sometimes seen are Barbary duck, Guinea duck, Cairo duck or musk duck (Mason, 1984).

Even if under natural conditions, the Muscovy duck is a tropical bird and lives in marshy forests; its robustness and hardiness have enabled it to adapt to different climates and habitats (CoE, 1999b).

Muscovy ducks have long claws on their webbed feet and a wide, flat tail. The wild Muscovy duck is blackish; its black feathers have a metallic green and purple iridescence with contrasting white wing patches. Domesticated birds may look similar; most are dark brown or black mixed with white, particularly on the head. Other colours such as lavender or all-white are also seen (Fig.10.4). Both sexes have a nude black and red or all-red face. Both sexes have a low erectile crest of feathers. The male Muscovy duck has a well-developed caruncle at the base of the beak, just behind the eyes, which is even more

Fig. 10.2. Mallards in snow close to a brook with partial open water. Wild ducks winter as far north as open water and food can be found (image I. Ekesbo).

Fig. 10.3. About 1000 Pekin ducks were raised for meat in this building. Intense indoor rearing of ducks on bedding in houses without windows and without access to water other than via nipples occurs in some countries (image I. Ekesbo).

Fig. 10.4. Wild and most domesticated Muscovy ducks are dark brown or black mixed with white. However, there are also all-white birds (image courtesy of Klaus Reiter).

prominent during the mating season. It is more pronounced in the domesticated than in the wild bird. The beak is richly innervated and very well supplied with sensory receptors (Clayton, 1984; CoE, 1999b).

Muscovy ducks are sexually dimorphic. The females weigh up to 1.5 kg and the males around 3 kg (Appleby *et al.*, 2004).

The Muscovy and domestic duck hybrid is obtained by crossing a female domestic duck and a male Muscovy. It is a sterile hybrid because of the difference in chromosome sizes between the two parents. It is hardier than the Muscovy. It shows little sexual dimorphism and is able to flourish in cooler conditions. Conversely, although crossing mallard drakes with Muscovy females is possible, the offspring are desirable neither for meat nor for egg production. The hybrid from the cross Muscovy by the common duck, the mulard, is raised for meat or for production of foie gras, especially in Taiwan, France, Belgium and Hungary (Mason, 1984; CoE, 1999b).

The term 'duck' is used for the adult female duck, 'drake' for the male and 'duckling' for young birds. For clear distinguishing between sexes, the terms 'female duck' and 'male duck' may be used, as the term 'duck' is used in a general sense to cover both males and females.

10.2   Innate or Learned Behaviour

Domestic duck and Muscovy duck

McKinney (1969), citing Lorenz's observations of imprinting in waterfowl, points out that, when ducklings are exposed to a conspicuous object, normally the mother, shortly after hatching, they become attached to it, following it if it moves, sitting beside it and giving contentment notes when it is stationary and expressing distress through lost peeping calls if it is removed. This behaviour can also be elicited by other things, anything from wooden duck decoys to humans. The most effective stimuli are those which move and make repetitive sounds similar to the calls of a broody female. Preening behaviour is another example of innate behaviour, as is swimming.

That ducks can be seen aggregating intensively when people appear in areas where feeding has occurred is an example of learned behaviour (Mason, 1984).

10.3   Different Types of Social Behaviour

Social behaviour

Domestic duck

Wild mallards show a well-defined seasonal cycle. Under wild conditions, the mallard lives socially in large flocks during autumn and winter, but disperses into pairs during the breeding season. Pair formation is reinforced by courtship displays and vocalizations. The pairs form during the winter and then, after the spring migration to the breeding ground, the flock spreads out and the pairs become socially isolated. At this time, aerial pursuits are frequent. As spring and summer progress, the males flock together and are joined later by the females when rearing is completed. Most mallard pair bonds break while the female is incubating and new bonds are established each year. Some adults re-pair with their old mates (McKinney, 1969, 1975).

Several behaviours are associated with the pairing and interactions between pairs, which are often combined with the emission of specific sounds. One is when the female walks or swims beside or behind a chosen male making ritualized threatening movements over one shoulder. Another is when the female swims rapidly among the drakes, stretching her head forward over the water's surface, a behaviour which often results in a burst of male displays. Males may perform a complex display involving the sudden raising of the closed wings to a position in which the scapulars almost touch the back of the head. As the wings and tail are lowered, the male turns his head so as to point his bill towards a female. The head is lowered and stretched forward along the water surface as the bird moves forward through the group. Another display involves sudden body actions in which the breast dips deeply in the water, the bill is jerked upward and outward, flipping up a column of water as it goes, and the tail is raised high out of the water (McKinney, 1969).

In the early part of the pairing season, bonds are temporary and it may be several weeks before firm pairs become apparent. Gradually, the two birds spend more and more time swimming close together, engaging in emitting sounds and becoming involved in hostile interactions with other birds (McKinney, 1969).

Muscovy duck

The Muscovy appears to be polygamous, unlike the common duck, which, at least in the wild or semi-natural state, form pair bonds (Mason, 1984). Far fewer studies have been published on the social behaviour of Muscovy ducks than on the mallard and the domesticated duck.

Communication
Domestic duck and Muscovy duck

Communication takes place mainly by signals provided by postures, display features associated with the head and neck and, for wild and domestic ducks, several different vocalizations. When the members of a wild or domestic duck pair become separated, the female will emit a series of quacks, which the male responds to by giving slow calls and both birds appear to recognize each other by voice (McKinney, 1969; Appleby et al., 2004). Far fewer studies have been published on the communicative behaviour of Muscovy ducks than on the mallard and the domesticated duck.

Behavioural synchronization
Domestic duck and Muscovy duck

Although ducks are flock animals during part of the year, behaviour synchronization is not general. In autumn and winter, when the birds are living in flocks, rest periods tend to be synchronized. In the breeding season, each pair follows its own routine and sleeping birds can be observed at all times of the day. Once the pair bond is strong between a male and a female, the two birds tend to synchronize their daily activities, bathing, preening and sleeping at the same time (McKinney, 1969).

Dominance features, agonistic behaviour
Domestic duck

When a pair is definitely established, the male behaves aggressively towards other males and the female makes inciting movements beside her mate when unpaired males approach. The threat posture is head lowered and stretched forward when swimming towards the adversary (McKinney, 1969).

Muscovy duck

Aggressive and sexual displays are simple and not well differentiated. In the male, these may take the form of crest raising, tail shaking and moving the head backwards and forwards. Muscovy ducks, especially the males, are more aggressive than the mallard breeds. In the breeding season, Muscovy drakes fight one another with great violence, using their strong, heavily clawed feet and wings (McKinney, 1969; Mason, 1984).

10.4 Behaviour in Case of Danger or Peril

Domestic duck

Domestic ducks have retained many antipredator responses such as freezing, alarm calling, attempts to take off or run rapidly away from danger, and vigorous struggling if caught. Such behavioural responses may be associated with, or replaced by, emergency physiological responses. Human approach or contact often elicits such responses (CoE, 1999a).

In the wild, mallards are agile in taking to the wing, being capable of a near vertical rise when alarmed. When approached on the nest, incubating mallards often remain frozen in a crouched posture, with feathers tightly sleeked and head lowered (McKinney, 1969).

Muscovy duck

Farmed Muscovy ducks have retained many anti-predator responses such as freezing, alarm calling, attempts to take off or run rapidly away from danger, and vigorous struggling if caught. Such behavioural responses may be associated with, or replaced by, emergency physiological responses. Human approach or contact often elicits such responses. Male Muscovy ducks and hybrids fight frequently using their claws, wings and beaks, particularly for chasing off intruders (CoE, 1999b).

10.5 Some Normal Physiological Frequency Values of Interest at Clinical Examination

Domestic duck and Muscovy duck

The body temperature of ducks is about the same as in other fowl. Bevan and Butler (1992) report it to be $41.5 \pm 0.2°C$. The body temperature is not influenced by diving (Caputa et al., 1998).

The heart rate values for individual birds swimming on the water surface range from 96 to 135 beats per min in the summer and from 143 to 167 beats per min in the winter (Bevan and Butler, 1992).

In ducks and geese, respiratory rates have been registered as being about 13–16 breaths per min (Funk et al., 1992). According to Fedde (1993), the respiratory rate for a Pekin duck of 2.0 kg is 11.9 breaths per min.

10.6 Active and Resting Behaviour Patterns

Activity pattern, circadian rhythm

Domestic duck

The domestic duck is, like its wild ancestor, a typical waterfowl and has the same need of permanent access to water for carrying out its normal behaviour repertoire during the day and night. The sequence of feeding, bathing, preening and sleeping may be repeated a number of times during the day (McKinney, 1969; Mason, 1984).

Muscovy duck

The diurnal rhythm of the Muscovy duck is comparable to that of the mallard duck. The sequence of feeding, bathing, preening and sleeping may be repeated a number of times during the day (McKinney, 1969; Rodenburg et al., 2005).

Exploratory behaviour

Domestic duck and Muscovy duck

Few studies seem to have been published on the explorative behaviour of ducks.

Diet, food searching, eating and eating postures (feeding behaviour)

Domestic duck

Mallards are omnivorous, feeding on seeds, plants, insects, worms and other invertebrates, depending on local conditions. They feed by foraging on land or by dabbling the beak along the water, which is then expelled through lamellae on each side of the beak, straining out planktonic organisms. Wild ducks perform about 70% of feeding activity in water. In deeper water, ducks perform 'upending' or even dive to seek food. Diving to feed from the bottom is a common practice, especially in ducklings. On land, crawling insects may be snapped up. The beak is richly innervated and very well supplied with sense organs (McKinney, 1969; Reiter, 1993; CoE, 1999a).

Muscovy duck

Wild Muscovy ducks are omnivorous, feeding on plants, worms, insects, fish, amphibians and reptiles.

Their diet consists of plant material obtained by grazing or dabbling in shallow water, with some small vertebrates and insects. Muscovy ducks search for food more on land than in water. They feed by dabbling, foraging and upending (CoE, 1999b). The Muscovy has a much more restricted appetite than the mallard (Mason, 1984).

Selection of resting areas, resting postures, sleep

Domestic duck and Muscovy duck

Ducks seldom keep their eyes closed for more than a few minutes at a time, but they do have long periods (30–60 min or longer) when they are concerned mainly with resting and sleeping. Mallards normally come out on land to rest. They sit down with the head sunk in the shoulders or turned back so that the bill rests in the scapular feathers. These postures may be maintained for long periods, but the eyes frequently blink open and sleeping birds can instantly become fully alert if disturbed. At night, Muscovy ducks often sleep on water, if available, so as to be able to flee quickly from predators if awoken (McKinney, 1969).

Walking, running, flying

Domestic duck

Wild mallards fly and walk efficiently. However, the heavier domestic birds, in particular those selected for meat production, may be unable to fly, have difficulty in walking and be subject to leg disorders (CoE, 1999a).

In wild ducks, pre-flight activities are well marked. The bird faces into the wind, adopts an erect stretched-neck posture and performs rapid, upward head-thrusting movements involving the forepart of the body. Repeated lateral headshakes and a variety of other comfort movements (wing flaps, stretches, etc.) are commonly seen when a bird in flying mood delays take-off for several minutes. These intention movements of flight appear to be important signals in all waterfowl ensuring synchronous take-off by the members of a pair, family or flock. Wild mallards are agile in taking wing, being capable of a near vertical rise when alarmed. Such a take-off from the water is achieved by a single vigorous downward stroke, the wings hitting the water. If the bird is on land, it jumps at the

same moment and the wings do not make contact with the ground (McKinney, 1969).

When descending, speed is decreased by braking; the head, tail and feet are brought forward, the long body axis becomes more vertical and the wings beat on a plane close to horizontal. At touchdown, tail and spread feet hit the water simultaneously. The feet are normally folded and lie under the tail during flight (McKinney, 1969).

Muscovy duck

Muscovy ducks fly and walk efficiently. Birds currently used for meat production have not undergone selection to the same extent as other poultry, but heavy birds may be unable to fly and have difficulty in walking (CoE, 1999b).

Swimming

Domestic duck

Mallards swim readily. When feeding in shallow water, a vigorous paddling movement is commonly seen. Swimming is accomplished by alternate foot movements, the webs being folded as the foot comes forward and expanded during the back stroke. At times, one foot may be shaken in the air and buried in the flank feathers as the other foot makes occasional paddling motions (McKinney, 1969).

Muscovy duck

Muscovy ducks swim readily.

10.7 Behaviour at Defecation and Urination (Eliminative Behaviour)

Domestic duck and Muscovy duck

Like all birds, ducks defecate randomly over their enclosure.

10.8 Body Care (Preening), Cleanliness

Domestic duck

Ducks spend considerable time performing complex preening behaviours. After feeding followed by bathing, ducks carry out a variety of shaking movements to remove water. Cleaning movements then remove foreign bodies and an elaborate sequence is carried out to distribute oil on the feathers from the uropygial gland above the tail. This is necessary for waterproofing and heat regulation. The oil on the duck feather, with its fully knitted barbules, does not allow water through and prevents wetting of the feather surfaces. Also, the down feathers of ducklings are oiled through contact with their mother's oily feathers, which helps to protect the ducklings when they come into contact with water. Preening is often followed by sleeping for a short period. Important elements of bathing are immersion of the head and wings, and shaking water from these over the body.

The recommendations of the Council of Europe (CoE, 1999b) prescribe that ducks should be able to dip their heads in water and spread water over their feathers. Studies by Jones *et al.* (2009) show that, without the opportunity at least to dip their heads and splash their feathers with water, ducks are unable to keep their eyes, nostrils and feathers fully clean.

Muscovy duck

The preening behaviour of the Muscovy duck is comparable to that of the mallard duck. However, Muscovy ducklings seem to perform their own oiling behaviour, starting from day one, and this behaviour may be stimulated by contact with water (Rodenburg *et al.*, 2005).

10.9 Temperature Regulation and Climate Requirements

Temperature regulation

Domestic duck and Muscovy duck

During chronic cold exposure, ducklings are able to develop some non-shivering, heat-producing mechanisms in skeletal muscles (Dividich *et al.*, 1992).

Domestic ducks use water for thermoregulation; heat stress can become a problem in husbandry systems with an inadequate water supply, especially under high temperatures (Rodenburg *et al.*, 2005).

Climate requirements

Domestic duck and Muscovy duck

Ducklings reared without being brooded do not get their down greased with oil from the uropygial gland. Such ducklings therefore are very sensitive to cold and wet weather.

Ducks also require access to open waters during wintertime. Farmed ducks tolerate cold if their resting areas are dry and free from draughts. The Muscovy duck is much more hardy against tropical climate influences than the domestic duck, which is less tolerant of hot dry conditions (Mason, 1984; CoE, 1999a,b).

10.10 Vision, Behaviour in Light and Darkness

Domestic duck and Muscovy duck

Vision in ducks, as for most birds, is an important sense and therefore the duck eye is a well-developed organ.

Ducks, and especially mallards, which naturally forage underwater, can change their refractive power by up to 50 dioptres, compared with the 16 dioptres and the 8 dioptres for fowl and humans, respectively. This is necessary because the powerful refraction resulting from the air–cornea boundary is lost underwater; for this same reason, non-aquatic mammals such as humans cannot focus well underwater (Prescott *et al.*, 2003).

Artificial sources of light usually produce little, if any, of the UV_A radiation which is biologically relevant in poultry species. There may also be a role for the inclusion of UV_A radiation in conventional lighting in duck production, given that the mallard is sensitive to wavelengths as low as 340 nm (Prescott *et al.*, 2003).

10.11 Acoustic Communication

Hearing

Domestic duck

The lowest frequency (number of waves that pass a fixed point in one unit of time) detected for mallard ducks is 100 Hz, the greatest sensitivity is 2 kHz and the highest frequency detected is 6 kHz (Heffner and Heffner, 1992b; Broom and Fraser, 2007).

Muscovy duck

There do not seem to be any results published regarding the hearing of Muscovy ducks.

Vocalization and acoustic communication

Domestic duck

Both sexes have a varied repertoire of sounds often combined with complex body actions. The female mallard makes quacking sounds with a variety of intensities and rhythms – producing several distinctive patterns which have different seasonal distributions and presumably serve different functions in communication. The male produces distinct calls, especially in situations of mild alarm and after take-off. During the mating season, wild female ducks emit particular calls, especially at twilight (Mason, 1984; Appleby *et al.*, 2004).

When members of a pair come together after a separation, another variant of sound is emitted by both sexes, the 'conversation call', according to Lorenz and cited by McKinney (1969).

Ducklings have two different kinds of vocalizations, one when the bird is cold, wet, hungry, alarmed or separated from the parent. The other is given when the distressing situation has been overcome (McKinney, 1969).

Muscovy duck

Muscovy ducks are not noisy birds and are seldom heard to vocalize. Even though they are not completely silent, they do not, in fact, quack. Vocalization is in the form of hissing. The cry of male Muscovy ducks is merely a sort of puffing and adult females are mute (Mason, 1984; CoE, 1999b; Appleby *et al.*, 2004).

10.12   Senses of Smell and Taste, Olfaction

Domestic duck and Muscovy duck

All domestic ducks and Muscovy ducks are very sensitive to water temperature, are reluctant to drink warm water above ambient temperature and can discriminate differences as small as 3°C (Appleby *et al.*, 2004).

The nasal cavity of many bird species has an olfactory epithelium that is structurally similar to that found in mammals. Experiments and observations of behavioural responses to olfactory cues provide evidence of a well-developed sense of smell in birds (Appleby *et al.*, 2004).

10.13   Tactile Sense, Sense of Feeling

Tactile touch

Domestic duck and Muscovy duck

The skin of birds is well supplied with sensory receptors, especially those areas of the body not covered by feathers, such as the beak. In the beak, there are also concentrations of touch receptors grouped to form special beak tip organs, which allow the bird to make very fine tactile discriminations (Appleby *et al.*, 2004). Ducks do not groom each other.

Sense of pain

Domestic duck and Muscovy duck

Birds possess pain receptors and show aversion to certain stimuli, which can be interpreted as being able to experience pain. They show fear behaviour and avoid frightening situations, implying that they experience fear. They show behaviour indicative of frustration. For example, Duncan and Filshie (1980) point out that, if suffering is defined as 'a wide range of unpleasant emotional states', then it is clear that domestic fowl are capable of both suffering and indicating fearfulness (Duncan and Filshie, 1980; Jones and Faure, 1982; Jones, 1986; Gentle and Hunter, 1991; Gentle and Wilson, 2004). There is no reason to assume that the same is not valid for waterfowl.

10.14   Perception of Electric and Magnetic Fields

Domestic duck and Muscovy duck

This does not seem to be described in ducks.

10.15   Heat and Mating Behaviour, Nesting, Egg Laying, Brooding

Domestic duck

Male ducks usually take the initiative in copulation (Appleby *et al.*, 1992). In wild ducks, mating takes place on water; in domestic breeds, it also takes place on land. Copulation is preceded by elaborate social courtship by both the male and female. The male often directs courtship at many females, pursuing and attempting to mate (McKinney, 1969).

Mounting is preceded by the female lowering her head forward along the water surface, spreading the folded wings slightly to each side, and the male mounts, grabbing her crown feathers in his bill. When he has mounted successfully, often after some paddling with his feet, presumably necessary to gain a well-balanced position on the female's back, the male passes his tail around the side of the female's cocked tail and, after some tail wagging, achieves intromission with a single thrust (Birkhead and Brennan, 2009). The act of copulation is brief, less than 30 s. On dismounting, the male immediately pulls his head back, often while still holding on to the female's head, and then lowers his head forward over the water and swings with a nodding head in a circle around the female. Copulation occurs once or several times each day and continues until incubation begins (McKinney, 1969).

The nest site is chosen by the female. The male accompanies the female during her visits to the site for egg laying. The incubation period is 23–26 days. Incubating birds leave the nest one to three times each day to feed, drink and bathe. The female duck leads her brood off the nest within 1–2 days of the emergence of the first duckling. She walks off slowly, calling constantly, and the ducklings follow closely in a tight group. Between spells of feeding in the water, broods come out on to land to sleep. Under normal circumstances, only one brood is raised annually. If the first clutch is destroyed, a second will be laid on a new site (McKinney, 1969).

Once incubation has begun, the male ranges more widely, the pair bond weakens and usually it breaks before mid-incubation and the drake departs to the moulting ground. The female rears the young until they can fly, the ducklings learning from the mother's actions. After rearing the young, the female duck also moves to the moulting area. In both sexes, the wing feathers are moulted at the same time and the birds are flightless for several weeks (McKinney, 1969).

The sitting female duck will leave the nest once a day for about 20 min to 1.5 h, and will then defecate, drink water, eat and sometimes bathe (McKinney, 1969).

Muscovy duck

Male ducks usually take the initiative in copulation. However, Muscovy ducks are an exception in this; females will approach males readily for mating (Appleby *et al.*, 1992).

In the wild, copulation occurs on water. After copulation, the female selects a nest site, usually in a tree hollow often high above the ground, or sometimes in rushes. The incubation period is about 35 days. Muscovy ducks are, unlike the mallard, efficient brooders. They show a preference for covered nests (Mason, 1984; Rauch *et al.*, 1993; CoE, 1999b).

In 23% of 1338 nests investigated, eggs were laid in the same nest by more than one duck, described as dump nests. Dump nesting appears to be a reproductive strategy used by the Muscovy duck to enhance duckling production. Thus, nesting behaviour of the domesticated Muscovy duck is similar to that of its wild ancestor (Harun *et al.*, 1998).

The male is polygamous and does not participate in nest-site selection or incubation. The female rears the young until they can fly and the ducklings learn from the mother's actions (CoE, 1999b).

When compared with domestic ducks, the embryonic development of Muscovy ducks is longer and the ducklings take longer to reach sexual maturity (McKinney, 1969). Domesticated Muscovy ducks can breed up to three times each year (CoE, 1999b).

10.16   During and After Hatching

Behaviour during and after hatching

Domestic duck and Muscovy duck

Newly hatched ducklings are relatively inactive until their down has dried off. A few hours later, they begin to move around, calling frequently and climbing over the duck. Shortly after hatching, they become behaviourally attached to the duck, sitting close beside her, giving contentment calls and expressing peeping calls if she disappears. The female leads her brood off the nest within 1–2 days of the emergence of the first duckling. She walks off slowly, calling constantly, and the ducklings follow closely in a tight group. Broods sleep on land (McKinney, 1969).

Litter size

Domestic duck

In the wild, the hen lays an average of 8–10 eggs per clutch (McKinney, 1969).

Muscovy duck

The Muscovy hen lays an average of 12–14 eggs per clutch (Mason, 1984).

Play behaviour

Domestic duck and Muscovy duck

There have been different interpretations of some duck behaviours as to whether or not they are play behaviours. According to McKinney (1969), however, play behaviours do not occur in ducks.

10.17 The Duckling

Domestic duck

Feeding by upending is first seen when the ducklings are about 4 weeks of age (McKinney, 1969).

If male ducklings are kept in captivity in groups without females during their sexual development, they may develop strong homosexual pair bonds. Drake ducks raised under such conditions show no interest in females, directing their courtship displays to males only and forming strong homosexual bonds (McKinney, 1969).

Muscovy duck

Muscovy ducklings are mostly yellow with buff-brown markings on the tail and wings. Some domesticated Muscovy ducklings have a dark head and blue eyes, others a light brown crown and dark markings on their nape. They are agile and active precocial birds. For the first few weeks of their lives, Muscovy ducklings feed on grains, corn, grass, insects and almost anything that moves. Their mother instructs them at an early age how to feed.

10.18 Assessment of Duck Health and Welfare

Domestic duck and Muscovy duck

For thorough overall inspection of the flock or group of birds, special attention should be paid to bodily condition, movements and other behaviour patterns, respiration, condition of plumage, eyes, skin, beak, legs and feet; attention should also be paid to the presence of external parasites, to the condition of droppings, to feed and water consumption and to growth. Where appropriate, the birds should be encouraged to walk or bathe.

The healthy duck

Domestic duck and Muscovy duck

The healthy bird has sounds and activity appropriate to its age, sex, breed or type, clear bright eyes, good posture, vigorous movements if unduly disturbed, clean healthy skin, good plumage, well-formed shanks and feet, effective walking, bathing and preening, and active feeding and drinking behaviour.

The sick duck

Symptoms

DOMESTIC DUCK AND MUSCOVY DUCK Sick birds show apathy or even lethargy and their plumage loses its lustre.

Examples of abnormal behaviour and stereotypies

DOMESTIC DUCK AND MUSCOVY DUCK Feather pecking may predispose to cannibalistic pecking activities and is often seen in Muscovy ducks when intensively housed as farm animals. The absence of open water can cause ducks to start showing a decrease in preening and abnormal behaviour, such as head shaking and stereotypic feather preening. Without open water, they often redirect their foraging behaviour to straw and their beak, nostrils and eyes may become dirty as they are unable to clean them if open water is absent (Rodenburg et al., 2005).

Examples of injuries and diseases caused by factors in housing and management, including breeding

DOMESTIC DUCK AND MUSCOVY DUCK Ammonia concentrations at 25 and 50 ppm induce eye lesions in young broiler chickens after 7 days of initial exposure. Light intensity of 0.2 and 20 lux for 14 days further exacerbates eye lesions. However, light intensity alone yields no significant eye lesions (Olanrewaju et al., 2007). There are reasons to suppose that this also is valid for ducks.

Pecking injuries have been reported in intensely housed Muscovy ducks. Footpad alterations are less severe in ducks with access to open water (Rodenburg et al., 2005).

Pecking damage and feet lesions have been reported in Mallard ducklings reared for game purposes (Wiberg and Gunnarsson, 2009).

Heavy domestic ducks, a result of genetic selection for muscular growth, and hybrids between Muscovy and domestic ducks are often subject to leg disorders.

Symptoms of pain

DOMESTIC DUCK AND MUSCOVY DUCK Lameness and other locomotion behaviour are the main symptoms of pain caused by joint disorders in heavy birds (CoE, 1999a).

De-beaking is carried out on domestic Muscovy ducks. Like other birds, the beak is abundantly innervated (Desserich et al., 1983, 1984). Beak trimming involves both cutting and cauterizing the beak and a significant amount of the remaining beak, including some of its nerves, are damaged by the cautery. After about 15 days, neuromas are present at the end of the nerves. It is clear that neuromas are formed as a result of the amputation and that these neuromas probably give rise to abnormal spontaneous nervous activity. Although most studies of the effects of de-beaking are made in domestic poultry, the results seem to be applicable to ducks also. The activation of specific nociceptors in humans and spontaneous discharges originating from stump neuromas are implicated in acute and chronic pain syndromes (Breward and Gentle, 1985; Gentle, 1986). Clinical studies of pain in amputated extremities in humans where neuromas have arisen show that tapping the neuromas accentuates not only local pain but also phantom pain from the originally amputated parts (Nyström and Hagbarth, 1981). There are reasons to believe that reactions in the amputated beak of ducks will be the same as in poultry (Gentle, 1986; Gentle and Wilson, 2004) during part of or the whole life of these animals.

Lameness and other locomotion behaviour are the main symptoms of pain caused by joint disorders in heavy Muscovy birds (CoE, 1999b).

10.19   Capturing, Fixation, Handling

Domestic duck and Muscovy duck

Birds should not be carried with the head hanging downwards or by the legs alone. Their weight should be supported by a hand placed under the body and an arm around the body to keep the wings in the closed position. Heavy birds should be carried individually.

10.20   The Importance of the Stockman

Domestic duck and Muscovy duck

Frequent calm and close approach and avoidance of unduly frightening domestic ducks from an early age facilitate the development of a positive relationship between humans and birds. Ducks, and young ducklings in particular, respond to being called or hearing human voices. If ducks are to be driven from one place to another, this should be done quietly and slowly (CoE, 1999a,b).

PART III
Non-Domesticated Farmed Animals

11 Deer: Fallow Deer (*Dama dama*) and Red Deer (*Cervus elaphus*)

11.1 Domestication, Changes in the Animals, their Environment and Management

Deer generally

It must first be said that neither fallow nor red deer are domesticated. They were taken from the wild after the middle of the 20th century and kept in captivity because of their meat.

The deer family (Cervuidae) has a subfamily of true deer (Ceroinae) to which the fallow, red and roe deer belong, together with some 13 other species and 60 or so subspecies. Antlers are characteristic features of deer; they are not integumentary derivatives and are shed and newly formed each year. In fallow and red deer, only the males have antlers. In calves from the age of 4 months, the so-called 'spikes', first year's antlers, grow from February from a club-like, thickening formation, knobs (pedicle). In older deer, the old hard antlers are shed in March–May. New antlers grow from then on. Unlike the rest of the body's arteries, the arteries of the knobs are formed from flexible muscle cells and can contract immediately when necessary. There are numerous nerve systems in the knobs, more than in other deer body organs. The connection between antler and knob is very strong. When a deer breaks an antler it never happens at the joint between knob and antler. The strength of this joint is intact until some hours before the antlers are shed. On shedding, the antler separates without great force from the deer. Usually, both antlers are shed within a short interval of time. During growth, the antlers are covered with velvet, which is a thick, deeply pigmented type of temporary skin and has many sebaceous and scent glands and a comprehensive artery system. When the antler is fully developed, the drier skin cracks and is rubbed off by the animal. Fully developed antlers are thus solid bone, without any epidermal covering. The rubbing-off of the velvet occurs between the end of August and the beginning of September. The reproductive lifespan of female deer is >15 years and for males is 9–11 years. A deer's lifespan can be more than 20 years (Johansson, 1989; Reinken *et al.*, 1990).

Fallow deer

The *Dama dama* species developed during the middle Pleistocene era and was recognizable as such from about 100,000 BC. The oldest remains are found in many caves in England, Germany, France, Poland, Spain and Italy. In the last Ice Age, *D. dama* was still found only in the southern regions of Mesopotamia and Asia Minor. Recent finds in eastern Macedonia in Greece point to the presence of fallow deer in the eastern Mediterranean in about 4000 BC. The European fallow deer was well known to the Phoenicians, Hittites, Assyrians, Babylonians and Sumerians, who kept them as sacrificial animals, probably in a partly domesticated state. Through the worship of Diana, Athena and Astarte by the Roman legions, fallow deer were spread to parts of Europe controlled by the Romans. A fallow deer antler from late Roman times was found in Trier. Sources show that during the 11th and 12th centuries there were fallow deer in England, Denmark and Hungary, and in Alsace in the 16th century. Literature from the end of the 17th century mentions fallow deer as being kept in Mediterranean countries in large game parks and town moats, and being quite tolerant of dry conditions. The *Handbook of the Science of Hunting* by Bechstein (1806) mentions that they were spread at that time over Italy, France, Spain, Palestine, Persia, China and Sweden. As a general rule, fallow deer do not coexist with red deer – the dominant wild species in Scotland. There is no wild population in North America, but in the USA fallow deer were introduced in about 1900. Fallow deer were introduced in New Zealand and Australia in the early

1980s, but in the UK it was the red not the fallow deer that formed the largest group of all the deer. In 1990, there were more than 50,000 fallow deer in Australia, farmed mainly for meat production. In several herds, the antlers are removed during the early velvet phase and sold for use in oriental medicine. In Texas, there are about 6000–7000 head on private hunting estates. There has been a lot more research into red deer farming than into fallow deer farming (Hamilton, 1986; Reinken et al., 1990; Mulley and English, 1991).

The skeleton of the fallow deer is largely similar to that of the red and roe deer, with prominent and strong neck vertebrae and elongated metatarsal bones. Shoulder height of fallow deer is 75–105 cm and body length 130–175 cm. The tail is 18–20 cm long. Body weight in females is about 40 kg, in males about 50 kg. A conspicuous body marking is the nose patch on the upper lip between the nostrils. Calves are born with 20 milk teeth, and there are 32 permanent teeth, with 8 front teeth biting against the callus in the maxilla. All permanent teeth are fully grown at about 3 years of age. There are neither incisors nor canine teeth in the upper gum, just three premolars and three molars. None of the native wild animals shows as many colour variations as the fallow deer. The skin pattern is individual and is unalterable from birth onwards, like human fingerprints. The normal colour of the wild species is rusty brown on the upper parts in summer, with lighter, cream-coloured spots and a light streak on the flanks. There is a darker, brownish streak over the back. The belly is pale cream, while the back of the upper part of the thigh is white, with dark surrounds on both sides. The winter coat is dark grey-brown on the upper parts, the spots are lost and the light belly areas become grey-brown. The moult starts about the beginning of May. By mid-June, depending on location and weather conditions, the summer coat has appeared. The moult is fairly slow and the animals often look unsightly during this time. The autumn change of coat begins at the start of September, lasting until the end of October with females and the beginning of November for males. This coat consists of soft whiskery and woolly hairs, the latter being the shortest. Fallow deer hides are particularly valued because of their flexibility and softness. In adult males, the antlers grow up to 75 cm long, a spread which is strongly palmate with increasing years, the central width is up to 60 cm and the weight up to 4 kg. Antlers in fallow deer are broad-bladed.

In older deer, the old hard antler is shed in April; the new antler grows from then on, with cleaning occurring between the end of August and the beginning of September. The tail of the fallow deer stands out clearly, with its dark, often black colour, against the white background of the anal region, the 'rump patch' (Fig. 11.1). Its skin is rich in glands. The tail shows the mood of the animal. In undisturbed animals, it normally hangs down loosely and is waved occasionally from side to side, which also serves to keep away annoying insects. This tail position is also held when on the move or feeding. When enemies are suspected, and when defending, the tail is held stiffly away from the body, on a level with the whole body posture. When jumping up and in flight, it is upright. It lies directly on the back when calves suck the female (Alexander, 1986; Johansson, 1989; Reinken et al., 1990; Alcock, 2007).

Fallow deer thrive everywhere, with the exception of high mountains, although they prefer a park-like landscape with small woodlands. Varying with the season, they seek out completely open fields or dense thicket. They are hardy and resistant to cold. Their great adaptability is demonstrated by their survival under a wide variety of environmental and climatic conditions. Annual home ranges for fallow deer are estimated at 2.1 km² for females and 9.75 km² for males. As a gregarious animal, the fallow deer is tied strongly to rutting and calving areas, which explains why their spread takes place very slowly. Fallow deer have been observed living with other species in the wild, like roe deer, red deer and sheep. Fallow deer will push roe deer aside. For choice of habitat, fallow deer prefer less dense woodland with larger trees than red and roe deer (Sinding-Larsen, 1979; Alexander, 1986; Reinken et al., 1990; Borkowski and Pudelko, 2007).

The female fallow deer is called a 'doe' or a 'hind', the male a 'buck' and the offspring a 'calf' or a 'fawn'.

Red deer

There is some international confusion of nomenclature regarding red deer (*Cervus elaphus*) and elk. In Europe, the word elk refers to *Alces alces*, whereas in the USA and Canada it refers to the wapiti, *C. canadensis* (Clutton-Brock et al., 1982). DNA studies conducted on hundreds of samples from red deer and wapiti show that there are no more than nine distinct subspecies and that they fall into two

Fig 11.1. Fallow deer observing something in their surroundings (image courtesy of Bengt Röken).

separate groups: the red deer from Europe, western Asia and North Africa, and the wapiti or elk from northern and eastern Asia and North America. From DNA evidence, the wapiti appear to be related more closely to sika deer and to Thorold's deer than to red deer (Ludt *et al.*, 2004).

The red deer were once found in the wild throughout much of the northern hemisphere, from Europe through northern Africa, Asia and North America. Body size apparently reached its peak during the last glaciation. Extensive hunting and habitat destruction have limited the red deer to a portion of their former range. Their traditional range extended over all of Europe up to 65° N in Scandinavia. In the 2000s, there are wild red deer in North Africa; they occur down to 33° N. Today, large populations of wild wapiti (elk, *C. canadensis*) in North America are found only in the western USA from Canada through the Eastern Rockies to New Mexico (Clutton-Brock *et al.*, 1982).

Red deer were taken from the wild after the middle of the 20th century and kept in captivity because of their meat. About 1970 deer farms were established in the UK and then spread to Australia and New Zealand. However, red deer from the UK had been introduced occasionally into Australia

and New Zealand during the 19th century. Since the mid-1980s, the farming of red deer has become firmly established in the UK and other European countries. In New Zealand, some of the by-products of deer farming increased the profitability of red deer farming, when compared with producing just venison alone. Very high prices are paid in the Far East for these by-products, for example antler velvet, tails and penises, for use as traditional remedies (Hamilton, 1986).

Red deer hinds (Fig. 11.2) are about twice as heavy as fallow deer, the adults weighing between 95 and 180 kg, with a height of 120–160 cm at the shoulders and a length of 185–215 cm. Scottish deer seem to be slightly smaller than those on the European continent. An old adult male (Fig. 11.3) can weigh 300 kg. Calves weigh 7–13 kg at birth. The summer coat is reddish brown, changing to grey-brown in winter. The rump patch is straw-coloured. Tail length is about 15 cm. The rutting period is from the beginning of October to December, while calving time is in June – earlier than for fallow deer. Red deer are calmer in behaviour than fallow deer and are not panicked so easily, but they seem to be somewhat more susceptible to illness. They are sometimes kept in very severe habitat and weather

Fig 11.2. Red deer hinds with calves (image courtesy of Bengt Röken).

conditions, for example in Scotland, where the forest coverage can be very slight and the rainfall is very high. The female range is concentrated, while the male range has three parts, one for reproduction, one for the winter period and one for the rest of the year. The stags can be very aggressive, particularly during the rutting season. The calves are born with 20 milk teeth and there are 32–34 permanent teeth with 8 front teeth biting against the callus in the maxilla. All permanent teeth are full-grown at about 3 years of age. Each red deer stag has specific antlers, like a human's fingerprints. Antlers are almost round in cross section. The loss of antlers during rutting fights seems to be less frequent than with fallow deer. The biggest antlers are found in males of 14 years of age (Sinding-Larsen, 1979; Alexander, 1986; Johansson, 1989; Reinken *et al.*, 1990; Pepin *et al.*, 2005).

In a study of red deer in Scotland, published by Clutton-Brock *et al.* (1982), the female range was 0.45 km² in August, 0.15–0.25 km² in September and October, increasing to 0.30 km² in November.

The female red deer is called a 'hind', the male a 'stag' or a 'hart' and the offspring a 'calf'.

11.2 Innate or Learned Behaviour

Fallow deer

Reinken *et al.* (1990) give examples of learned behaviour, such as begging for food from visitors in fallow deer kept in estate herds and even that certain individuals react to the names given to them.

Red deer

Innate behaviour in red deer, as seen in connection with parturition, i.e. the avoidance of other deer and the removal to high ground before calving, appears in hinds calving for the first time (Clutton-Brock *et al.*, 1982).

11.3 Different Types of Social Behaviour

Social behaviour

Deer in general

The female herds consist of adult females, young hinds, calves and some young males younger than 2 years. By 1 year of age, the young males are already leaving the female flock occasionally, and by 2 years of age they have left the female herd

Fig 11.3. Red deer stag (image courtesy of Bengt Röken).

once and for all. These young males usually go to herds far away from the one they have left, and young red deer males can make very long journeys, especially in connection with the rut. In the beginning, they are at the periphery of this new male herd, which will consist mainly of individuals older than 2 years (Johansson, 1989).

Fallow deer

Social grouping is largely similar to that of many animals which live in herds in open country. There is much flexibility in the formation of groups, so fallow deer can thrive in constantly changing surroundings. A female group may range from two to several hundred animals. According to Reinken *et al.* (1990), larger groups are found mainly in winter, smaller ones in summer and male groups are largely similar to female ones in size and seasonal

changes. However, according to Clutton-Brock *et al.* (1982) and Johansson (1989), groups are usually larger in the summer, when competition for food is far less than in the winter.

Mixed groupings can be observed in fallow deer from October to April, while from May to September the sexes stay separate. Young males must be considered as a separate entity. Calves and yearlings of both sexes at times come together in groups of young animals. A complete separation of the calves from the females can be particularly noticed during the rutting period. The next period of isolation comes before the birth of new calves (Reinken *et al.*, 1990).

Male groups are largely similar to female groups in size and seasonal changes. The rutting period in autumn brings about a break-up of grouping. The younger animals go about singly or in small groups between mating, or they stay close to the females.

At the rut, the groups split up, with bucks collecting a 'harem' of fertile does. After mating, larger groups are formed again, which stay together until early spring, according to the availability of food and weather conditions. Female groups are formed after mating comprising adult does and calves born during the summer, and also unmated yearling does. The mother has a close relationship with her young. Calves and yearlings keep close to the mother when there is any disturbance. Splitting up of the winter groups usually occurs at the time of antler shed. Variations in group size are more clearly pronounced with males than with females (Reinken *et al.*, 1990).

Red deer

Hinds and stags are segregated for most of the year. Individuals of both sexes are most commonly found in parties of four to seven animals, excluding calves. The frequency with which calves associate with their mothers peaks during their first autumn and subsequently declines. Daughters usually adopt their mother's ranges. Mothers and daughters are associated with each other and it is possible to identify and define long-lasting matrilineal groups within the hind population. Hinds regularly associate with the same individuals, who are usually matrilineal relatives, whereas the frequency of association between individual stags is lower and members of the same male party are seldom closely related. Male calves associate with their mothers less than do female calves, and calves whose mothers become pregnant in the rut following their birth associate with them less than calves whose mothers fail to conceive again. Most hinds spend 80–90% of their time in parties that do not include any stags over 3 years old. Stags disperse from their mother's ranges between the ages of before 2 and up to 3 years and join stag groups. Members of stag groups share overlapping ranges, but individuals associate with each other less regularly than hinds. Stags spend most of the year in bachelor groups, which break up in September when older individuals move separately to their traditional rutting areas. Fighting ability and reproductive success are related to body size and condition. Size is affected by early development and stags that show advanced development at 1 year old will be the most successful breeders at the age of 6 (Clutton-Brock *et al.*, 1982).

Communication

Deer in general

For both fallow and red deer, communication on special occasions takes place with vocalization, but most communication is performed by body behaviour, e.g. the posture of the head, neck and tail, and the way of locomotion (Johansson, 1989).

Behavioural synchronization

Fallow deer

The female groups are more synchronized during feeding and rumination periods than the male groups (Reinken *et al.*, 1990).

Red deer

All individuals in a group of deer mainly feed or ruminate synchronously (Clutton-Brock *et al.*, 1982). Alarm postures induce social facilitation, imitative behaviour. The act performed by one individual elicits new alarm postures in the remaining members of the group (Recuerda *et al.*, 1986).

Dominance features, agonistic behaviour

Deer in general

In the male groups, there is a pronounced dominance order between the older bigger animals and the dominant male usually keeps his established position in the hierarchy for several years in succession. When the stags have shed their antlers and when new antlers are growing, their threats are similar to those of the hinds. They avoid using their antlers completely. Instead, conflicts are solved by rearing on the hind legs and direct kicks against each other's head, neck and shoulder (Johansson, 1989).

Fallow deer

In most species of deer, group leadership is undertaken by an adult female, but with fallow deer, the leader can be an adult female or male, or even, in moments of danger, the nearest available animal. In groups of females and in mixed groups, the leading position is usually taken by an adult female. She is particularly alert, gives warning of danger and indicates the possible route of flight. Within groups of males, one of the elders always takes the lead. At the time of antler shedding, his leadership may or may not be retained when other stags still have their antlers (Johansson, 1989; Reinken et al., 1990).

When pushing, as an occasional precursor of threatening behaviour, the buck walks in a peculiarly stilted way, stiff-legged and with the neck held horizontal. When threatening, the buck approaches his opponent with head slightly raised. He then drops his neck below the horizontal and lays his ears back, opens his mouth and steps towards his opponent. With further threatening, the antlers are extended forward. A smaller buck will turn away. One of equal weight will face up to the stag, threatening him and they will exchange blows with their antlers. During fighting, antlers clash and often become entangled with one another. After separating, the stags will make another attack. In serious battles, the effectiveness of the antler thrust is strengthened by turning the head and attempting to knock the opponent off balance. The fights last from 5 to 12 min. The defeated animal usually takes flight but generally he is not followed, apart from a few steps. Younger bucks may show shadow fighting, scraping with antlers and forelegs, bucking, striding around and marking trees and bushes (Reinken et al., 1990).

Red deer

Social relationships between members of hind and stag groups differ in a variety of ways. Among both hinds and stags, the frequency of threats is highest during times of food shortage. Threat rates are higher in winter than in summer (Clutton-Brock et al., 1982).

Hinds threaten each other less frequently than stags and their dominance relationships are less predictable. This is apparently because hinds seldom threaten their matrilineal relatives and commonly associate with them. Nearest-neighbour distances are shorter among hinds and they sometimes groom each other, whereas stags rarely do so. Dominance rank is age-related among hinds, but less regular and less consistent among hinds than among stags. Apart from age, there are no obvious correlates of dominance and, among hinds, the commonest forms of agonistic interaction are nose and ear threats. One hind may poke her head towards another, usually at the neck, shoulder or rump, and often exhale loudly. Another hind may instead lay back her ears while moving towards another animal. Displacements, where one hind walks steadily towards another, are also common.

More intense threats include kicking, usually with a single foreleg, but sometimes with both, and boxing, when two hinds rear on their hind legs and slap at each other's heads or shoulders with their hooves. Bites are also common and are usually directed at the head, neck or shoulders: they often occur as the finale of a gradually escalating series of threats and are followed frequently by a chase or an attempt to butt the opponent. Oestrous females in harems show elevated aggression rates, compared to when they are in harems but not in oestrus, and also when they are in foraging groups outside the breeding season (Clutton-Brock et al., 1982; Bebie and McElligott, 2006).

Dominance rank between stags is age-related among animals that have not yet reached adult body size. After this, a difference in body size is apparently more important than age. Situations where one stag walks directly towards another individual that moves away are commoner than all other categories of 'threats'. The dominant animal sometimes uses a stiff, pronounced gait, often accompanied by a concentrated sideways glance. In more intense threats, stags raise their heads and point their chins at their opponents, sometimes curling back the upper lip and hissing or grinding their teeth at the same time. When a dominant stag approaches a subordinate and the latter fails to move away, the dominant usually responds with a nod of his head, directing his antlers towards the subordinate, often flattening his ears or rolling his eyes at the same time. Stags sometimes kick at other individuals, with either one or both forelegs, and occasionally chase them, attempting to horn them in the flank. However, such events are rare. Unlike hinds, stags rarely try to bite each other. Older and more dominant stags cast their antlers earlier than younger stags and thereby in some cases become subordinate to them, a situation that will reverse when the older stag gets his new antlers (Clutton-Brock et al., 1982).

11.4 Behaviour in Case of Danger or Peril

Deer in general

The principal deer defence strategy is to avoid everything that might seem dangerous. Their defence is dependent on an efficient and continual supervision of the environment by at least one member of the group and a thorough familiarity with the ground where they are staying. Each individual takes a turn, one after the other, to keep watch over the surrounding area (Johansson, 1989).

Fallow deer

Hearing strange sounds provokes a reaction, but even sounds occurring naturally in woodland, like the rustle of leaves or moving twigs, can cause restlessness. Generally, though, sounds bring about flight relatively slowly. One exception is when an adult deer crosses another deer's path. Then, audible stimuli, in combination with visual or olfactory awareness, lead to flight relatively quickly. During attacks and danger periods, the females with fawns at foot run at a gallop or trot, head erect, ears alert and tail raised. Later, the neck is outstretched, the ears laid back and the mouth opened, snarling. Pawing with the front hooves may be noticed, and a blow might even be turned on the aggressor. If dogs and cats approach farmed fallow deer, the hinds are observed to goose-step up to them and drive them off by kicking (Reinken et al., 1990).

Red deer

Deer are reported to be more nervous and frequently more alert when feeding in small groups than when feeding in large ones, and hinds, which mostly have calves with them, are frequently more alert than stags when in small groups. Hinds sometimes give a short, staccato alarm bark when they are disturbed. These are almost always given by adults, though not necessarily by older animals, and instantly alert other members of the flock. Males rarely give alarm barks, though young stags occasionally do so (Clutton-Brock et al., 1982).

In a study, the percentage of animals that were vigilant was higher in sites disturbed by humans than in less-disturbed sites, and higher in disturbed grassland with poor cover and in heather with intermediate cover than in disturbed woodland with good cover. The majority of the vigilant animals in disturbed heather and woodland habitats and in all the less-disturbed habitats were standing. In disturbed grassland, however, lying was the main posture while vigilant. In both disturbed grassland and heather, the percentage of vigilant animals that were moving was higher than in woodland or the less-disturbed habitats. In disturbed sites, the deer were more likely to aggregate when vigilance levels were high. Thus,

red deer respond to disturbance from human activities by increasing their level of vigilance, but the nature of their response varies with the level of cover available. Red deer thus may lie down when keeping vigil in grasslands, because lying animals are less conspicuous and the low cover will still allow animals to scan their surroundings (Jayakody et al., 2008).

11.5 Some Normal Physiological Frequency Values of Interest at Clinical Examination

Fallow deer

The normal rectal temperature of adult animals is between 38 and 39°C (Johansson, 1989).

In adult animals, the pulse rate varies between 65 and 100 beats per min and in adult animals at rest the respiratory rate is 13–18 breaths per min (Johansson, 1989). In a telemetric study, Hoffmeister (1979) registered a pulse rate of 60 beats per min in resting animals and up to 80 beats per min in animals either standing or walking. Visual disturbances from one person or a person with a dog cause increases in the pulse rate of between 18 and 70 beats per min. Auditory disturbances like the barking of dogs or noise from an engine cause an increase between 11 and 73 beats per min.

Red deer

The normal rectal temperature of adult animals varies, according to Adam (1986), between 38.5 and 38.9°C and, according to Johansson (1989), between 38 and 39°C.

The heart rate is reported to be 60–80 beats per min (Adam, 1986). In restrained 10-month-old stags, the pulse rate was 60–80 beats per min (Pollard and Littlejohn, 1995). In 3-year-old hinds, it varied from about 60 when the animal was lying to about 85 when the animal was walking (Price et al., 1993). If it changed from idling to any other behaviour, then the heart rate increased further, by 1.7 beats per min if it started ruminating, by 4.0 beats per min if it started foraging, by 5.5 beats per min if it started grooming and by 12.8 beats per min for neck-up. These changes associated with head movements are similar in magnitude to those observed in ewes. By this reckoning, heart rate while walking foraging appears low, but this is probably because foraging reduces walking speed.

The high heart rate during walking neck-up may reflect the 'alarm' responses which would sometimes have been associated with this behaviour (Price *et al.*, 1993).

The respiratory rate varies between 10 and 14 breaths per min (Adam, 1986; Johansson, 1989).

11.6 Active and Resting Behaviour Patterns

Activity pattern, circadian rhythm

Fallow deer

For adult fallow deer, the first feeding period starts at dawn and finishes about 2–2.5 h later. There is then a ruminating period of about 2 h, in which there are periods of sleep of about 20 min. During the summer, up to the start of the rut, five to six grazing periods alternate with periods of ruminating. During the rut, there is an irregular daily rhythm, while in the winter there may be only two grazing and ruminating periods. The daily rhythm of male fallow deer is not as clearly organized as with the females. With young animals there is a much more regular daily rhythm. Observations during daylight from mid-May to the end of September show that adults graze for 27.5% of the time; 7.2% is taken up in lying down ruminating, 5.3% in resting without ruminating, 14.4% in moving and standing and 2.2% in other activities (Reinken *et al.*, 1990).

Red deer

Because of their shyness and vigilance, red deer are mostly on the move during the dark part of the day and night. If wetlands exist within their area, they try to go there some time each day and night. They avoid grazing on open surfaces (Sinding-Larsen, 1979).

When they start to hold harems, the activity budgets of harem-holders show dramatic changes. The proportion of daytime spent grazing falls from 44% during the summer to less than 5% in the rut, while the proportion of time spent moving and standing inactive increases from 3 and 4% to 15 and 33%, respectively. Night-time budgets show similar changes. Young stags and stags without harems do not show as marked a decline in grazing time, though they graze less than at other times of the year. The decline in grazing associated with harem holding is reversed quickly

when the stag ceases to hold a harem (Clutton-Brock *et al.*, 1982).

Exploratory behaviour

Deer in general

Deer seem to react with curiosity and hesitation to even small changes, e.g. a newly fallen tree, or foreign objects in their environment. They seem to feel a compulsion to investigate each strange and unfamiliar object and to show a behaviour characterized by a combination of curiosity and fear. They approach the object with great caution, all the time prepared to withdraw. When after many advances and retreats the investigation is performed, they lose interest rapidly. A similar behaviour occurs when a non-group member is let into an established group of farmed deer; it will be subject to intense investigation (Johansson, 1989).

Diet, food searching, eating and eating postures (feeding behaviour)

Fallow deer

Where the fallow deer and roe deer occur within same area, they do not appear in the same biotope. This depends probably on the fact that the two species have different food preferences. Roe deer prefer herbs and leaves. The fallow deer is the least particular of the deer in the choice of food. Fallow deer are not as selective as roe deer. They graze on the plants most commonly found on site and are indifferent to a variety of food plants. During most of the year, they graze on herbs and grass. Agricultural crops might also be included in the diet (Sinding-Larsen, 1979).

While eating, fallow deer take short, quick bites of plants, which are bitten off or ripped out by closing the incisors on to the gums. The lips often are not a part of the movement. Rest periods between bites are not regular. Deer can take leaves or fruit from trees by rearing up on their hind legs to reach heights of about 2 m. In winter, snow is scraped away with the forelegs and pushed aside with the snout and forehead so that food can be reached. Ruminating takes place mainly when deer are lying in the resting place, only occasionally when standing. There are on average about 40 grinding movements per bolus, with each grinding lasting about 0.7 s. A fallow deer's drinking needs are low in the summer due to the moisture content of the feed. In the winter, the daily need is estimated to be 1.5 l (Johansson, 1989; Reinken et al., 1990).

During the rut, fallow bucks reduce their feeding and increase the time and energy spent on vocalizing and fighting to gain matings, and consequently their body condition declines greatly (Vannoni and McElligott, 2009).

Red deer

During the summer, red deer graze mainly on herbs and fresh grass. On the fields, barley, wheat and oats are eaten. In the forest, a lot of reindeer moss, berry sprigs and heather are eaten. Red deer also eat a lot of leaves and twigs of different trees and bushes such as ash, mountain ash, beech and osier and willow. During winters with much snow, they eat mainly juniper and pine, but can also eat a lot of bark from especially young spruces (Sinding-Larsen, 1979).

Grazing is aggregated into bouts varying from 10 to over 100 min. The duration of daytime grazing bouts is shorter in summer than in winter and does not differ significantly between the sexes. Night-time bouts in winter are considerably shorter than daytime bouts. In summer, both stags and hinds usually fit four to six grazing bouts in the day and between one and three bouts at night. In winter, as grazing bout length increases, the number of daytime bouts drops to two or three. Bite rate is between 50 and 60 bites per min. Red deer chew bones and cast antlers (Clutton-Brock et al., 1982).

In winter, red deer might feed partly by bark stripping on spruce (Jerina et al., 2008). In 36 sites examined, the rate of bark stripping was highly variable (from 0 to 84% of susceptible trees debarked), with less damage in Scotland than in other European sites, for which bark-stripping rates were higher at high red deer density. Bark sometimes made up a large proportion of red deer diet (>10%), especially in areas with severe winters (high levels of snow), whereas in study sites with mild winters, bark was practically not eaten at all. These results suggest that severe bark stripping could be related to a reduction in food resource availability (Verheyden et al., 2006).

A red deer's drinking needs are low in the summer due to the moisture content of the feed.

However, during long warm and dry periods, this is not enough. In the winter, the daily need is estimated to be 3 l (Johansson, 1989).

Selection of lying areas, lying down and getting up behaviour, resting postures, sleep

Fallow deer

The fallow deer is not very choosy in its choice of resting place. Only areas with no growth or very damp places are avoided, the preference being for sheltered spots. The resting place is usually scraped with the forelegs, so that the greater part of the body can lie on the cleared area. The chosen resting places of the males are larger than those of the females. Even during rest periods, the animals often stand up to change position or to groom themselves. The look-out is undertaken in turn by the animals. Before lying down, deer drop their head and neck, bend the forelegs and shake and turn the body. The most common resting position is the ruminating one, with forelegs drawn up underneath the body. The neck is usually held vertically, but sometimes it is laid down on the ground. In the lying position, the head leans against or is supported on the flanks. When the sun is shining, fallow deer often take up a position on their side, with fore- and hind legs stretched out from the body. When getting up, the head and neck are lifted and first the forelegs and then the back legs are raised (Reinken et al., 1990).

Red deer

The red deer select an undisturbed environment as a resting area. Therefore, lying areas during the day are found in dense deciduous or coniferous forests. In the summer, they might select fields where the crop has grown high (Johansson, 1989).

Studies indicate that resting bouts are shorter during the night than during the day during summer and do not vary between sexes. Resting place visibility is lower during the day. Use of cover prevailed during the daytime, whereas night resting place characteristics were more variable, indicating less constrained behaviour. Thus, cover as well as food is an important factor in red deer habitat use (Adrados et al., 2008).

It has been shown by means of satellite image analysis, field observations and measuring 'deer beds' in snow that grazing and resting red and roe deer (n = 2974 at 241 localities) across the globe align their body axes in roughly a north–south direction. The same has been shown for domestic cattle (Begall *et al.*, 2008).

Locomotion (walking, running)

Fallow deer

Normal movement is at a walk, singly or in groups, with the body at ease and relaxed. The alarm position sees the body taut, head and neck held high, front legs close together and ears pricked. When mildly excited, movement is at a trot. If excitement rises, the fallow deer executes its typical spring, taking 2–3 m leaps, with front and hind legs close together, tail raised and ears pricked. This movement is a means of conveying messages to other members of the group. Finally, the quickest motion is a gallop-like step with long strides, each covering several metres (Reinken *et al.*, 1990).

Fallow deer can run at up to 65 km per h (Johansson, 1989).

Red deer

Few wild, even-toed hoofed mammals exhibit so graceful locomotion as the red deer. Their movements are extraordinarily graceful, whether they are moving at a light, springy walking pace, in a floating trot or at a gallop (Sinding-Larsen, 1979). However, there is one exception, the parallel walk shown by competing stags during the rut. Then they show a slow, regular and stiff gait (Clutton-Brock *et al.*, 1982).

At top speed, a red deer stag can run at 70 km per h (Johansson, 1989).

Swimming

Fallow deer

Like all deer, fallow deer can swim.

Red deer

Red deer are good swimmers (Sinding-Larsen, 1979).

11.7 Behaviour at Defecation and Urination (Eliminative Behaviour)

Deer in general

In deer, both sexes remain standing while urinating. The male deer generally retains his usual body posture or spreads out his hind legs. The female's position is with hind legs spread or a slight bending down of the body. Defecating can also take place while moving and walking. Body posture is unchanged, with the tail being held upwards or to the side (Reinken *et al.*, 1990).

11.8 Body Care, Cleanliness

Fallow deer

Signs of contentment include scratching with the hind leg, rubbing, grooming with the mouth, snorting, stretching, yawning, shaking and tail swishing. Social grooming in females is observed as mainly licking another member of the group. Intentional bodily contact between males is rare, with the exception of 'pecking order' fights (Reinken *et al.*, 1990).

Red deer

Nearest-neighbour distances are shorter among hinds, and they sometimes groom each other, whereas stags rarely do so. Hinds groom each other by licking and nibbling the hair of their counterparts, particularly around the neck, head and ears. Grooming by hinds is confined to other hinds and their own dependent offspring and, with the exception of those in oestrus during the rut, hinds rarely groom stags over 2 years old (Clutton-Brock *et al.*, 1982).

Both sexes of red deer create pits, by treading in moist ground where water and soil are mixed, in which they perform wallowing (Sinding-Larsen, 1979). During the rut, the stag often urinates into the wallows (Clutton-Brock *et al.*, 1982).

11.9 Temperature Regulation and Climate Requirements

Temperature regulation

There do not seem to be any results published regarding the temperature regulation of deer.

Climate requirements

Fallow deer

Long, snowy winters do not suit fallow deer but, at the same time, they are hardy and resistant to cold: no fallow deer died in the harsh winter of 1939/40 in East Prussia, but red and roe deer suffered losses (Reinken *et al.*, 1990).

Red deer

Stags spend more time in shelter during adverse weather than hinds, maybe because of increased heat loss. An additional reason for increased heat loss may be that their poor body condition after the rut delays the growth of their winter coat, which is often thinner than that of hinds (Clutton-Brock *et al.*, 1982).

A study of red deer behaviour in dry and wet weather indicated a need for shelter from the rain during wet days for stags at pasture in New Zealand (Pollard and Littlejohn, 1999).

11.10   Vision, Behaviour in Light and Darkness

Deer in general

As regards perceptive senses, the good eyesight of the fallow deer and the red deer is especially important. Wild deer are shown to shy when 700 m away from a man moving in open country. They do not seem to observe immobile beings as well, although fallow deer seem to be the red deer's superior in this respect. Their sight in twilight and at night is similar to that of humans. In red deer, optical disturbances generally induce a more intensive reaction than acoustic ones (Sinding Larsen, 1979; Johansson, 1989; Reinken *et al.*, 1990; Herbold *et al.*, 1992).

11.11   Acoustic Communication

Hearing

Deer in general

Both fallow and red deer have relatively large funnel-shaped ears which can be moved independently of each other and thereby be slanted around a lateral arc of about 150° in order to catch where a sound comes from (Johansson, 1989). There do not seem to be any results published regarding the hearing range of deer.

Vocalization and acoustic communication

Fallow deer

Fallow deer communicate by calls, i.e. warning cries, or the doe's bleating sound when calling her calf to suckle or in order to keep contact with it when on the move. The males give rutting calls, a belching noise similar to snoring, in order to attract females. The calls are divided into: lowing, bleating, mewing, wailing, screeching and the rutting bellow. Bleating is similar to the sound made by sheep and is emitted only by females with fawns. The mothers bleat to call their young to suckle and to keep contact when on the move. The young are alone in making a short piping or peep-peep noise. This is a sound which shows awareness but also creates a bond between mother and fawn. It can be heard throughout the year. Mewing, a more choked 'mi-mi-mi', is heard from both sexes and all ages and is always linked with a soothing gesture. Wailing is a long drawn-out, high-pitched bleating and is heard from fawns in need, when they are in bodily pain and sometimes as a death cry. The rutting call is not a typical bellowing, as with the red deer, but a snoring or belching sound made with the mouth partly open, head slightly raised and the tail lifted. Short bursts follow in quick succession. This call is used mainly to attract females. Fallow deer differ completely from red deer in their rutting call. The buck usually walks about restlessly when calling, but he can also emit his call when lying down. A particular territorial position is not evident, as with the red deer stag. Sometimes, rutting calls are also emitted after a successful fight with a rival (Reinken *et al.*, 1990).

During the breeding season, physiological and social factors change and these can have strong effects on the structure of calls and calling. Groaning rates follow a clear curvilinear trend, reaching their highest levels at the peak of the rut when most matings occur, and are lower during the early and late rut. The rapid decrease in groaning rate at the end of the rut may be explained partly by the decline in male body condition. Short-term changes in groaning rates mainly represent a threat signal directed to other males (Vannoni and McElligott, 2009).

Red deer

During rutting, the mature stags give deep guttural roars, often in several bouts, as signals to the hinds, but also to competing stags. It is described as a mixture of the boom of church bells and the roaring of a lion (Sinding-Larsen, 1979). The roars are aggregated into bouts of one to ten that are often given on the same exhalation. During the days before and after parturition, hinds sometimes bellow, raising their noses in the air and giving a loud deep moan similar to a stag's roar but softer in tone. Another sound is barking, which is a series of short barks typically directed from a harem holding older stags or young stags, after they have been chased away. Hinds sometimes give a short, staccato alarm bark when they are disturbed. Hinds when frightened or disturbed give barking sounds as a warning for the calf or for other members of the flock. Young stags sometimes run through a harem held by a senior stag giving alarm barks, and thus causing the hinds to scatter, in order to get an opportunity to mate with any of the hinds. Does and calves both emit sounds during the first weeks after delivery. The doe's sound is a low, hoarse bleating and the calf's a piping sound. Older calves give out piping and whickering noises (Clutton-Brock *et al.*, 1982).

11.12 Senses of Smell and Taste, Olfaction

Deer in general

Scent glands are similar in red and fallow deer. Both sexes have scent glands between the hooves, on the lateral parts of the hocks and below the front corner of the eye sockets (antorbital glands). The males have scent glands in the preputium and the hinds excrete scents from glands in the urinary tract (Johansson, 1989).

Fallow deer

A fallow deer's sense of smell is poorer than that of red deer and reindeer. It can smell humans from about 200 m, while red deer, depending on humidity, react at up to 1200 m. Potential sexual partners occasionally find each other by scent; during the rut, bucks seek out females with their noses lowered, and the smell of urine has a clearly sexual connotation. However, the sense of smell is not used in the initial approach, prior to pairing-off for mating. Compared with optical signals, warnings of enemies through scent have little importance. As a rule, after picking up a scent, fallow deer try to find their enemy with the aid of eyes and ears and, if they cannot do this, then further restlessness is a normal reaction after a certain length of time. The distance at which fallow deer show a reaction when picking up a scent is less than if they become aware of an enemy by sight. Also, scent alone plays less part in picking up human presence than does sight (Reinken *et al.*, 1990).

Red deer

Olfaction plays an important role in the red deer rut period. During the rut, the stag often urinates into the wallow pool and wipes his antorbital glands and antlers on nearby vegetation. At the onset of the rut, the hinds begin to give off a characteristic scent from the region of the vulva that grows in intensity during the following weeks and stags sniff and lick around the base of the tail of the hinds (Clutton-Brock *et al.*, 1982).

11.13 Tactile Sense, Sense of Feeling

Tactile touch

Deer in general

There do not seem to be any results published regarding the tactile sense of deer.

Sense of pain

Deer in general

The fact that deer as prey animals do not show obvious signs of pain does not mean that they do not suffer if exposed to pain (Arney, 2009). Antler removal, which apparently causes considerable pain, is not legally forbidden in some countries. In a study comparing behaviour in two groups of red deer, one just restrained and the other restrained together with antlers removed under analgesia, it was obvious that behavioural indications of pain increased following antler removal even with analgesia (Webster and Matthews, 2006).

11.14 Perception of Electric and Magnetic Fields

Deer in general

This does not seem to be described in deer.

11.15 Heat and Mating Behaviour, Pregnancy

Deer in general

Deer are seasonally polyoestrous. Females can still conceive when aged 20 or older, but the probability of this is small (Mulley *et al.*, 1990; Reinken *et al.*, 1990).

Fallow deer

Sexual maturity occurs at 16–20 months. The rutting season is September–October. The oestrus cycle is 22 ± 1.3 days. The fertility of fallow does is high in the wild. In farmed fallow deer, the annual conception rate varies between 95 and 98% (Fletcher, 1986).

The breeding season for fallow deer is longer than the rut. The breeding season is a short period for many species of deer but with fallow deer it can last from October to March. Ovulation is normally seen from September to January. The greatest number of conceptions takes place during the main rutting period, in October. Gestation period in fallow deer is 234 days. Unlike red deer, fallow deer that conceive late can rear their calves successfully. The antlers are cast in late April–May and the new ones cleaned in late August–September (Fletcher, 1986; Reinken *et al.*, 1990).

The start of antler rubbing usually coincides with shedding the velvet and the start of the rut. During the rut, secondary sexual signs appear, such as the call, strong rut smell, increased growth of the neck and also stretching and colour changes to the skin at the end of the penis. This skin becomes dark to almost black in colour, and marked papillae are formed which give out a strong rut scent. Sexual behaviour is marked by restlessness in the bucks at the start of the rut, when they make for the areas where the females gather. From the end of August to September, the young bucks starting earlier than the older ones, these animals, who are normally very wary and keep to cover, now become bold and lose their fear of humans. Many older stags return every year to the same rutting area, marking it by scraping holes in the ground with their antlers and hooves. These rutting scrapes are 30–300 cm across, width and depth varying according to the hardness of the ground, and they are often sprayed with sharp-smelling urine, the tufts of hair at the tip of the penis helping to scatter this. Marking of the area is done by the bucks rubbing and knocking against bushes and trees, using their antlers on the branches and trunks, often stripping off large areas of bark. Whole branches are often broken off and strewn about, and even large trees are not safe from antler attack (Reinken *et al.*, 1990).

Mating behaviour consists of courtship and mounting. Females may display rejection signals, by means of which they show their unwillingness to mate. These include lying down or chewing, with the head and neck stretched well forward and the ears close to the head. Finally, mating animals rub their lower jaws, flanks or cruppers or lick each other's flanks. This behaviour is noted particularly in the main rut and during the phase of mounting by the male. The male attempts to arouse the female by touching her head and body with his mouth, licking her and finally rubbing his head on her flanks. This is usually followed by pushing, nose-wrinkling, smelling, then the rutting call and the mounting attempt. The male's head and neck are laid on the cruppers of the female, while several attempts at mounting are made. The female withdraws often, stepping forward or turning away. After renewed attempts, there will be pauses during which the female in turn chases and tries to arouse the male. If the doe is willing to mate, she stands with legs spread wide apart, pointing backwards. The coupling is of short duration. The stag rises on his hind legs, his head and neck are pushed along the female's back and his forelegs hang down by the cruppers or belly of the female. She lowers her head and neck and, after mating, leaves the male at the rutting place (Reinken *et al.*, 1990).

The gestation period in fallow deer is 234 days (Fletcher, 1986).

Red deer

Sexual maturity occurs at 16–20 months. The rutting season is September–October. The oestrus cycle is 18.8 ± 1.7 days and the duration 12–24 h. The antlers are cast in March–April and the replacements cleaned in August–September (Fletcher, 1986).

In late August, mature stags over 5 years old become progressively intolerant of each other and, after the middle of September, stag groups begin to fragment. One by one, the stags leave their usual ranges, some moving directly to their traditional rutting grounds, others spending several days on their own in peripheral areas before moving to their rutting areas. Mature stags seldom attempt to form harems immediately and, during the first few days after arriving in their rutting area, their tolerance of other stags varies. As mature stags spend more of their time associating with hind groups, they move less and less, remaining with their harems in one particular area of the rutting grounds and moving only as their harems move to new feeding areas or to sheltered positions. At the end of September, hinds gather in traditional rutting areas, where they are joined by the stags. The stags begin to display regularly and interact frequently with their hinds. Since hinds aggregate in groups, it is possible for a stag to monopolize access to a considerable number of hinds and there is intense competition for harems between stags during the autumn rut. A stag's breeding success is related closely to his ability to control the behaviour of young stags that attempt to abduct hinds from his harem and, of course, of other mature males. Direct competition for hinds is common between mature stags and fights are frequent and dangerous. However, it seems as if stags assess their opponents carefully, fighting only in cases where they are likely to win. Only stags that have reached full adult weight are able to hold large harems during the period of peak conception, though young and old stags may hold harems before and after this period. Harem size may vary between stags from one to more than 20 hinds. As far as possible, stags control the movement of hinds, preventing individuals from leaving wherever possible by herding them back to the group and driving in extra hinds that approach. Rutting stags feed little and display regularly. Both the frequency and the intensity of many rutting displays are related to the probability that other stags would attack them or try to infiltrate their harems. Stags displace young stags or rivals that approach their harems by walking steadily towards them. Stags chase yearlings or young stags away from their harems by running directly at them, often pursuing them until they are more than 100 m from the harem. Chases terminate when the harem-holder stops and turns back to his harem. Stags often end a chase with a scissor kick of the forelegs followed by a bout of roaring. On some occasions, stags also chase calves and even yearling females out of the harem. Removal of calves from the harem must be of dubious advantage to the stag, since their mothers often follow them when they leave. Apart from avoiding mating with immature stags, there is no evidence that hinds consistently select any particular category of stags as breeding partners. Female–female competition over mates could play a role in the mating behaviour of red deer (Clutton-Brock et al., 1982; Bebie and McElligott, 2006).

After sniffing a hind or the place where a hind has urinated, stags sometimes show flehmen, raising their heads and curling back their upper lips. Stags frequently approach lying hinds and lick the back of their heads and necks, gradually working over the head towards the preorbital region. This may be followed by an attempt to sniff and lick around the base of the tail. Bouts of sniffing and licking, which may last several minutes, are usually terminated by the hind getting to her feet and moving away. After this, the stag commonly sniffs and licks the grass where the hind has rested. Hinds often urinate while lying and this may permit the stag to scent pheromones released in the urine. When a hind is in oestrus, the harem-holder directs most of his attentions towards her, interacting more frequently with her than with other hinds. Stags take an intense interest in oestrous hinds, standing close to them, licking their preorbital glands or vulvas and intermittently chasing them through the harem. A variety of behavioural changes occur during the 12–24 h a hind is in oestrus. The proportion of time she spends feeding tends to decline and the proportion of time she spends standing inactively or moving tends to increase. In the early stages of oestrus, hinds run when the stag moves towards them or tries to place his chin on their backs, and only as oestrus progress are they willing to stand and allow him to mount. At this stage, hinds also show interest in the stag, licking him and sometimes mounting him. Stags usually mount several times before ejaculating, and after ejaculation their interest in the hind apparently wanes, though several successive sequences of chasing, mounting and ejaculation may occur sometimes with the same hind. Stags often head-off hinds that are attempting to leave their harems by walking outside them, head held high, with a stiff, prancing gait. The stag's head is typically at an angle to the direction of movement, with the eyes half closed. If the hind persists in attempts to leave the harem, the stag may threaten her with lowered antlers or by giving a brief scissor kick with his front legs and may bark at her. Hinds that are herded are almost always on the periphery of the harem. In over 90% of observed cases, hinds move back towards the centre of the harem when herded. Stags frequently chase hinds over short distances within their harems.

During these chases, the stag trots after the hind with neck outstretched, sometimes extending his tongue. Chases end when the stag stops, apparently losing interest, and they are often followed by a bout of roaring. Only hinds in oestrus allow the stag to mount. Mating sequences usually involve several mountings over a period of up to an hour or more. Ejaculation can be identified easily by a sudden thrust that jerks the stag's body upright, often throwing the hind forward several paces (Clutton-Brock et al., 1982).

As a result of the high energy costs of rutting and reduced food intake, the body weight and condition of rutting stags show a rapid decline during the rut. Some individuals lose as much as 20% of their body weight over this period. It is presumably declining condition and the associated changes in hormone levels that cause stags to terminate their rutting activities. This typically occurs suddenly, and stags could switch from a high level of activity one day to an almost total cessation of rutting activities combined with intense feeding on the subsequent day (Clutton-Brock et al., 1982).

The gestation period in red deer is 231 days, with a range of 226–238 days (Fletcher, 1986).

11.16 Before and After Parturition

Behaviour before and during parturition

Deer in general

Calving problems are rare in wild deer (Fletcher, 1986).

Fallow deer

Calving time in Europe is from May to the end of August, with the peak period in June. Some days before parturition, the doe leaves the flock and searches out a protected place for the delivery. For several hours before giving birth, the doe is very restless and takes almost no food. She tries to keep a distance of at least 50 m between herself and the other members of her group. She prefers sites with long grass and tall weeds where she strides to and fro, head against flanks or cruppers. If disturbed, birth will often be delayed, although severe disturbances caused, for example, by fire, low-flying aircraft, etc., can lead to premature calving. The presence of the mate, on the other hand, can delay the birth (Reinken et al., 1990).

The birth itself is usually in the lying position, with labour taking anywhere between 20 and 340 min, normally taking less than an hour. With breech presentation or with weak elderly females, the birth can be even longer. Sounds of pain, such as moaning, piping, groaning and panting, can be heard. During labour, the female licks the vagina and protruding amniotic sac. When the calf emerges, the mother often stands up, thereby severing the umbilical cord (Reinken et al., 1990).

Red deer

Calving time in Europe is from May to the end of August, with the peak period in June (Fletcher, 1986).

Hinds about to calve typically move away from their matrilineal groups and their usual home ranges. Isolation of nursing mothers from their usual social groups and separation of the calf from the mother throughout most of the day probably help to make both of them inconspicuous to predators hunting by visual or olfactory cues (Clutton-Brock et al., 1982).

The first sign that a hind is about to calve is a marked swelling of the udder, usually 1–2 days before parturition. The hind searches out a protected spot for the delivery, often on high ground. A few hours before birth, the perineal area becomes red and slightly swollen. During labour, which usually lasts 30–120 min, the hind becomes more and more restless, frequently nuzzling or grooming her flanks, udder and perineal area. In the later stages of labour, the hind normally lies on her side, straining as the fetus begins to emerge. Once it is partly exposed, the hind frequently stands, allowing it to fall out. Afterwards, the hind licks the calf dry and eats the afterbirth (Clutton-Brock et al., 1982; Johansson, 1989).

Number of calves born

Fallow deer

The number of twin births is few; Reinken et al. (1990) report 0.5% over 11 years.

Red deer

There is generally only one calf born and hardly any twin births among red deer (Fletcher, 1986; Reinken et al., 1990).

After the birth

Fallow deer

The mother immediately begins to tend her calf, licking it and removing and eating the birth sac. The calf's rear quarters are licked particularly well, apparently to remove dirt from the skin. A weakened mother will lick the calf while she is lying down. After about 20 min, the afterbirth is passed and is almost always eaten by the mother. She will also eat up all traces of the birth, such as stained grass, earth and leaves from the site – this is important for the survival of the newborn calf in the wild in that it removes signs which could alert potential predators. The calf's initial attempts to stand are usually unsuccessful, but it should be able to stand for the first time within 30 min. First attempts to suck take place 25 min to 1 h after birth, and suckling may be done either standing up or lying down. After a few hours, the mother and calf leave the birth site and look for a resting place with cover. Here, the calf will lie down low on the ground between suckling periods. The mother usually moves the calf from cover for suckling. The calf nuzzles up to the belly and flanks of its mother, sucking from the side between intervals of 2–3 h. After that, the suckling time becomes shorter and the intervals longer. The doe will move away, lie down or turn away from her calf after suckling. There is also some pushing away of the suckling calf. There is close rapport between mother and calf in the first days of life. Does usually keep an eye on the calf's resting place, calling the calf from its cover by bleating and piping. Smelling and touching are important factors in the mother–calf bond: they will lick each other on the head and neck while lying down side by side. When the calf is about 1 week old, it follows the hind and other members of the herd. If mother and calf become separated, they call each other, the doe by a low, hoarse bleating and the calf by a piping sound. Older calves give out piping and whickering noises. Fallow deer does are more tolerant towards foreign calves than red deer hinds. When her own calf is sucking, the doe usually accepts that a foreign calf may at the same time, suck a teat from behind. The calves will pull grass stalks and suck them as early as the 3rd or 4th day, while the first eating of grass usually occurs from the 10th day onwards (Johansson, 1989; Reinken *et al.*, 1990).

Red deer

Immediately after the birth, the mother licks the calf clean. Calves are usually able to stand within half an hour of birth, and the first suckling bout typically occurs within 40 min of the end of labour. The placenta is expelled 1–1.5 h after the birth and is eaten immediately by the hind, which licks up the amniotic fluid from the ground and the grass where it fell, until the ground is cleaned. The calf starts trying to get up and usually is on its feet within 1 h. After cleaning her calf, and after having suckled the calf for the first time, a hind commonly rests close to it for a short period. Subsequently, another bout of licking and suckling occurs and, at the end of this, the mother normally moves away slowly from the place of birth, but seldom goes any further than 300 m. She encourages the calf to follow her and, at some stage, the calf leaves the mother, selects a well-protected hiding place in the neighbourhood of the calving area and eventually lies down, adopting a characteristic hunched posture with head held low in a typically curled up position in a patch of deep grass or heather. During the next few hours, the mother grazes or rests within sight of the calf, but rarely approaches it closely. After a further 2–3 h, she usually visits the calf and a third bout of licking and suckling occurs – the mother eats the calf's faeces and urine as the calf sucks. Subsequently, the calf lies down again and the mother moves further away. It stays there for 2–3 days to 1 week. During this time, the hind is in the neighbourhood but goes to see the calf only in connection with suckling. During these occasions, she licks the calf. When their calves are standing, and therefore visible, hinds are intensely vigilant. If hinds with visible young calves are suddenly disturbed, they commonly bark and the calf hides quickly. When the calf is about 1 week old, it follows the hind. If they get separated, they call each other, the hind by a low mooing and the calf by a piping sound. During the first day or two after birth, calves suck every 2–3 h while their mothers are with them. When they begin to accompany their mothers after a week or so, suckling frequency is about eight times over 24 h, then it declines to 4–5 times per day by the time they are 3 months old. Suckling frequency differs little between day and night, though it peaks at dawn and dusk. Red deer hinds only rarely suckle calves that are not their own and they often react aggressively to approaches by other calves. The hiding response seems to be uncommon among calves over 1 week old. Instead, hinds with calves over 1 week old flee quickly when disturbed. As calves grow older, both their own behaviour and that of their mother change rapidly. During their first 3 weeks, calves spend over 80% of the time within 3 m of their mothers. However, they rarely lie down close to their mothers and usually move at least 20 m away before doing so. This appears to represent a preference for lying away from the mother, since they frequently ignore lying places close to her that are apparently suitable. After the third week of life, though calves spend less time close to their mothers, they more often lie down closer to them (Clutton-Brock *et al.*, 1982; Johansson, 1989).

Play behaviour

Deer in general

Play is an important part in the development of young deer and accordingly it is mainly calves that perform play behaviour. This behaviour contains moments of fight, flight and pursuit. During play, the calves learn to combine different parts of their behaviour to patterns of behaviour that might be necessary for survival later in life (Johansson, 1989).

Play among calves consists of trotting and galloping close to the mother, bucking, chasing, sometimes with sideways pushing, head-to-head pushing, kicking out with the hind legs, bounding and rubbing. The younger females already show greater alertness than the males. As the calves grow older – at about 4 months – play lessens and feeding, ruminating and herd activities fill up the day (Reinken *et al.*, 1990).

Weaning

Deer in general

Weaning occurs usually when the calf is about 6 months old (Johansson, 1989).

Red deer

As calves grow older, their mothers are progressively more likely to frustrate attempts to suck by walking forward as soon as the calves begin to search for the nipple. The proportion of sucking attempts that are rejected by the mother increases steadily over the first 6 months of the calf's life.

Pregnant mothers often wean their calves by the time they are 5–7 months old. Non-pregnant mothers usually do not wean their calves until the following summer or autumn (Clutton-Brock *et al.*, 1982).

11.17   The Fawn or Calf and the Young Deer

Deer in general

With the exception of a short period during the rut, deer calves constantly follow their mothers from a few days up to 1 week after the birth, until they are about 1 year old or more. Then the male calves leave the hind flock and join a flock of males in the neighbourhood. The female calves usually stay in the neighbourhood of their mother during their whole life (Johansson, 1989).

11.18   Assessment of Deer Health and Welfare

Deer in general

In order to determine whether or not a deer is healthy, special attention should be paid to bodily condition, movements and posture, rumination, condition of hair, skin, eyes, ears, tail, legs and feet.

The healthy deer

Healthy deer have sounds, activity, movements and posture appropriate to their age, sex, breed or physiological condition. These include: clean and shiny coat, clear bright eyes, alert and keen watchfulness and vigilance to the environment; posture with straight, sound feet and legs, straight spine and elevated head; normal sucking or suckling, feeding, ruminating, drinking, defecation and urination behaviours; normal getting up, lying down and resting behaviour; and otherwise normal movements and behaviour. The appearance of the faeces varies according to the nature of the feed. Oestrus appears regularly and in a normal way.

The sick deer

Symptoms

DEER IN GENERAL   In wild deer, disease problems are mainly in relation to starvation. In both wild and farmed deer, the sick animal moves aside or stays at the periphery of the flock. The sick animal gets a lustreless coat and might show apathy, inactivity or disinclination to move. The changed behaviour may lead other deer to show agonistic behaviour towards them.

In farmed deer, diseases are caused mainly if they are exposed to muddy or wet ground, dirty feeding places or bad feed hygiene. Keeping many deer on too small an area increases the risk of parasitic diseases. In some countries, infectious diseases occur in farmed deer, e.g. tuberculosis, Johne's disease (paratuberculosis), salmonella.

RED DEER Studies of post-mortems of calves that died during winter indicated that they had all suffered from malnutrition. Other important causes of calf mortality can be attacks by eagles. Observations in Scotland have shown that eagles can kill calves weighing up to 20 kg (Clutton-Brock *et al.*, 1982).

Examples of abnormal behaviour and stereotypies

DEER IN GENERAL Stereotypies as described in domestic animals are not described in wild deer and do not seem to be described in farmed deer.

Symptoms of pain

DEER IN GENERAL Prey animals do not show obvious signs of pain; any individual that does so would be selected out of the population by a sharp-eyed predator. So, although they can suffer from feelings of pain, they might not show it (Arney, 2009).

11.19 Capturing, Fixation, Handling

Deer in general

Treatment of farmed deer must be made bearing in mind that they must be regarded as wild animals. A handling set-up with pens, etc., must therefore be available that makes it possible to catch, separate, treat or gather the deer for sale or slaughter without immobilizing them. The animals should be given the opportunity to acquaint themselves with the handling set-up by letting them pass through it without any treatment. In order to avoid trample injuries, agonistic behaviour and attacks from adult males, these must always be handled apart from females and calves.

11.20 The Importance of the Stockman

Deer in general

In connection with handling deer, the stockman must avoid touching the animals more than is absolutely necessary. They should not be patted as they react negatively to human touch. Stockmen must perform all deer handling in a calm and firm manner.

12 Ratites: Ostrich (*Struthio camelus*), Rhea or Nandu (*Rhea americana*) and Emu (*Dromaius novaehollandiae*)

12.1 Domestication, Changes in the Animals, their Environment and Management

Ratites generally

First it must be said that ratites are not domesticated. They were taken from the wild after the mid-19th century and kept in captivity, first because of their feathers and later at the end of the 20th century for their meat.

Ratites are large, long-necked, terrestrial birds with powerful, heavily muscled legs adapted for running. All species of ratites are unable to fly. Their sternum is smooth or raft-like because, unlike other flightless birds, it lacks a keel to which flight muscles could be anchored. The latter property has given the ratites their name. *Ratis* in Latin is a vessel without a keel, a raft. All ratites use their legs and feet by thrusting forward in both defence and offence. The skin on the plantar surface of the digits is modified for arid environments. The ostrich and rhea have relatively large wings, while those of the emu are rudimentary. The crop is absent in all ratites. The cloaca in ratites has three chambers. The large intestine runs into the first chamber. The urine and the sperm tract run into the second chamber. The third chamber holds the extensible penis in the males (Kreibich and Sommer, 1995; Fowler, 1996; CoE, 1997).

The male is called a 'cock' or 'rooster', the female a 'hen' and the young 'chicks'.

Ostrich

Besides several places in Africa, fossil remains of ostriches have been found on the west and north coast of the Black Sea, in China, Mongolia and India. Images of ostriches have been discovered in paintings and carvings in the Sahara dating between 5000 and 10,000 years BC. The ostrich is referred to in the Bible on several occasions. Solomon saw the symmetrical ostrich feather as a sign of justice. Tutankhamun had a long-handled golden fan, which held ostrich feathers and was decorated with an image of the Pharaoh hunting ostriches. In Africa, ostriches have been hunted for their meat for centuries by Bushmen, who often adorned themselves in the feathered skin of the bird. The eggshells have been used as storage and drinking vessels in Africa and Arabia (Drenowatz *et al.*, 1995; Kreibich and Sommer, 1995; Deeming, 1999).

Ostriches have been kept on farms in the South African Cape region since after the mid-19th century. Serious interest in ostrich farming arose in about 1860 following the demand for ostrich feathers in Europe, which reached its peak just before World War I. The first artificial hatching of eggs from wild ostrich nests took place in 1857 in Algeria. The first breeding in a zoo took place at the end of the 19th century in Marseilles, France. The introduction of ostriches into Germany by Carl Hagenbeck in 1906 showed that it was possible to keep ostriches for several years in northern Germany, thus proving the capability of ostriches to adapt to strange habitats. In the last decades of the 20th century, ostrich farms were established in the USA, Australia and in Central and Western Europe. At the end of the 20th century, more than 80% of all ostriches lived on farms or in zoos (Batty, 1994; Kreibich and Sommer, 1995).

The ostrich is the largest living bird, males measuring up to 2.75 m in height and females up to about 2 m, and they weigh up to 160 kg and 110 kg, respectively (Fig. 12.1). The adult male bird is mainly black with white wing primaries and tail feathers and a grey-coloured neck. The female is dull brown-grey all over with light grey to white wing primaries and tail feathers. Juvenile birds resemble the females, whereas young chicks

Fig. 12.1. Male ostrich on a farm in Sweden. In the background, other males and females (image I. Ekesbo).

are mottled brown, yellow, orange and cream with black quills on the back. Ostrich feathers are fluffy and symmetrical. The massive thighs are devoid of feathers. The wings of the ostrich are poorly developed and there are no substantial pectoral muscles. The sternum is large and bowl-shaped without a keel. Ostriches have no significant breast muscles. Due to the flying ability of their prehistoric predecessors, ostrich bones are light and break easily. They comprise a thin outer wall, either hollow or filled by a porous bone substance. The ostrich is the only bird in the world which has retained only two of the original four toes, namely the third and fourth (Fig. 12.2). The third toe, the smaller one located on the outside, is involved in an ongoing decline and provides balance only. The fourth

inner toe is large and has a wide, resilient cushioning callus underneath, ending in a claw approximately 7 cm long. Together with the powerful legs, this gives the ostrich a most dangerous kick. A small nail may be present on the distal part of the fourth digit. Single, large footpads consisting of thickened skin with long papillae cover the plantar surfaces of the third and fourth digits. In place of a crop, ostriches have a bag-like oesophagus where food accumulates. When the bird lifts its head, the food is passed to the proventriculus. The proventriculus leads into the gizzard. The content of the gizzard of an adult bird comprises food and approx 1.5 kg of pebbles, which grind the food to a pulp. These pebbles, due to wear, remain in the gizzard for a limited time and have to be replaced

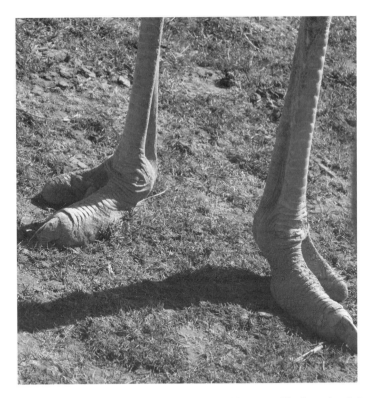

Fig. 12.2. The ostrich has retained the third and fourth of the original four toes. The large fourth inner toe has a wide, resilient cushioning callus underneath, ending in a long claw of approximately 7 cm. The smaller third toe located on the outside provides balance only (image I. Ekesbo).

constantly. A pyloric opening with a sphincter type muscle leads to the small intestine. In adult birds, the small intestine is approximately 6 m long, the large intestine is about 16 m and tapers down towards the end. Food needs a long time to pass through the system, even in chicks it takes more than 36 h, achieving optimal absorption of fibrous food (Kreibich and Sommer, 1995; CoE, 1997; Bezuidenhout, 1999; Deeming, 1999).

The preferred habitat is open, short-grass plains and semi-desert, although ostriches are found in hot, fringing desert steppes. They avoid areas of tall grass, bush lands and dense woodland, but will occupy or cross more open plains and woodland. The birds tend to keep to lowland areas. Breeding groups of ostriches usually consist of one territorial male to two to four hens, one being the major hen. Out of the breeding season, the ostrich in the wild is a gregarious species, forming groups of birds of mixed gender and age, particularly around waterholes. They avoid close contact with other animals (Kreibich and Sommer,

1995; CoE, 1997; Deeming, 1999; Deeming and Bubier, 1999).

Rhea

The greater rhea, *Rhea americana*, from South America, also called nandu (Fig. 12.3), is a fast-running bird. In the wild, it is distributed through Argentina, Bolivia, Brazil, Paraguay and Uruguay. It has been hunted since the earliest pre-Hispanic times. It is smaller than the ostrich, reaching a height of up to 1.5 m. Rheas reach sexual maturity between 20 and 24 months of age. As with other ratites, rheas grow at exceptionally slow rates, with a final mature live weight of around 25 kg, depending on gender. The great rhea has grey-brown upper parts and whitish under parts. Unlike ostriches, the legs of rheas are feathered down to the tibiotarso-metatarsal joint. Their wings are large for a flightless bird and are spread while running, to act like sails. All the three digits of rheas carry nails. The birds frequent open, treeless country. During the non-breeding season, they may form mixed flocks of between 10 and 100 birds, breaking up in the breeding season when males generally become solitary, females form small cohesive groups and yearlings remain as a flock. Its lifespan in the wild is 6–10 years. It is, in the wild, an endangered species as it damages crops and is therefore actively hunted. Rheas have many uses in South America. The feathers are used for feather dusters, skins for cloaks or leather and their meat is a staple to many people. The two most common schemes for farming adult

Fig. 12.3. Rhea in a Slovenian zoological park (image courtesy of Peter Dollinger).

rheas are intensive in pens or semi-intensive in large breeding paddocks (Drenowatz *et al.*, 1995; Fowler, 1996; CoE, 1997; Bezuidenhout, 1999; Navarro and Martella, 2002; Sales, 2006b).

Emu

Widely distributed throughout the Australian continent, the emu has been hunted by humans for thousands of years. Emus are smaller than ostriches, reaching heights of up to 1.9 m and weights up to 50 kg. Like the ostrich, it is a fast-running, powerful bird. Both sexes are brownish, with dark grey head and neck. The neck of the emu is pale blue and shows through its sparse feathers. Emu eyes are golden brown to black. They have a soft bill, adapted for grazing. The wings are small and vestigial. Their ability to run at high speeds is due to their highly specialized pelvic limb musculature and they are the only birds with gastrocnemius muscles in the back of the lower legs. Unlike ostriches, the legs of the emu are feathered down to the tibiotarso-metatarsal joint (Fig. 12.4). All the three digits of emus carry nails (Fig. 12.5). Emus in the wild may be found alone, in pairs or in small groups. Emus are most common in the evergreen forests occurring in areas of seasonal water shortage and savannah woodland, and least common in dense forest and in very arid areas, except during wet periods. In the wild, emus live for between 10 and 20 years; captive birds can live longer than those in the wild. Emus are farmed for oil, meat and skins. Commercial emu

Fig. 12.4. Emus on a private Swiss farm (image courtesy of Andreas Steiger).

Fig. 12.5. Emus have nails on three toes (image courtesy of Andreas Steiger).

farming started during the 1980s in Australia, but later also in the USA, Canada, China and some other countries. Emus breed well in captivity, if kept in large, open pens to avoid leg and digestive problems, which arise with inactivity. They are typically fed on grain supplemented by grazing, and are slaughtered at 50–70 weeks of age (Fowler, 1996; CoE, 1997; Bezuidenhout, 1999; Sales, 2007).

12.2 Innate or Learned Behaviour

Ratites in general

In ratites, as in other animals, reproductive and many other behaviours are innate. However, finding food might be learned, as the male ratites teach the chicks how to do this.

12.3 Different Types of Social Behaviour

Social behaviour

Ratites in general

In the wild, ratites are typically gregarious animals and their groups have a complex social structure. Group size depends on the natural resources available and the breeding status (CoE, 1997).

Ostrich

The social behaviour of ostriches is both complex and variable and has been compared to that of animals belonging to the most developed and complicated social orders. Areas around waterholes usually serve as communal pastures for mixed flocks of adult and immature males and females. Observations have been made of mixed groups in the wild of 14 or more adult birds and of herds with up to 100 1- and 2-year-old birds. Through social contact with other groups at waterholes, where so-called super-herds can be formed, new family groupings are constantly being established. In such herds, the family groups remain intact. Before the first rains arrive in their African habitats, the male birds usually establish or select their old breeding sites and territories, with or without their mates. Although socially aggregative among their own kind, ostriches tend to avoid close contact with other types of animals, preferring to maintain a certain social distance. There is a social hierarchy within ostrich groupings, headed and maintained by an adult male or major female. The ostrich has a low interspecific ranking: zebras, jackals, baboons, smaller gazelles and even crows all have precedence over ostriches at watering holes (Hallam, 1992; Kreibich and Sommer, 1995; Deeming and Bubier, 1999).

Communication

Ostrich

Communication consists mainly of warning sounds from the males (Kreibich and Sommer, 1995).

Behavioural synchronization

Ostrich

An ostrich that starts cleaning its feathers usually sets off a chain reaction in the group, with the result that all members clean simultaneously.

Dominance features, agonistic behaviour

Ratites in general

Both male and female ratites demonstrate considerable aggression (CoE, 1997).

Ostrich

Aggression can be demonstrated between hens, between cocks and between a cock and a hen. Shortly before attacking, the male raises both wings high. A submissive attitude is assumed by a courting hen in front of the cock, and also by a cock who wishes to show a hen a place to nest. In submissive individuals, both wings are spread but hang downwards, shaking periodically, and it hangs its head, neck and tail. As a rule, an aggressive cock holds his head and neck high, stretches the front half of his body upwards and raises his tail feathers. A dominant hen might drive other hens away. To do so, she holds her wings out to the side and approaches her rivals with her beak open, uttering a hoarse noise. In reaction to this threatening gesture, the victimized flee and in turn vent their aggression on other hens. Aggression begins with the highest ranking bird and ends with the lowest in the pecking order. Among the male birds, dominant cocks in particular can be observed driving other cocks out of their territory with threatening gestures and warning noises. In farmed animals, hefty arguments and fights can be seen between two cocks on neighbouring pastures, even if separated by a wire mesh fence. After initially walking up and down the fence line together, they stop and stand opposite each other, face to face. Then, with wings spread and beaks open, they kick with one of their legs. The purpose of this well-aimed kick, accompanied by a muffled noise, through the fence against the breast or thigh of the opposition is to assert its own position (Kreibich and Sommer, 1995).

Rhea

Most aggression is observed between males and occurs mainly during winter and spring, suggesting strong intrasexual competition (Carro and Fernandez, 2008).

12.4   Behaviour in Case of Danger or Peril

Ratites in general

In ratites, defence against predators depends on high vigilance based on acute vision and avoidance behaviour, and includes running away at high speeds.

Ostrich

Feeding and vigilance behaviours are mutually exclusive because once the head is lowered to feed, the ability of the bird to look around is lost. Males are more vigilant than females in all group sizes and this is presumed to be related to searching behaviour for predators and for potential mates or rivals. Male ostriches kept in semi-natural conditions patrol boundary fences (Deeming and Bubier, 1999).

Ostriches tend to panic if frightened and run as a group against the fences and corners of an enclosure, or try to climb the fences and gates, causing severe damage to themselves or others. In farmed ostriches, a simple action such as a person entering an enclosure suddenly and hastily, the moving of birds from one enclosure to another, or attempts to separate one or more birds from the flock may, on many occasions, cause panic that could end with injured or dead birds (Perelman, 1999).

A loud tractor or aeroplane can cause the whole herd to panic and take flight across the field. Sometimes, in conjunction with this response, the ostriches perform what is commonly known as the ostrich waltz. To do this, they spin themselves on their own axis and lift their wings. Like some other animals, ostriches play dead when caught unawares, prone on the ground and with extended neck (Kreibich and Sommer, 1995).

Rhea

When in danger, rheas flee in a zigzag course, using first one wing, then the other, similar to a rudder (Kreibich and Sommer, 1995).

Emu

Studies by Boland (2003) showed that emus in larger groups spent less time in vigilance and more time foraging. None the less, the combined vigilance of group members ensured that emus detected a predator sooner as group size increased. After detecting the predator, larger groups waited longer until opting to flee, and then spent less time and energy doing so. Thus, emus benefit from grouping in terms of both the 'many-eyes effect' and the 'dilution effect'.

12.5 Some Normal Physiological Frequency Values of Interest at Clinical Examination

Ratites in general

Body temperature is lower in ratites than in other birds (Maloney, 2008).

Ostrich

Normal cloaca body temperature is slightly lower in the ostrich than in other avian species. Normal body temperature of the ostrich varies from 37.8 to 38.9°C.

The respiratory rate under normal conditions is 3–5 breaths per min in the adult and 12–60 breaths per min in the young. Duke (1977) states a respiration rate of 5 breaths per min for an ostrich weighing 100 kg. Their respiration rate is the slowest of any bird. However, respiration rates of ostriches fall within two ranges, either within a low range of 3–5 breaths per min or a high range of 40–60 breaths per min in response to heat stress.

The heart rate in the adult is 28–36 beats per min and 80–164 beats per min in the young. Respiratory and heart rates vary with body mass (Huchzermeyer, 1994; Skadhauge and Dawson, 1999).

Rhea

The body temperature of rheas at rest is reported to be 39.5°C (Taylor et al., 1971).

Emu

The normal body temperature of the emu varies from 37.8 to 39.8°C. The heart rate is 42–76 beats per min. The respiratory rate is 4–17 breaths per min (Cornick-Seahorn, 1996).

12.6 Active and Resting Behaviour Patterns

Activity pattern, circadian rhythm

Ratites in general

Ratites are diurnal. Daily activity is divided mostly between feeding, walking, running, standing alert, preening and dust bathing (CoE, 1997).

Ostrich

Under natural conditions, ostriches will range daily over an area of an average radius of up to 20 km in search of food. The ostrich spends its time walking around its environment, only running if threatened. Ostriches do not move during the night, when they are presumed to be sleeping. Gathering food is time-consuming and implies that the birds spend a large part of the day walking around. Usually, pastoral feeding is done while standing or walking. During daylight hours, approximately 60% of the time is spent walking, 20% pecking and 16% standing. Preening and sitting each occupy less than 3% of the time budget and all other behaviours combined account for less than 1%. Ostriches spend 33% of their time during daylight hours feeding, 29% travelling, 18% being alert and 9% preening (Deeming and Bubier, 1999). Feeding is carried out mainly in the morning and in the evening. Of course, the time spent feeding and searching for food depends on the available food source (Kreibich and Sommer, 1995; Deeming, 1999).

In a study of farmed ostriches, it was observed that they spent about 27% of their time resting. This was the largest fraction of total behaviour in one day. At the beginning of twilight, almost 80% of the herd was already resting. Generally, the birds were lying close to one another, with only a few ostriches still eating or standing watch. These observations were made during a rainy period. It was conspicuous that the birds preferred to sit in the rain than to seek shelter under the covered area (50 m²) provided. The shelter was visited in the mornings at feeding time only. In August, when the weather was very hot, the birds rested less than in the damper months of May and June. When standing, ostriches are constantly on the watch. Their eyes are positioned so that they can even watch what is happening behind their backs. They can hold the same head-and-body position for a long time. If an alarming noise is heard, the head turns immediately in the direction of the noise. On a daily average, the herd spends 18% of its time standing. Apart from being alert – which, despite a daily percentage of only 18%, is doubtless more important in their natural habitat – feeding, drinking, cleaning, courting, feather pecking and calling are all activities performed while standing (Kreibich and Sommer, 1995).

Yawning has been observed in ostriches. At the beginning of the resting phase, it calms the birds, at the end it promotes simultaneous, synchronized activity, e.g. getting up, in the group (Kreibich and Sommer, 1995).

Larger groups can identify the enemy faster and the individual bird does not need to be so alert, so it can devote more time to eating. Accordingly, the time a bird grazing alone keeps watch is about 34.9%. For a pair, it is only 22.9% and, for groups of three or four birds, it is a mere 14% of the daytime (Kreibich and Sommer, 1995).

Exploratory behaviour

Ratites in general

Like most animals, ratites perform exploratory behaviour.

Ostrich

The often quoted 'burying your head in the sand' when danger threatens is a myth. It is true that ostriches put their heads into openings and holes in the ground, but they do so from curiosity (Kreibich and Sommer, 1995).

Diet, food searching, eating and eating postures (feeding behaviour)

Ratites in general

Ratites are mainly herbivorous, occasionally supplementing their diet with insects and small vertebrates. They need a certain amount of grit (small stones) to aid digestion of coarse material. Ratites readily swallow all kinds of foreign bodies (CoE, 1997).

Ostrich

Ostriches are herbivores but also eat small mammals, lizards and beetles. The wild ostrich is a selective grazer. Food selection is primarily a visual choice, with chicks being attracted to bright green foliage, and a secondary response apparently being taste. Ultimately, selection is for selected forbs and new grasses, and mature grass leaves are not consumed. Adult and subadult ostriches tend to select low-growing plants and rarely eat woody materials. Green annual grasses and forbs (pasture herbs) are preferred, although leaves, flowers and fruit from succulent and woody plants are also consumed. Feeding is a plucking process whereby the grass is ripped up with the beak. After several plucking motions, the ostrich wobbles its head. This procedure is repeated several times until enough food has been gathered in the throat. The following swallowing movements can be observed easily on the outside. A fist-sized ball of food is swallowed slowly down the oesophagus and can be recognized clearly as a large bulge in the throat. Because the head and neck are usually raised during this sequence, the ostrich can keep watch simultaneously by turning its head. It is, however, also possible for an ostrich to swallow food with its head and neck lowered. Adult ostriches strip, not bite, leaves from shrubs and woody plants. They have a significant preference for forbs low in phenolics and high in fibre. Ostriches kept in semi-natural conditions forage from the veld even when provided with supplementary rations (Kreibich and Sommer, 1995; Angel *et al.*, 1996; Cilliers and Angel, 1999; Deeming and Bubier, 1999).

In order to digest their food, ostriches must ingest small pebbles and other marble-sized objects, which facilitate grinding down their fibrous food. With their long, nimble necks and sharp eyes, ostriches are ideally suited to finding little invertebrates, which supplement their otherwise vegetarian diet (Kreibich and Sommer, 1995).

Ostriches drink at watering places, but can do without water for considerable lengths of time. When drinking, with the beak open they let the lower movable beak repeatedly scoop water into the oesophagus. The neck is then lifted in an 'S' form and the water swallowed (Kreibich and Sommer, 1995).

When farmed ostriches are likely to be exposed to heat stress, a fresh water supply is essential (Mitchell, 1999).

Rhea

Rheas are omnivorous, preferring broad-leafed plants, but also eat seeds, roots, fruit, lizards, beetles and grasshoppers (CoE, 1997).

Emu

Emus are opportunistically nomadic and may travel long distances to find food; they feed on a variety of plants and insects. Emus forage in a diurnal pattern. They eat a variety of native and introduced plant species; the type of plants eaten depends on seasonal availability. They also eat insects, including grasshoppers and crickets, ladybirds, soldier and saltbush caterpillars, moth larvae and ants. Like other ratites, emus require pebbles and stones to assist in the digestion of plant material. Individual stones may weigh 45 g and they may have as much as 745 g in their gizzard at one time (Angel *et al.*, 1996; Fowler, 1996; CoE, 1997).

Selection of lying areas, lying down and getting up behaviour, resting postures, sleep

Ostrich

Ostriches sleep prone on the ground and with extended neck at night, but only for short periods of time and never all birds in a flock at the same time. Yawning often precedes sleep and, together with stretching, takes place after waking (Kreibich and Sommer, 1995; Deeming and Bubier, 1999).

Locomotion (walking, running)

Ostrich

Although ostriches are adapted to running (cursorial), they were reported either to walk or run during only 15% of the time they were under observation. Movements occur usually in association with the search for food or with social behaviour. When walking, the birds have a pace length of about 1 m. This can increase to more than 3 m when the bird is running. Hens usually run in small groups along a fence or in search of food. They usually let their tail feathers hang, unless they are in an aggressive mood. In contrast, the dominant cocks are always found alone, with their tail feathers up, when pacing their territory. Ostriches can run at up to 70 km per h and can sustain this pace for more than 10 min (Kreibich and Sommer, 1995; CoE, 1997).

Rhea

Their wings are large for a flightless bird and are spread while running, to act like sails. Rheas are able to turn at right angles to alter course suddenly (CoE, 1997).

Emu

Emus can travel great distances at a fast, economical trot and, if necessary, can sprint at 50 km per h for some distance at a time (CoE, 1997).

Swimming

Ratites in general

Ostriches will bathe and emus and rheas will bathe and swim in water if it is provided (CoE, 1997).

12.7 Behaviour at Defecation and Urination (Eliminative Behaviour)

Ratites in general

Differing from other birds, ratites excrete urine and faeces separately. The excretion is performed standing (Kreibich and Sommer, 1995).

Ostrich

The flaccid phallus in the male has to be partly protruded to allow for defecation and urination (Fowler, 1996).

12.8 Body Care, Cleanliness

Ratites in general

Ratites lack a preen gland and so do not have the ability to oil their feathers to provide a waterproof plumage. Embryos have such a gland, but it regresses as the embryos grow. For ratites, body care behaviour includes plumage care and dust bathing (Kreibich and Sommer, 1995; CoE, 1997).

Ostrich

Observations of ostriches show that on a daily average 5% of the time is spent on cleaning. When cleaning itself, an ostrich pushes its beak into its plumage and pulls the feathers through its beak. In this manner, loose feathers can be removed. Cleaning, however, also includes the head, neck, legs and toes. Usually, an ostrich that starts cleaning sets off a chain reaction in the group, with the result that all members clean simultaneously. Fellow ostriches are not, however, included in mutual bodily care. In order to sand bathe, the ostriches lie down in a dry, sandy place. The neck moves like a snake in the sand and the spread wings shovel sand on to the body. While plumage care is independent of climate and can be done all year round, sand bathing requires a dry, covered location, even in winter (Kreibich and Sommer, 1995).

12.9 Temperature Regulation and Climate Requirements

Temperature regulation

Ratites in general

Ratites, like other birds, lack sweat glands and under heat stress they rely on increased evaporation from the respiratory system for heat transfer.

Ostrich

Ostriches are adept at behavioural thermoregulation. They have various facilities to control their heat exchange. The feathers offer protection from the sun's harmful rays by day and in the cold desert nights they stop the loss of body heat. During hot weather, they lose heat by panting and fluffing up their feathers to expose bare skin on their thorax and upper legs. Under desert conditions, heat loss by radiation and by convective cooling can be augmented by behavioural adjustments, which are also affected by the wind speed. At high ambient temperature and with a wind blowing, the birds react with feather erection. An increase in the feather layer from about 3 to some 10 cm has been registered, and wing drooping thereby exposes bare skin on the thorax. The birds put their heads up at an angle with their beaks open and pull their haw (nictitating membrane) partly over their eyeball. Respiration rates of ostriches fall within two ranges, either within a low range of 3–5 breaths per min or a high range of 40–60 breaths per min. The increase in respiration rate from the low range to the high range is sudden and occurs in response to heat stress. At the same time, the tidal volume doubles, resulting in a 16-fold increase in ventilation. The increase in respiration rate is not necessarily associated with an increase in the rate of oxygen consumption. The feathers are flattened during low ambient temperatures at night (Kreibich and Sommer, 1995; Deeming and Bubier, 1999; Skadhauge and Dawson, 1999).

Rhea

In studies of heat balance of the running rhea, Taylor et al. (1971) found that at higher running speeds the rheas stored, rather than dissipated, large amounts of heat and the body temperature exceeded 45°C. The highest cloacal temperature that was recorded at the end of a run was 46.4°C, which caused no apparent ill effects to the bird. However, the rheas became very reluctant to run and the experiments were discontinued.

Emu

On very hot days, emus pant to maintain their body temperature; their lungs work as evaporative coolers and, unlike some other species, the resulting low levels of carbon dioxide in the blood do not appear to cause alkalosis. For normal breathing in cooler weather, they have large, multifolded nasal passages. Cool air warms as it passes through into the lungs, extracting heat from the nasal region (Kreibich and Sommer, 1995).

Climate requirements

Ostrich

The colder, wetter climates of Europe and parts of North America should be regarded as potentially deleterious to ostriches unless there is proper management of the birds. The birds show a propensity to sit out in rainy weather rather than seeking shelter and could thus be affected adversely by wetting during cold weather as ostriches have poor waterproofing of the feather cover. This may lead to a major disruption of the insulative properties of the plumage during wetting and increase heat loss markedly, causing profound hypothermia, especially in conjunction with air movement. Ostriches should be brought into shelter during prolonged periods of rain, particularly in cold European and North American winters. If heat stress is likely, for example in indoor pens, they should have the freedom to use wing spreading and postural changes to increase heat loss (Mitchell, 1999).

12.10 Vision, Behaviour in Light and Darkness

Ostrich

The exceptionally large eyes with their 50 mm diameter are the largest of any vertebrate. They are widely separated on the head and positioned to produce an image from in front of and below the eye. The large blind spot behind and above the head is considered to shade the eye. Ostriches have excellent eyesight. Using their periscope-like neck, they can discern movement up to a distance of 3.5 km. Ostriches can also perceive objects directly under their beaks. Each eye has a mass of approximately 60 g, four times the weight of a human eye. Apart from an upper and lower lid, the ostrich eye is equipped with a nictitating membrane which runs sideways across the eye and helps to keep the eye clean. Through their keen eyesight, they may act as guards for other animals, in the wild as well as on large farms (Kreibich and Sommer, 1995; Deeming, 1999).

12.11 Acoustic Communication

Hearing

Ratites in general

Ratites are very sensitive to sound. Constant noise, even at low levels, must therefore be avoided and animals should be protected against sudden noise (CoE, 1997).

Ostrich

The ear flaps face to the rear of the bird. Their hearing is well developed (Deeming, 1999).

Vocalization and acoustic communication

Ostrich

Although the ostrich is considered a silent bird, limited vocalization has been identified. Pumping up the neck is obviously related to the strange sounds the cock produces when courting a hen or challenging another male; these sounds are particular to males, mainly in the mating season, but sometimes, usually at night, without discernible reason. This particular emission of sound is called 'brooming'. Ostriches can broom only when stationary, not when running. Brooming is a peculiar,

soft murmuring tone and it is difficult to determine where it comes from. It gives the impression that the bird could increase this sound to a powerful roar if it wanted to. Each individual broom comprises three individual notes, two short and one long. The beak remains closed and there is no exhalation of air; the neck swells up like a balloon at each tone, especially on the third. This behaviour is used during the day to mark out a territory and generally causes the hens to pause in their feeding activities. They raise their heads and stand still, listening carefully. This sound – it cannot be classed as a call or cry – is uttered and sounds uncanny and ferocious on a quiet night. Other sounds are produced by both sexes: hissing when enraged, low gurgling sounds when in fear and a short, sharp cry warning of danger. The hens use noises predominantly in conjunction with aggressive behaviour towards other hens. To do so, they open their beaks and utter a hoarse noise. Hens who feel threatened also make such a noise, as do birds in a herd which is panicking. Cooing, pigeon-type noises can also be heard now and again. The growing, non-adult birds of all ages produce a reverberating, penetrating and plaintive call (Kreibich and Sommer, 1995; Deeming, 1999).

Rhea

Rheas tend to be silent birds, the exception being when they are chicks or when the male is seeking a mate (Kreibich and Sommer, 1995).

Emu

The peculiar structure of the trachea of the emu is correlated with the loud booming note of the bird during the breeding season. Their calls consist of loud booming, drumming and grunting sounds that can be heard up to 2 km away. The booming sound is created in an inflatable neck sac that is 30 cm long and thin-walled (Kreibich and Sommer, 1995).

12.12 Senses of Smell and Taste, Olfaction

Ratites in general

In the ostrich, the faculties of smell and taste are retarded (Kreibich and Sommer, 1995). However, in preference studies of different shrubs, Pollock *et al.*

(2007) showed that *Aristotelia fruticosa* was preferred by ostriches. And smell is used by the nocturnal kiwi to locate food (Sales, 2006a).

12.13 Tactile Sense, Sense of Feeling

Tactile touch

Ratites in general

Ratites do not mutually touch each other.

Sense of pain

Ratites in general

All birds possess pain receptors and show aversion to certain stimuli, which can be interpreted that they experience pain.

12.14 Perception of Electric and Magnetic Fields

Ratites in general

This does not seem to be described in ratites.

12.15 Heat and Mating Behaviour, Nesting, Egg Laying, Brooding

Ostrich

A hen in the wild reaches sexual maturity at the age of about 3 years and the male a year later (Hallam, 1992). Ostriches are photoperiod dependent, coming into season during periods of increasing daylight (Hicks-Alldredge, 1996).

Day length regulates the breeding season. In the beginning of the breeding period, males start to establish territories. Each territorial male digs a number of nest scrapes, which he shows to any female that enters his territory. The nest is a simple shallow hole in the ground. Defence of the territory usually involves parallel walking, chasing and kantling (striking the back of the head on either side of the back) by the males. In contrast, females cover a mean range of at least 25 km² and they enter territories held by a variety of males, although a few males are actively avoided by some females (Deeming and Bubier, 1999).

As the cock comes into breeding condition, his beak and shins become a bright scarlet colour. While the cock generally demonstrates his urge to mate by a loud mating song, the hen reacts to these noises by standing still for a moment and stopping her feeding activities. Cocks differ widely in their temperament and this is particularly noticeable during the mating season. Female birds show prenuptial courtship displays towards male companions involving posturing in front of potential mates. A dominant female exhibits aggressive behaviour against other females and any yearlings in the group, which often adopt a characteristic submissive posture (head down, neck in an 'S' shape and tail down) to appease the aggressor. Males appear to develop courtship behaviours later than females, with a slow development of the characteristic red flush of the beak, neck, thigh and shin skin. Dominance by males in mixed groups is established by posturing, usually with the tail held erect, and aggression against companions. The erect phallus is also distended from the cloaca and displayed to other birds. The phallus doubles in size during the breeding season. The flaccid phallus is about 20 cm long, bright red in colour and lies in a phallic pocket in the ventral wall of the proctodeum. When erect, the phallus is about 40 cm long and projects from the cloaca in a ventrocranial curve. The female clitoris is about 3 cm long. Many males approach the hen gently and court her patiently. Other cocks act like brutes, assaulting and mauling the hen to the point of exhaustion. Part of the courtship play consists of 'rolling'; the cock goes down on his haunches before the hen, raises his wings slightly and, while moving them to and fro, he hits his head alternately against the right and left arch of his back with a thudding sound. The hen crouches when she is receptive and flutters her wings, drops to the ground with her head extended and makes a snapping motion with her beak. Just prior to mounting, the male stamps his feet on the ground several times. The cock then mounts her, putting his left foot next to her on the ground at her left side and placing his right foot on her back slightly to the right side. She pushes her rump to the right, thereby making the so-called cloacal kiss possible. He drops to his hocks and intromission occurs. The curved penis passes down on the left side between the body and tail into the cloaca of the hen. Repeated thrusts of the phallus are often required before intromission. During intromission, he rolls from side to side, performs a kantle display to climax with his head being brought forward and a deep guttural grunt emitted, while the hen snaps her beak and shakes her head. When ejaculation occurs, the male extends his head forward, making a guttural sound. During the 30–60 s of mating, the female usually remains impassive, holding her head forward, then shakes her head and claps her beak. There is very little post-copulation behaviour (Hallam, 1992; Fowler, 1996; Hicks-Alldredge, 1996; Deeming and Bubier, 1999; Soley and Groenewald, 1999).

In each territory, a major female pairs with the male and lays most of her eggs in the nest site she chooses. This hen demonstrates her domination to minor hens and her mate. She always lays in the nest first, and she will be the one who later guards the nest and incubates the eggs with her mate. In addition, other minor females visit the territory and may lay an egg in an already established nest. To lay an egg, the hen briefly lifts her tail high shortly before laying and then sits down abruptly. The egg is laid in the short period the hen is sitting. Directly afterwards, she stands up again and rolls the egg with her beak. The minor females may be major females in another male's territory. Each major hen usually contributes about 11 eggs to her nest and the total number of eggs in the clutch, an average of 26 (Deeming and Bubier, 1999), depends on the number of 'minor' hens laying in the nest. In the wild, it has been found that hens lay 8–20 eggs per nest (Hallam, 1992). The surplus eggs are pushed out of the nest by the dominant hen. She does not push out her own eggs, but retains these as she is able to recognize them. The deciding factors are probably differences in size and shape, colour, the character of the pores and the surface of the shell. If she requires any balance to be made up, then she will keep eggs from the other hens. This breeding system is reported throughout the natural range of the ostrich and hence is considered to be typical of the species. There is usually a lull in laying during protracted wet periods. A major difference between wild and domesticated ostriches is that wild birds are very aware of danger and disturbance. If all the eggs are removed from the nest, then birds in the wild may desert the site and, after 2–3 weeks, start laying in another nest some distance from the original site. Eggs are laid in the late afternoon or early evening, and the clutch builds up over a period of up to 30 days. During this period, the nest is attended by both male and female, although full incubation does not proceed until the clutch is complete. The birds take turns sitting on the eggs. Usually, a bird remains sitting on the eggs for between 15 and 90 min at a time, before getting up and turning the eggs. In the beginning phase of brooding, the cock and the hen are often at the nest together. During the daytime, the hen, with her unobtrusive, earth-coloured feathers sits on the nest; the black-coloured male takes the night shift. On average, it takes 42 days until the eggs hatch. However, not all eggs will hatch. Hatching takes place over a period of 2–3 days, during which time the hatchlings remain brooded by an adult, although they start to consume small stones during periods of activity. A sitting ostrich can cover and warm about 19 eggs. The rest lie in a casual ring around the edge and, as they cannot develop, either rot or are broken. At the first sign of danger, the birds rely on camouflage to conceal them from predators, although they perform distraction displays or attack potential predators if necessary (Hallam, 1992; Kreibich and Sommer, 1995; Hicks-Alldredge, 1996; Deeming and Bubier, 1999).

Rhea

Rheas are not sexually mature and do not breed until they reach 2 years of age. Rheas are polygamous, with males courting between 2 and 12 females. During the nesting season, males compete for territories on the plains. Once established, each male tries to attract groups of females by running quickly towards them with outspread wings. Once sufficient females are assembled, he displays with voice and wing-shaking. The flaccid phallus of the rhea males is about 1 cm long and has a spiral conformation. Female rheas have no clitoris. After copulation, the male leads the female to a nest which he has prepared previously, and there she will lay her eggs. Only 5–6% of the male population breed successfully each year in the wild. The nest consists of a simple scrape in the ground, lined with grass and leaves. He tries to get as many females as possible to lay in this group nest and accumulates 13–30 eggs. The male alone incubates. Some of the females, meanwhile, may move on and mate with other males and lay eggs in other nests. Females may lay eggs for up to 12 different males during the laying season. The male will use a decoy system and place some eggs outside the nest and sacrifice these to predators, so that they will not attempt to get inside the nest. However, the male may use another subordinate male to incubate his eggs, while he finds another harem to start a second nest. The incubating period is around 42 days. All eggs hatch synchronously within a 24–28 h period. While caring for the young, the males will charge at any perceived threat that approaches the chicks, including female rheas and humans (Fowler, 1996; Hicks-Alldredge, 1996; Labaque *et al.*, 1999; Sales, 2006b).

Emu

Emus are not sexually mature and do not breed until they reach 2 years of age. Emus form breeding pairs during the Australian summer months of December and January, and may remain together for about 5 months. Mating occurs in the cooler months of May and June. In the northern hemisphere, farmed emus reproduce from November to March in response to a decreasing photoperiod. In the southern hemisphere – their normal range – they reproduce during the corresponding opposite season. The emu has been regarded as mating for life. However, it occurs that the female mates with other males and may lay in multiple clutches; thus, as many as half the chicks in a brood may be fathered by others, or by neither parent, as emus also exhibit brood parasitism. Those males commencing incubation earliest in the season tend to have the highest levels of paternity in their own nests. In a study, over 70% of males and females were classified as socially monogamous, but 7% of males and 3% of females were classified as socially polygamous. Also, 15% of the females were engaged in sequential polyandry. Before they began incubating, the males were prompt to court females other than their mates. It can be concluded that the social mating system of the emu is of a monogamous type, but a few individuals are promiscuous – the males before they start incubating and the females after their mate has started incubating. In addition, there are significant numbers of extra-pair copulations. During the breeding season, males lose their appetite and construct a rough nest in a semi-sheltered hollow on the ground from bark, grass, sticks and leaves. The pair mates every day or two, and every second or third day, the female lays an egg. During copulation, the female sits on the ground with her cloaca extended. The male drops to his hocks behind her and intromission occurs. The male pecks the back of the hen during mating and makes guttural noises during ejaculation. Emu males have a spiral conformation of the phallus, the females have a slight clitoris. Emu hens are dominant and vocalize by making a drumming sound during the breeding season. The male emu makes a growling sound rather than booming. In the wild, male emus gather and incubate from 7 to 10 dark-green eggs in the ground nest for about 60 days. The male becomes broody after his mate starts laying and begins to incubate the eggs before the laying period is complete. From this time on, he does not eat, drink or defecate, and stands only to turn the eggs, which he does about 10 times a day. Over 8 weeks of incubation, he will lose a third of his weight and will survive only on stored body fat and on any morning dew that he can reach from the nest. Some females stay and defend the nest until the chicks start hatching, but most leave the nesting area completely to nest again. In a good season, a female emu may nest three times. The male stops incubating the eggs shortly before they hatch. Newly hatched chicks are active and can leave the nest within a few days (Kreibich and Sommer, 1995; Fowler, 1996; Hicks-Alldredge, 1996: Blache *et al.*, 2000; Taylor *et al.*, 2000).

12.16   During and After Hatching

Behaviour during and after hatching

Ratites in general

Chicks are precocious and leave the nest within a few days after hatching (CoE, 1997).

Ostrich

Ostrich chicks are accompanied and protected by an adult for up to 9 months (CoE, 1997).

Rhea

The male alone cares for the striped young for the next 6 months. Lost youngsters are thus sometimes adopted by other males. Males frequently accept chicks from other groups, including them in their own brood. The young reach full adult size in about 6 months, but they remain in their groups of siblings. They breed when they reach 2–3 years of age (Kreibich and Sommer, 1995; CoE, 1997).

Emu

The male stays with the growing chicks for up to 7 months, defending them and teaching them how to find food (Kreibich and Sommer, 1995; CoE, 1997).

Litter size

Ratites in general

In the wild, there can be great losses of eggs and chicks from predators. The litter sizes thereby vary.

Play behaviour

Ratites in general

Play behaviour as it occurs in mammals does not seem to be described in ratites.

12.17   The Ratite Chick

Ratites in general

Chicks are precocious and leave the nest within a few days (CoE, 1997).

Ostrich

Little is known about the behaviour of ostrich chicks under natural conditions. Once the chicks leave the nest, they are difficult to track in grassland but the parents of young birds brood them as protection against the elements and predators. Adults will feign a 'mock injury' to distract a potential predator from the chicks. Families of chicks are combined into creches, which may number up to 300 birds and are overseen by a single pair of adult birds. When groups of chicks meet, there is a vigorous behavioural contest between the guardians of each group over which adults will take charge of the enlarged creche. Usually, younger chicks are accepted into groups of older chicks, but not vice versa. By the time the chicks are 1 year old, they have been abandoned by the adults and spend their time in a compact peer group (Deeming and Bubier, 1999).

Rhea

When the young are half-grown, they can wander on their own, but generally they remain in their groups of siblings until 2–3 years old (Kreibich and Sommer, 1995; CoE, 1997).

Emu

Chicks grow very quickly and are full-grown in 5–6 months; they may remain with their family group for another 6 months or so before they split

up to breed in their second season. They are protected by an adult, possibly for up to 18 months (Kreibich and Sommer, 1995; CoE, 1997).

12.18   Assessment of Ratite Health and Welfare

The healthy ratite

Ratites in general

Physical examination of farmed ratites should start with watching the birds in their enclosure, their locomotion, feeding, resting, social and other behaviour.

Physical examination of the single bird is based mainly on visual and manual examination and requires good physical restraint of the bird. The condition of the eyes, nostrils, mouth and throat, as well as the colour and texture of the faeces or the colour of the urine, is of great importance in the diagnosis of nutritional deficiencies, as well as of infectious diseases (Perelman, 1999).

The sick ratite

Symptoms

OSTRICH   Sick, weak or stressed birds usually stay apart from the rest of the flock, walking slowly with an 'S' shaped neck and low head and spending most of the day near the fences and away from other birds. Reluctance to move may indicate locomotive and neurological problems, and could be indicative of nutritional, toxicological or infectious disorders (Perelman, 1999).

Examples of abnormal behaviour and stereotypies

OSTRICH   One example of abnormal behaviour is pecking directed at conspecifics. In contrast to normal body care behaviour in ostriches, which never involves other individuals, pecking involves the seeming collaboration of another ostrich. This abnormal behaviour is carried out predominantly by hens on other hens, and only seldom on cocks. Feather pecking is usually carried out on the torso of another either resting or standing ostrich. Without any apparent motive, the feathers are pulled out with the beak. The pecked bird shows no inclination to aggression or to flee, although the pecked areas are often bloody. Cocks are generally pecked less than hens. This is probably due to the higher social status held by the male birds, which at the same time leads to greater distances between the birds, especially between cocks. Possibly, hens may fear sexual contact from cocks and therefore prefer pecking other females. Pecking can also occur as a sort of normal feeding behaviour, directed against objects in the environment, fences, fence posts, the ground, etc. During this process, a bird pushes its shut beak down on the object in question, then opens it and snaps it shut again, for example, to pick up or play with little stones which may be lying around on the ground. A swallowing procedure follows (Kreibich and Sommer, 1995).

These symptoms of behavioural aberrations may be caused by disorientation stress, desertion stress or frustration. Disorientation may take place when ostrich chicks or juveniles are moved from one place to another, from one pen to another or even from the night accommodation to the outside run, and particularly from farm to farm. Frustration can be caused by non-recognition of feed after a change to a different feed. Stressed birds compensate by abnormal pecking or feeding behaviour. Frequently, this leads to the ingestion of long grass, litter, foreign bodies, sand or gravel, which might accumulate and thereby lead to an occlusion of the oesophagus or the intestines. Access to such foreign bodies can cause problems. This behaviour can be associated with whole groups of birds, but it can also affect single birds (Kreibich and Sommer, 1995; Huchzermeyer, 1994).

One explanation for abnormal pecking behaviour among ostriches might be that feeding no longer requires the extensive time and movement it once did in the wild. If the birds are given a very limited time, let us say 1 h per day, to select and consume their food, the bird is still programmed for searching and eating food during one-third of its time during daylight hours. This might explain this abnormal feeding behaviour in an environment without grazing possibility. Similar abnormal feeding behaviour is known from cattle kept indoors and given concentrated feed during limited periods without access to roughage the rest of the day (see Chapter 4 on cattle).

Examples of injuries and diseases caused by factors in housing and management

RATITES IN GENERAL Farmed ratite chicks which are not encouraged to exercise outdoors are at risk of retardation of their normal behavioural and physical development and the emergence of skeleton muscle disorders or abnormal behaviour.

Ratites, and especially ratite chicks, are subject to respiratory and alimentary diseases if exposed to wet and otherwise adverse weather conditions. All farmed ratites older than 5 days are at risk of alimentary diseases if grit is not available. Farmed ostriches kept indoors may get injured when jumping if the height of the ceiling in buildings or shelters is lower than 2.5 m (CoE, 1997).

OSTRICH In the wild, the reasons for mortality of ostrich chicks have been poorly investigated, but predation may be a key factor (Verwoerd *et al.*, 1999).

Environmental conditions, particularly cold, can predispose farmed chicks to enteritis. Ostriches lack mesenteric lymph nodes and often are unable to limit the spread of infection. Clinically, affected chicks are very depressed, and the condition can spread rapidly through the whole group. There may not always be visible diarrhoea (Huchzermeyer, 1994).

Injuries are the most common cause of fatalities in farmed adult ostriches. Startled ostriches may run off in a panic and try to burst through any obstacle low enough for them to see over. This makes them prone to injuries, especially broken legs. Broken bones mend only reluctantly and birds injured in this way usually have to be put down (Kreibich and Sommer, 1995).

Symptoms of pain

RATITES IN GENERAL Impaired walking ability caused by chronic pain can be a result of footpad disorders caused by wet and dirty surfaces or the wrong technique of toe trimming in farmed ratites (Glatz and Miao, 2008).

12.19 Capturing, Fixation, Handling

Ratites in general

The preferred method of lifting young chicks (from newly hatched to about 10 weeks) is with one hand under the body, gathering the legs with the other hand to stop kicking. When older and larger, the bird can be restrained by standing astride it, holding its body, with the handler's legs placed behind the wings and the hands around the chest or base of the neck area. Ratites must not be restrained by the legs, the feathers, or a single wing as this can cause dislocation. The wings and tail can be used to guide the bird but care must be taken never to use a single wing. After catching, hooding is recommended, taking care not to obstruct the nostrils. Hoods should not be left on for longer than is absolutely necessary. Care should be taken when releasing birds to protect them from aggressive behaviour and injury from conspecifics (CoE, 1997).

Ostrich

Small chicks are easy to catch and handle, as they tend to sit when threatened. Very young birds can be grabbed gently by the base of the neck and then held by placing one hand under the chest bone and the other below the pelvic bone; ostrich chicks held in this way tend to paddle with the legs using strong movements. At no time should ostrich chicks be restrained by the legs only or held upside down. While holding ostrich chicks, it is necessary to hold both legs of the bird tightly. A young chick should be supported in a sternal position, with the legs tucked under the bird in the arms of the handler, who maintains gentle downward pressure on the back. Cradling the bird in the handler's arms limits struggling (Kreibich and Sommer, 1995; Wade, 1996; Raines, 1998; Perelman, 1999).

Capture and restraint of an adult require two to three experienced, physically fit handlers. The response of ostriches to handling and restraint can be unexpected and dangerous to both birds and handlers. As ostriches kick forward and down, the side and back are the safest positions when tackling them. A hook should be used to catch adult ostriches and this should be done in the gentlest possible way. The hook should be placed about the upper third of the neck, and the neck should be pulled carefully downwards, following the bird's movement. Once the bird has been hooked, the hook should not be turned sideways. After the ostrich is caught, one should pull gently to reach the head in order to apply the hood. The bird's head should be held below the level of its chest until the hood is in position. Most ostriches become quieter and easier to handle when hooded (Kreibich and Sommer, 1995; Tully and Shane, 1996; Perelman, 1999).

Emu

Emus are best handled in open areas, because they are capable of jumping nearly 2 m high and will bounce off walls in an attempt to evade capture. Close confinement can cause injury to both birds and handlers. Due to their small size and the relative ease in capture, chicks up to 3 months of age present no problems in restraint. It is important to remember that rough handling may lead to serious injury or death. Emu chicks are restrained easily by cradling them on their back with a hand restraining their legs in a tucked position. Juvenile emus from 3 to 8 months of age can be restrained easily for most examinations and treatment procedures. A young emu normally can be caught by approaching from behind and scooping the bird up with one arm between the legs, with the hand on the sternum. Applying gentle pressure on the caudal sternum seems to calm these birds. The other hand is used to restrain and immobilize the head and neck. It is important to remember that emus may struggle briefly, sometimes violently, when caught. Care should be taken to keep the neck and head of the bird away from its thrashing feet. The same is true for the face and arms of handlers or bystanders, who should be aware of the possibility of injury. Because emus very often defecate when captured, the vent should be directed away from either handler. Most treatments and sample collections on young birds may be accomplished with light but firm manual restraint. Older juvenile and adult emus present more difficulty in handling because of their well-developed legs and considerable body mass. Some individuals may be extremely fractious or aggressive. Adult emus can be straddled between the legs of the handler while applying upward pressure caudally to the sternum or breastplate in a bear hug. The body of the bird is pulled back against the handler and the bird is tilted slightly more upright than normal. The handler's other hand can be used to help restrain the bird or to administer oral or injectable medication. Birds can be walked into a trailer or new pen by grasping the wings, straddling them or walking beside them, and steering them in the desired direction (Mouser, 1996; Raines, 1998).

12.20 The Importance of the Stockman

Ratites in general

It is essential for the stockman to have close contact with ratites, particularly when the birds are young, to train them to become accustomed to people (CoE, 1997).

Ostrich

A typical trait of ostriches is their ferocity. Even farmed young ostriches may become completely unmanageable if not in early and continuous contact with people (Kreibich and Sommer, 1995).

Glossary

Abomasum   The fourth compartment of the ruminant stomach. It is an elongated sac, comparable in structure and function to the stomach of nonruminants. It lies in the right half of the abdominal cavity, largely on the abdominal floor.

Afterbirth   Discharged placenta.

Agonistic behaviour   Each behaviour which is connected with conflict or fight between two individuals. It comprises behaviour patterns which express attack, flight or passivity.

AI   Artificial insemination.

Allogrooming   The licking cattle perform on another animal. It is a component of social behaviour.

Antorbital   In front of the eyes.

Aviary   A housing system with several floors where domestic fowl are kept loose.

Bruxism   Grinding of teeth.

Caecotrophs   Pellets of soft caecal contents expelled periodically from the anus in rabbits and reingested as a source of nutrients. They have an outer greenish membrane of mucus that encloses semi-liquid caecal ingesta.

Carpus   Foreknee.

Conditioned reflex (or **conditioned response**)   A response that does not occur naturally in an animal but which may be developed by regular association of some physiological function with an unrelated outside event, such as ringing of a bell or flashing of a light. The physiological function starts whenever the outside event occurs.

Congenital behaviour (also called **innate behaviour** or **fixed movement pattern**)   Various behaviours found in different species, for example getting up or lying down behaviour, the way in which the young suck, or the position at urination.

Conspecifics   Animals of the same species.

Cranial   Pertaining to the cranium or to the head of the body. The word is also used to refer to a point or part of an organ situated at or directed towards the front; opposite to posterior.

Cubicle   See **free stall**.

Dominant behaviour   An individual is regarded as being dominant over another when it has precedence in feeding or sexual behaviour and is superior to others in aggressiveness and group control. Dominance is shown by superiority in fighting capacity over conspecifics.

Dystocia   Difficult birth (delivery).

Eliminative behaviour   Urination and defecation.

Epidemiology   The science of the distribution of disease in animal (and human) populations and the factors determining that distribution.

Ethogram   A catalogue of the discrete behaviours typically employed by a species or a single animal. Can be described as a behaviour catalogue.

Ethology   The science of animal behaviour and its causes.

Farrowing hysteria   Affected gilts restless at farrowing savage piglets without cannibalizing them: likely to recur at the next farrowing.

Feather pecking   The pecking and pulling of feathers from another hen. Feather pecks are directed towards body regions like the areas near the vent and preen gland, and the wings, back and tail.

Fixed movement pattern   See **congenital behaviour**.

Flight reaction   A characteristic reaction in order to avoid a danger, especially a particular enemy or a particular situation. The reaction occurs as soon as the enemy or the situation exists.

Fluorescent light   Is emitted from fluorescent lamps, which are gas-discharge lamps that use electricity to excite mercury vapour. The excited mercury atoms produce shortwave ultraviolet light that then causes a phosphor to fluoresce, producing visible light. A fluorescent lamp converts electrical power into useful light more efficiently than an incandescent lamp. Lower energy cost typically offsets the higher initial cost of the lamp. The lamp is more costly because it requires a ballast to regulate the flow of current through the lamp.

Free stall (or **cubicle**) Individual cow bed spaces (stalls) separated by partitions. Located in loose-housing cow accommodation in which the cow is free to wander at will.

Grooming The licking that some animals perform on themselves or on others.

Haw A thin membrane in the corner of the eye behind the eyelids in reptiles, birds and some mammals. It takes part in the distribution of the tear fluid on the cornea. It is also called the **nictitating membrane** or the **third eyelid**.

Health The state when the organs and the organ systems function in harmony with each other and with the environment. **Disease** is consequently a state when this harmony no longer exists.

Hierarchy Each social order of precedence which has been established through direct fight, threat, passive submission or through a combination of these behaviours.

Imprinting A process of learning in the very young during a very limited and early period of the development of the young individual. The classic example is of newly hatched ducklings that have a following behaviour by which the very sight of a moving object functions as a key stimulus or as an unconditioned stimulus. It is thus possible to get such ducklings to follow almost any object, e.g. a balloon. If they continue to do so for an hour or so, they will henceforth follow only balloons and no other objects, not even a living female duck.

Incandescent light Is emitted from a source of electric light that works by incandescence (a general term for heat-driven light emissions). An electric current passes through a thin filament, heating it to a temperature that produces light. The enclosing glass bulb contains either a vacuum or an inert gas to prevent oxidation of the hot filament. Incandescent bulbs are also sometimes called **electric lamps**, a term also applied to the original arc lamps. Incandescent bulbs are electric lamps, arc lamps, producing light with arcs.

Individual distance The distance that an animal assumes as its private domain and inside which it will attack any intruder.

Innate behaviour See **congenital behaviour**.

Instinct handling A biological relevant movement pattern which is genetically conditioned. One example is the following behaviour in newborn ducklings.

Key stimulus Very specific stimuli on the different senses, sight, touch, hearing, that trigger fixed movement patterns. The capacity to respond to a key stimulus can vary depending on the animal's hormonal status and other similar conditions. A thirsty cow reacts to drink when offered water; a non-thirsty cow does not show a reaction. A sow in oestrus stands for the boar and the boar shows interest in her; a non-oestrous sow shows no interest in the boar and the boar does not show any interest in her.

Loose housing There are closed and open loose-housing systems. In the closed system, the cattle are kept loose in one or several groups in a barn. For each group there are separate feeding and lying areas connected by walkways and the animals can go freely between these. If the barn is built for dairy cows, there is a section for milking.

Mastitis–metritis–agalactia syndrome (MMA) A syndrome that occurs in sows that have farrowed for 12–48 h. It is manifested by an indeterminate set of signs including anorexia, lethargy, fever, agalactia, swelling of mammary glands, constipation and a marked disinterest in the piglets. Is also called farrowing fever.

Meconium A dark greenish mass that accumulates in the bowel during fetal life and is discharged shortly after birth.

Micron, μm (or **micrometre**) Metric unit of measure for length equal to 0.001 mm. Commonly employed to measure the thickness or diameter of microscopic objects, such as microorganisms and colloidal particles.

Milliampere, mA 1 mA = 0.001 ampere.

Nanometre, nm Unit of length used chiefly in measuring wavelengths of light, equal to 10^{-10} m.

Oedema An abnormal accumulation of fluid in the cavities and intercellular spaces of the body.

Oesophagus Gullet.

Oestrus Heat.

Omasum The third and smallest compartment of the forestomachs of the ruminant. Connects with the reticulum through the reticulo-omasal orifice and with the abomasum through the omaso-abomasal orifice.

Osteoporosis A reduced mineralization of the main structural bones of the skeleton which weakens them, leading to fractures.

Pheromone A substance secreted to the outside of the body and perceived (e.g. by smell) by other individuals of the same species, releasing specific behaviour in the percipient.

Philtrum Divided upper lip.

Pinna (pinnae) External ear(s).

Placenta An organ that connects the developing fetus to the uterine wall to allow nutrient uptake, waste elimination and gas exchange via the mother's blood supply.

Plantar Of or relating to the sole of the foot.

Pododermatitis Inflammation of the plantar surface and connective tissue of the foot/feet.

Polyoestrous Polyoestrous species are those that, in the absence of mating, have several oestrous cycles during the year.

Reticulum The smallest, most cranial section of the compound stomach of ruminants, lined with mucous membrane folded in a hexagonal pattern. It is also called honeycomb. It communicates cranially with the oesophagus and caudally with the rumen.

Rumen The largest of the compartments of the forestomachs of ruminant animals, which serves as a fermenting vat. It is lined with a keratinized epithelium bearing numerous absorptive papillae; it is partly subdivided by folds (pillars). These include dorsal and ventral sacs and a caudo-dorsal blind sac and a caudo-ventral blind sac. It communicates directly with the reticulum cranially and has no other exit.

Rumination Rumination includes regurgitation, remastication, ensalivation and reswallowing. The regurgitation of fluid reticular contents occurs as a result of a positive lowering of intrathoracic pressure, the arrival of a ruminal contraction at the cardiac sphincter at the appropriate time, relaxation of the cardiac sphincter and reverse peristalsis in the lower oesophagus. The regurgitus is compressed at the back of the tongue and the fluid reswallowed immediately. The solid material is chewed for about 1 min and reswallowed. The cycle is then ready to recommence. Rumination requires a positive approach by the cow and it is easily dissuaded by fright. It is also called 'chewing the cud' or 'cudding'.

SCAHAW Scientific Committee on Animal Health and Animal Welfare in the EU.

Scotopic vision Twilight or night vision.

Sleep A period of rest during which volition and consciousness are in partial or complete abeyance and the bodily functions partially suspended. It is a behavioural state marked by characteristic immobile posture and diminished but readily reversible sensitivity to external stimuli. Sleep is divided into two main types, REM (rapid eye movement) and NREM (non-REM); each recurs cyclically several times during a normal period of sleep. REM sleep is characterized by increased neuronal activity of the forebrain and midbrain, by depressed muscle tone and by dreaming, rapid eye movements and vascular congestion of the sex organs. Slow-wave sleep (SWS), often referred to as deep sleep, consists of stages 3 and 4 of non-rapid eye movement sleep (NREM). As of 2008, the American Academy of Sleep Medicine (AASM) has discontinued the use of stage 4, such that the previous stages 3 and 4 are now combined as stage 3. REM sleep shows EEG low voltage and fast activity similar to wakefulness, but the individual shows very little muscle activity. An animal is aroused more easily during SWS sleep than during REM sleep.

Social behaviour Behaviour of an animal to others in its social group of herd, flock or neighbours. Includes establishment of the dominance order.

Stereotypy Frequent, almost mechanical, repetition of the same movement. Such repetitive behaviour is usually associated with restricted environments and constraints that prevent or frustrate the performance of highly motivated behaviours. It is also seen in connection with lack of stimuli in barren environments, sometimes in combination with overstimulation caused by exposure to one environmental factor, e.g. constant noise.

Sternum Breastbone.

Tachycardia Abnormally rapid heart rate.

Tachypnoea Very rapid respiration.

Tarsus Hock.

Thermoneutral zone The range of ambient temperatures at which an animal does not have to regulate its body temperature actively (the range of temperatures animals can withstand without elevating their metabolic rate).

Twitch A loop of rope or strap which is tightened over a horse's lip as a restraining device.

Unconditioned reflex A reflected involuntary reaction or movement of an organism to a stimulus. This reaction always follows the same pattern. The knee-joint reflex in humans is an example of an unconditioned reflex. A reflex is built into the nervous system and does not need the intervention of conscious thought to take effect.

UV$_A$ light Ultraviolet light with a longer wavelength ($320 < l < 400$ nm). Many bird species use UV$_A$ light for a variety of functions, including mate choice. Humans do not perceive UV$_A$ light.

UV light Ultraviolet light, electromagnetic radiation with a wavelength shorter than that of visible light 60 (100) to 380 (400) nm.

UV radiation Electromagnetic radiation with wavelengths between X-rays and visible light. The lower boundary is set at 60–100 nm, the upper at 380 or 400 nm.

Waste management Handling urine, dung and spill water from an animal house or plant. Sometimes waste refers to just urine and manure.

Definitions from several dictionaries have been used to elaborate this glossary, among them:

Blood, D.C. and Studdert, V.P. (2000) *Saunders Comprehensive Veterinary Dictionary,* 2nd edn. W.B. Saunders, London, 1380 pp.

Concise Veterinary Dictionary (1988) Oxford University Press, Oxford, 890 pp.

Dorlands Illustrated Medical Dictionary (2003) 30th edn. Saunders, Philadelphia, Pennsylvania, 2190 pp.

Gilpin, A. (1976) *Dictionary of Environmental Terms.* Routledge and Kegan Paul, London, 183 pp.

Last, J.M. (1983) *A Dictionary of Epidemiology.* Oxford University Press, Oxford, 114 pp.

Lindskog, B.I. and Zetterberg, B.L. (1981) *Medicinsk Terminologi Lexikon.* Nordiska Bokhandelns Förlag, Stockholm, 630 pp.

Mack, R. (1988) *Wörterbuch für Veterinärmedizin und Biowissenschaften.* Verlag Paul Parey, Berlin and Hamburg, 321 pp.

The Merck Veterinary Manual (1979) 5th edn. Merck and Co., Inc., Rahway, New Jersey, 1672 pp.

References

Abourachid, A. (1991) Comparative gait analysis of two strains of turkey, *Meleagris gallopavo*. *British Poultry Science* 32, 271–277.

Adam, C.L. (1986) Physiological values – red deer. In: Alexander, T.L. (ed.) *Management and Diseases of Deer*. Veterinary Deer Society, BVA, London, p. 227.

Adrados, C., Baltzinger, C., Janeau, G. and Pepin, D. (2008) Red deer *Cervus elaphus* resting place characteristics obtained from differential GPS data in a forest habitat. *European Journal of Wildlife Research* 54, 487–494.

Aland, A., Lidfors, L. and Ekesbo, I. (2002) Diurnal distribution of dairy cow defecation and urination. *Applied Animal Behaviour Science* 78, 43–54.

Albright, J.L. and Arave, C.W. (1997) *The Behaviour of Cattle*. CAB International, Wallingford, UK, 306 pp.

Alcock, I. (2007) Hair today, gone tomorrow? Do red deer moult twice a year? *Journal of the British Deer Society, London* 14, 42–45.

Alcock, M.B. (1992) Role in landscape and wildlife conservation. In: Phillips, C.J.C. and Piggins, D. (eds) *Farm Animals and the Environment*. CAB International, Wallingford, UK, pp. 383–410.

Alexander, T.L. (1986) Identification of free-loving and farmed deer. In: Alexander T.L. (ed.) *Management and Diseases of Deer*. Veterinary Deer Society, BVA, London, pp. 233–235.

Algers, B. (1984a) Animal health in flat deck rearing of weaned piglets. *Zentralblatt Veterinary Medicine A* 31, 1–13.

Algers, B. (1984b) Early weaning and cage rearing of piglets: influence on behaviour. *Zentralblatt Veterinary Medicine A* 31, 14–24.

Algers, B. (1989) Vocal and tactile communication during suckling in pigs. PhD thesis, Report 25, Swedish University of Agricultural Sciences (SLU), Department of Animal Hygiene, Skara, Sweden, 162 pp.

Algers, B. and Jensen, P. (1985) Communication during suckling in the domestic pig. II. Effects of continuous noise. *Applied Animal Behaviour Science* 1, 49–61.

Algers, B. and Jensen, P. (1990) Thermal microclimate in winter farrowing nests of free-ranging domestic pigs. *Livestock Production Science* 25, 177–181.

Algers, B. and Jensen, P. (1991) Teat stimulation and milk production during early lactation in sows: effects of continuous noise. *Canadian Journal of Animal Science* 71, 51–60.

Algers, B., Ekesbo, I. and Strömberg, S. (1978a) The impact of continuous noise on animal health. *Acta Veterinaria Scandinavica* Suppl. 67, 26 pp.

Algers, B., Ekesbo, I. and Strömberg, S. (1978b) Noise measurements in farm animal environments. *Acta Veterinaria Scandinavica* Suppl. 68, 19 pp.

Allen, T. and Clarke, J.A. (2005) Social learning of food preferences by white-tailed ptarmigan chicks. *Animal Behaviour* 70, 305–310.

Al-Rawi, B. and Craig, J.V. (1975) Agonistic behaviour of caged chickens related to group size and area per bird. *Applied Animal Ethology* 2, 69–80.

Andersson, B.E. (1977) Temperature regulation and environmental physiology. In: Swenson, M.J. (ed.) *Dukes' Physiology of Domestic Animals*, 9th edn. Cornell University Press, Ithaca, New York, pp. 686–695.

Andersson, B.E. and Jonasson, H. (1993) Temperature regulation and environmental physiology. In: Swenson, M.J. and Reece, W.O. (eds) *Dukes' Physiology of Domestic Animals*, 11th edn. Cornell University Press, Ithaca, New York, pp. 886–895.

Andrews, A.H., Blowey, R.W., Boyd, H. and Eddy, R.G. (2004) *Bovine Medicine: Diseases and Husbandry of Cattle*, 2nd edn. Wiley-Blackwell, Ames, Iowa, 1218 pp.

Angel, C.R., Scheidler, S.E. and Sell, J.L. (1996) Ratite nutrition. In: Tully, T.N. Jr and Shane,

S.M. (eds) *Ratite Management, Medicine and Surgery*. Krieger Publishing Co., Malabar, Florida, pp. 11–30.

Anthony, D.W. (1986) The 'Kurgan culture', Indo-European origins, and the domestication of the horse: a reconsideration. *Current Anthropology* 27, 291–311.

Appleby, M.C. and Duncan, I.J.H. (1989) Development of perching in hens. *Biology of Behaviour* 14, 157–168.

Appleby, M.C., McRae, H.E. and Duncan, I.J.H. (1983) Nesting and floor laying by domestic hens: effects of individual variation in perching behaviour. *Behaviour Analysis Letters* 3, 345–352.

Appleby, M.C., Hughes, B.O. and Elson, A. (1992) *Poultry Production Systems. Behaviour Management and Welfare*. CAB International, Wallingford, UK, 238 pp.

Appleby, M.C., Mench, M.A. and Hughes, B.O. (2004) *Poultry Behaviour and Welfare*. CAB International, Wallingford, UK, 276 pp.

Arave, C.W. and Albright, J.L. (1981) Cattle behaviour. *Journal of Dairy Science* 64, 1318–1329.

Archer, J. (1988) *The Behavioural Biology of Aggression*. Cambridge University Press, Cambridge, UK, 257 pp.

Arellano, P.E., Pijoan, C., Jacobson, L.D. and Algers, B. (1992) Stereotyped behaviour, social interactions and suckling pattern of pigs housed in groups or in single crates. *Applied Animal Behavioural Science* 35, 157–166.

Arney, D.R. (2009) Welfare of large animals in scientific research. *Scandinavian Journal of Laboratory Animal Science* 36, 97–101.

Ashworth, C.J. and Pickard, A.R. (1998) Embryo survival and prolificacy. In: Wiseman, J., Varley, M.A. and Chadwick, J.P. (eds) *Progress in Pig Science*. Nottingham University Press, Nottingham, UK, pp. 303–312.

Austin, A.R. (1996) Travel sickness in pigs and sheep. *Veterinary Record* 139(23), 575.

Avellaneda, Y., Rodriguez, F., Grajales, H., Martinez, R. and and Vasquez, R. (2006) Puberty determination in rams according to body characteristics, and evaluation of quality of ejaculate and testosterone. *Livestock Research for Rural Development* 18(10) October, Article 138.

Ayala-Guerrero, F., Mexicano, G. and Ramos, J.I. (2003) Sleep characteristics in the turkey *Meleagris gallopavo*. *Physiology and Behaviour* 78, 435–440.

Bachmann, I. and Stauffacher, M. (2002) Prävalenz von Verhaltensstörungen in der Schweizer Pferdepopulation. *Schweizer Archiv für Tierheilkunde* 144, 356–368.

Bäckström, L. (1973) Environment and animal health in piglet production. A field study of incidences and correlations. Thesis. *Acta Veterinaria Scandinavica* Suppl. 41, 240 pp.

Baldock, N.M. and Sibly, R.M. (1990) Effects of handling and transportation on the heart rate and behaviour of sheep. *Applied Animal Behaviour Science* 28, 15–39.

Barber, C.L., Prescott, N.B., Wathes, C.M., Le Sueur, C. and Perry, G. (2004) Preferences of growing ducklings and turkey poults for illuminance. *Animal Welfare* 13, 211–224.

Barnett, J.L. and Hemsworth, P.H. (1986) The impact of handling and environmental factors on the stress response and its consequences in swine. *Laboratory Animal Science* 36, 366–369.

Barnett, J.L., Winfield, C.G., Cronin, G.M., Hemsworth, P.H. and Dewar, A.M. (1985) The effect of individual and group housing on behavioural and physiological responses related to the welfare of pregnant pigs. *Applied Animal Behaviour Science* 14, 149–161.

Batty, J. (1994) *Ostrich Farming*. Beech Publishing House, Midhurst, UK, 110 pp.

Baumann, P., Oester, H. and Stauffacher, M. (2005) The influence of pup odour on the nest related behaviour of rabbit does (*Oryctolagus cuniculus*). *Applied Animal Behaviour Science* 93, 123–133.

Baxter, M.R. (1983) Environmental determinants of excretory and lying areas in domestic pigs. *Applied Animal Ethology* 9, 195–200.

Baxter, M.R. (1989) Intensive housing: the last straw for pigs? *Journal of Animal Science* 67, 2433–2440.

Baymann, U., Langbein, J., Siebert, K., Nurnberg, G., Manteuffel, G. and Mohr, E. (2007) Cognitive enrichment in farm animals – the impact of social rank and social environment on learning behaviour of dwarf goats. *Berliner und Münchener Tierärztliche Wochenschrift* 120, 89–97.

Beausoleil, N.J., Stafford, K.J. and Mellor, D.J. (2005) Sheep show more aversion to a dog than to a human in an arena test. *Applied Animal Behaviour Science* 91, 219–232.

Beaver, J.M. and Olson, B.E. (1997) Winter range use by cattle of different ages in southwestern Montana. *Applied Animal Behaviour Science* 51, 1–13.

Bebie, N. and McElligott, A.G. (2006) Female aggression in red deer: does it indicate competition for mates? *Mammalian Biology* 71, 347–355.

Begall, S., Červený, J., Neef, J., Vojtěch, O. and Burda, H. (2008) Magnetic alignment in grazing and resting cattle and deer. *Proceedings of the National Academy of Sciences of the United States of America* 105, 13451–13455.

Bell, D.J. (1983) Mate choice in the European rabbit. In: Bateson, P.P.G. (ed.) *Mate Choice*. Cambridge University Press, Cambridge, UK, pp. 211–223.

Bell, D.J. (1999) The European wild rabbit. In: Poole, T. (ed.) *The UFAW Handbook on the Care and Management of Laboratory Animals*, 7th edn. Blackwell Science, Oxford, UK, pp. 389–394.

Bendixen, P., Vilson, B., Ekesbo, I. and Åstrand, D.B. (1986) Disease frequencies of tied zero-grazing dairy cows and of dairy cows on pasture during summer and tied during winter. *Preventive Veterinary Medicine* 4, 291–306.

Bendixen, P., Vilson, B., Ekesbo, I. and Åstrand, D.B. (1987) Disease frequencies in dairy cows in Sweden, III. Parturient paresis. *Preventive Veterinary Medicine* 5, 87–97.

Bendixen, P., Vilson, B., Ekesbo, I. and Åstrand, D.B. (1988) Disease frequencies in dairy cows in Sweden, V. Mastitis. *Preventive Veterinary Medicine* 5, 263–274.

Bendixen, P., Vilson, B., Ekesbo, I. and Åstrand, D.B. (1989) Disease frequencies in dairy cows in Sweden, VI. Tramped teat. *Preventive Veterinary Medicine* 6, 17–25.

Bengtsson, G., Ekesbo, I. and Jacobsson, S.-O. (1965) Ett presumtivt fall av gödselgasförgiftning. *Svensk VeterinärTidning* 17, 248–254.

Bennett, B. (2001) *Storey's Guide to Raising Rabbits*, 3rd edn. Storey Communications Inc., Pownal, Vermont, 288 pp.

Berg, C. and Sanotra, G.S. (2001) Survey of the prevalence of leg weakness in Swedish broiler chickens – a pilot study. *Svensk Veterinär Tidning* 53, 5–13.

Berg, L. (1998) Foot-pad dermatitis in broilers and turkeys – prevalence, risk factors and prevention. Thesis. *Acta Universitatis Agriculturae Sueciae, Veterinaria* 36, 85 pp.

Berman, A., Folman, Y.M., Kaim, M., Mamen, Z., Herz, D., Wolfenson, A. and Graber, Y. (1985) Upper critical temperatures and forced ventilation effects for high-yielding dairy cows in a tropical climate. *Journal of Dairy Science* 68, 488–495.

Bessei, W. (1999) The behaviour of fattening turkeys – a literature review. *Archiv für Geflügelkunde* 63, 45–51.

Bessei, W. (2006) Welfare of broilers: a review. *World's Poultry Science Journal* 62, 455–466.

Bevan, M. and Butler, P.J. (1992) The effects of temperature on the oxygen consumption, heart rate and deep body temperature during diving in the tufted duck. *Journal of Experimental Biology* 163, 139–151.

Bezuidenhout, A.J. (1999) Anatomy. In: Deeming, D.C. (ed.) *The Ostrich, Biology, Production and Health*. CAB International, Wallingford, UK, pp. 13–49.

Birkhead, T.R. and Brennan, P. (2009) Elaborate vaginas and long phalli: post-copulatory sexual selection in birds. *Biologist* 56, 33–38.

Blache, D., Barrett, C.D. and Martin, G.B. (2000) Social mating system and sexual behaviour in captive emus *Dromaius novaehollandiae*. *EMU* 100, 161–168.

Blokhuis, H.J. (1984) Rest in poultry. *Applied Animal Behaviour Science* 12, 289–303.

Blokhuis, H.J. (1986) Feather pecking in poultry: its relation with groundpecking. *Applied Animal Behaviour Science* 12, 289–303.

Blokhuis, H.J. and van der Haar, J.W. (1992) Effects of pecking incentive during rearing on feather pecking of laying hens. *British Poultry Science* 33, 17–24.

Blood, D.C. and Studdert, V.P. (2000) *Saunders Comprehensive Veterinary Dictionary*, 2nd edn. W.B. Saunders, London, 1380 pp.

Bøe, K.E. (2007) Flooring preferences in dairy goats at moderate and low ambient temperature. *Applied Animal Behaviour Science* 108, 45–57.

Boivin, X. and Braastad, B.O. (1996) Effects of handling during temporary isolation after early weaning on goat kids' later response to humans. *Applied Animal Behaviour Science* 48, 61–71.

Boland, R.J. (2003) An experimental test of predator detection rates using groups of free-living emus. *Ethology* 109, 209–222.

Borkowski, J. and Pudelko, M. (2007) Forest habitat use and home-range size in radio-collared fallow deer. *Annales Zoologici Fennici* 44, 107–114.

Bradshaw, R.H. and Hall, S.J.G. (1996) Incidence of travel sickness in pigs. *Veterinary Record* 139(20), 503.

Bradshaw, R.H., Kirkden, R.D. and Broom, D.M. (2002) A review of the aetiology and pathology of leg weakness in broilers in relation to welfare. *Avian and Poultry Biology Review* 13, 45–103.

Brant, A.W. (1998) A brief history of the turkey. *World's Poultry Science Journal* 54, 365–373.

Breuer, K., Hemsworth, P.H., Barnett, J.L., Matthews, L.R. and Coleman, G.J. (2000) Behavioural response to humans and the productivity of commercial dairy cows. *Applied Animal Behaviour Science* 66, 273–288.

Breward, J. and Gentle, M.J. (1985) Neuroma formation and abnormal afferent nerve discharges after partial beak amputation in poultry. *Experientia* 41, 1132–1134.

Broad, K.D., Mimmack, M.L., Keverne, E.B. and Kendrick, K.M. (2002) Increased BDNF and trk-B mRNA expression in cortical and limbic regions following formation of a social recognition memory. *European Journal of Neuroscience* 16, 2166–2174.

Brooks, D.L. (2004) Nutrition and gastrointestinal physiology. In: Quesenberry, K.E. and Carpenter, J.W. (eds) *Ferrets, Rabbits, and Rodents, Clinical Medicine and Surgery*, 2nd edn. Saunders, St Louis, Missouri, pp. 155–160.

Broom, D.M. and Fraser, A.F. (2007) *Domestic Animal Behaviour and Welfare*. CAB International, Wallingford, UK, 438 pp.

Broom, D.M. and Kennedy, M.J. (1993) Stereotypies in horses: their relevance to welfare and causation. *Equine Veterinary Education* 5, 151–154.

Broom, D.M., Mendl, M.T. and Zanella, A.J. (1995) A comparison of the welfare of sows in different housing conditions. *Animal Science* 61, 369–385.

Buchenauer, D. von, and Fritsch, B. (1980) Zum Farbsehvermögen von Hausziegen (*Capra hircus* L.). *Zeitschrift für Tierpsychologie* 53, 225–230.

Buchholz, R. (1995) Female choice, parasite load and male ornamentation in wild turkeys. *Animal Behaviour* 50, 929–943.

Buchwalder, T. and Huber-Eicher, B. (2003) A brief report on aggressive interactions within and between groups of domestic turkeys (*Meleagris gallopavo*). *Applied Animal Behaviour Science* 84, 75–80.

Buchwalder, T. and Huber-Eicher, B. (2004) Effect of increased floor space on aggressive behaviour in male turkeys (*Meleagris gallopavo*). *Applied Animal Behaviour Science* 89, 207–214.

Buchwalder, T. and Huber-Eicher, B. (2005) A brief report on aggressive interactions within and between groups of domestic turkeys (*Meleagris gallopavo*). *Applied Animal Behaviour Science* 93, 251–258.

Budiansky, S. (1997) *The Nature of Horses: Exploring Equine Evolution, Intelligence, and Behaviour*. Free Press, New York, 290 pp.

Canali, E., Ferrante, V., Mattiello, S., Sacerdote, P., Panerai, A.E., Lebelt, D. and Zanella, A. (1996) Plasma levels of 3-endorphin and *in vitro* lymphocyte proliferation as indicators of welfare in horses in normal or restrained conditions. *Pferdeheilkunde* 12, 415–418.

Capello, V. and Gracis, M. (2005) *Rabbit and Rodent Dentistry Handbook*. Zoological Education Network, Florida, 274 pp.

Caputa, M., Folkow, L. and Blix, A.S. (1998) Rapid brain cooling in diving ducks. *American Journal of Physiology – Regulatory, Integrative and Comparative Physiology* 275, 363–371.

Carro, M.E. and Fernandez, G.J. (2008) Seasonal variation in social organisation and diurnal activity budget of the Greater Rhea (*Rhea Americana*) in the Argentinian pampas. *EMU* 108, 167–173.

Cashman, P.J., Nicol, C.J. and Jones, R.B. (1988) The effect of transportation on tonic immobility fear reactions and selected meat quality characteristics in broiler chickens. In: Unshelm, J., van Putten, G., Zeeb, K. and Ekesbo, I. (eds) *Proceedings of the International Congress on Applied Ethology in Farm Animals*. Department of Animal Hygiene, Swedish University of Agricultural Sciences, Skara, Sweden, pp. 369–373.

Castrén, H., Algers, B. and Jensen, P. (1989a) Occurrence of unsuccessful sucklings in newborn piglets in a semi-natural environment. *Applied Animal Behaviour Science* 23, 61–73.

Castrén, H., Algers, B., Jensen, P. and Saloniemi, H. (1989b) Suckling behaviour and milk consumption in newborn piglets as a response to sow grunting. *Applied Animal Behaviour Science* 24, 227–238.

Cena, K. and Monteith, J.L. (1975) Transfer processes in animal coats, II. Conduction and convection. *Proceedings Royal Society London, B* 188, 395–411.

Chamberlain, M.J., Leopold, B.D. and Burger, L.W. (2000) Characteristics of roost sites of adult wild turkey females. *Journal of Wildlife Management* 64, 1025–1032.

Cherry, P. and Morris, T. (2008) *Domestic Duck Production. Science and Practice.* CAB International, Wallingford, UK, 239 pp.

Christensen, J.W., Malmkvist, J., Nielsen, B.L. and Keeling, L.J. (2008) Effects of a calm companion on fear reactions in naive test horses. *Equine Veterinary Journal* 40, 46–50.

Christopherson, R.J. and Young, B.A. (1986) Effects of cold environments on domestic animals. In: Gudmundson, O. (ed.) *Grazing Research at Northern Latitudes.* Plenum Publications, London, pp. 247–257.

Chu, L.R., Garner, J.P. and Mench, J.A. (2004) A behavioural comparison of New Zealand white rabbits (*Oryctolagus cuniculus*) housed individually or in pairs in conventional laboratory cages. *Applied Animal Behaviour Science* 85, 121–139.

Cilliers, S.C. and Angel, C.R. (1999) Basic concepts and recent advances in digestion and nutrition. In: Deeming, D.C. (ed.) *The Ostrich, Biology, Production and Health.* CAB International, Wallingford, UK, pp. 105–128.

Clark, J.A. and McArthur, A.J. (1994) Thermal exchanges. In: Wathes, C.M. and Charles, D.R (eds) *Livestock Housing.* CAB International, Wallingford, UK, pp. 97–122.

Classen, H.L., Riddel, C., Robinson, F.E., Shand, P.J. and McCurdy, A.R. (1994) Effect of lighting treatment on the productivity, health, behaviour, and sexual maturity of heavy male turkeys. *British Poultry Science* 35, 215–225.

Clayton, G.A. (1984) Muscovy duck. In: Mason, I.L. (ed.) *Evolution of Domesticated Animals.* Longman, London, pp. 340–344.

Clegg, H.A., Buckley, P., Friend, M.A. and McGreevy, P.D. (2008) The ethological and physiological characteristics of cribbing and weaving horses. *Applied Animal Behaviour Science* 109, 68–76.

Close, W.H. (1992) Thermoregulation in piglets: environmental and metabolic consequences. Neonatal survival growth. *Occasional Publications, British Society Animal Production* 15, 25–33.

Cloutier, S., Newberry, R.C., Honda, K. and Alldredge, J.R. (2002) Cannibalistic behaviour spread by social learning. *Animal Behaviour* 63, 1153–1162.

Clutton-Brock, J. (1999) *A Natural History of Domesticated Mammals,* 2nd edn. Cambridge University Press, Cambridge, UK, 238 pp.

Clutton-Brock, T.H., Guinness, F.E. and Albon, S.D. (1982) *Red Deer: Behaviour and Ecology of Two Sexes.* Edinburgh University Press, Edinburgh, UK, 378 pp.

CoE (Council of Europe) (1992) Standing Committee of the European Convention for the Protection of Animals Kept for Farming Purposes (T-AP), Recommendation Concerning Sheep. CoE, Strasbourg, France.

CoE (Council of Europe) (1997) Standing Committee of the European Convention for the Protection of Animals Kept for Farming Purposes (T-AP), Recommendation Concerning Ratites (Ostriches, Emus and Rheas), adopted by the Standing Committee on 22 April 1997, T-AP (94) 1. CoE, Strasbourg, France.

CoE (Council of Europe) (1999a) Standing Committee of the European Convention for the Protection of Animals Kept for Farming Purposes (T-AP), Recommendation Concerning Domestic Ducks (*Anas platyrhynchos*), adopted by the Standing Committee on 22 June 1999, T-AP (94) 3. CoE, Strasbourg, France.

CoE (Council of Europe) (1999b) Standing Committee of the European Convention for the Protection of Animals Kept for Farming Purposes (T-AP), Recommendation Concerning Ducks (*Cairina Moschata*) and Hybrids of Muscovy and Domestic Ducks (*Anas platyrhynchos*), adopted by the Standing Committee on 22 June 1999, T-AP (95) 20. CoE, Strasbourg, France.

CoE (Council of Europe) (1999c) Standing Committee of the European Convention for the Protection of Animals Kept for Farming Purposes (T-AP), Recommendation Concerning Domestic Geese (*Anser anser f. domesticus, Anser cygnoides f. domesticus*), adopted by the Standing Committee on 22 June 1999, T-AP (95) 5. CoE, Strasbourg, France.

CoE (Council of Europe) (2001) Standing Committee of the European Convention for the Protection of Animals Kept for Farming Purposes (T-AP), Recommendation Concerning Turkeys (*Meleagris gallopavo* ssp.), adopted by the Standing Committee on 21 June 2001, T-AP (95) 15. CoE, Strasbourg, France.

Collier, R.J., Dahl, G.E. and VanBaale, M.J. (2006) Major advances associated with environmental effects on dairy cattle. *Journal of Dairy Science* 89, 1244–1253.

Cooper, J. and Mason, G.J. (1998) The identification of abnormal behaviour and behavioural problems in stabled horses and their relationship to horse welfare, a comparative review. *Equine Veterinary Journal* Suppl. 27, 5–9.

Cooper, J.J., McDonald, L. and Mills, D.S. (2000) The effect of increasing visual horizons on stereotypic weaving: implications for the social housing of stabled horses. *Applied Animal Behaviour Science* 69, 67–83.

Coppock, C.E., Woelfel, C.G. and Belyear, L. (1981) Forage and feed testing programs – problems and opportunities. *Journal of Dairy Science* 64, 1625–1633.

Cornelison, J.M., Hancock, A.G., Williams, A.G, Davis, L.B., Allen, N.L. and Watkins, S.E. (2005) Evaluation of nipple drinkers and the lott system for determining appropriate water flow for broilers. *Avian Advice* 7, 1–4.

Cornick-Seahorn, J.L. (1996) Anaesthesiology of ratites. In: Tully, T.N. Jr and Shane, S.M. (eds) *Ratite Management, Medicine and Surgery*. Krieger Publishing Co., Malabar, Florida, pp. 79–94.

Coureaud, G., Schaal, B., Coudert, P., Hudson, R., Rideaud, P. and Orgeur, P. (2000) Mimicking natural nursing conditions promotes early pup survival in domestic rabbits. *Ethology* 106, 207–225.

Coureaud, G., Fortun-Lamothe, L., Rodel, H.G., Monclus, R. and Schaal, B. (2008) The developing rabbit: some data related to the behaviour, feeding and sensory capacities between birth and weaning. *Productions Animales* 21, 231–237.

Cowan, D.P. (1987) Aspects of the social organisation of the European wild rabbit (*Oryctolagus cuniculus*). *Ethology* 75, 197–210.

Crawford, R.D. (1984) Goose. In: Mason, I.L. (ed.) *Evolution of Domesticated Animals*. Longman, London, pp. 345–351.

Crawford, R.D. (1990) Origin and history of poultry species. In: Crawford, R.D. (ed.) *Poultry Breeding and Genetics*. Elsevier, Amsterdam, pp. 1–41.

Cronin, G.M., Leeson, E. and Dunmore, B.W.N. (1986) The effects of farrowing nest orientation and size on sow behaviour and piglet survival. In: *Proceedings Advances in Animal Behaviour and Welfare in Australasia and Africa*. International Society for Applied Ethology, Christchurch, New Zealand, 23 June 1996, p. 16.

Crowell-Davies, S.L. (1994) Daytime rest behaviour of the Welsh pony (*Equus caballus*) mare and foal. *Applied Animal Behaviour Science* 40, 197–210.

Crowell-Davis, S. and Houpt, K.A. (1985) The ontogeny of flehmen in horses. *Animal Behaviour* 33, 739–745.

Cutler, R.S., Fahy, V.A., Spicer, E.M. and Cronin, G.M. (1999) Preweaning mortality. In: Straw, B.E., D'Allaire, S., Mengeling, W.L. and Taylor, D.J. (eds) *Diseases of Swine*, 8th edn. Iowa State University Press, Ames, Iowa, pp. 985–1001.

Cuyler, C. and Øritsland, N.A. (1999) Effect of wind on Svalbard reindeer fur insulation. *Rangifer* 22, 93–99.

Cuyler, C. and Øritsland, N.A. (2004) Rain more important than windchill for insulation loss in Svalbard reindeer fur. *Rangifer* 24, 7–14.

Cymbaluk, N.F. (1990) Cold housing effects on growth and nutrient demand of young horses. *Journal of Animal Science* 68, 3152–3162.

Cymbaluk, N.F. and Christison, G.I. (1989) Effects of diet and climate on growing horses. *Journal of Animal Science* 67, 48–59.

Dallaire, A. (1986) Rest behaviour. *Veterinary Clinics of North America, Equine Practice* 2, 591–607.

D'Allaire, S., Drolet, R. and Brodeur, D. (1996) Sow mortality associated with high ambient temperatures. *Canadian Veterinary Journal* 37, 237–239.

Dane, B. (1967) Social behaviour of mountain goat. *American Zoologist* 7, 797.

Dannenmann, K., Buchenauer, D. and Fliegner, H. (1985) The behaviour of calves under four levels of lighting. *Applied Animal Behaviour Science* 13, 243–258.

Dawkins, M.S. (1989) Time budgets in red jungle fowl as a baseline for the assessment of welfare in domestic fowl. *Applied Animal Behaviour Science* 24, 77–80.

Dawkins, M.S. (1995) How do hens view other hens? The use of lateral and binocular visual fields in social recognition. *Behaviour* 132, 591–606.

Dawkins, M.S. (2002) What are birds looking at? Head movements and eye use in chickens. *Animal Behaviour* 63, 991–998.

Day, J.E.L., de Weerd, H.A.V. and Edwards, S.A. (2008) The effect of varying lengths of straw bedding on the behaviour of growing pigs. *Applied Animal Behaviour Science* 109, 249–260.

Deeb, B. (2000) Digestive system and disorders. In: Flecknell, P. (ed.) *Manual of Rabbit Medicine and Surgery*. British Small Animal Veterinary Association, London, pp. 39–46.

Deeming, D.C. (ed.) (1999) *The Ostrich, Biology, Production and Health*. CAB International, Wallingford, UK, 358 pp.

Deeming, D.C. and Bubier, N.E. (1999) Behaviour in natural and captive environments. In: Deeming,

D.C. (ed.) *The Ostrich, Biology, Production and Health*. CAB International, Wallingford, UK, pp. 83–104.

de Jong, I.C., Wolthuis-Fillerup, M. and Reenen, C.G. van (2007) Strength of preference for dustbathing and foraging substrates in laying hens. *Applied Animal Behaviour Science* 104, 24–36.

Delgadillo, J.A., De Santiago-Miramontes, M.A. and Carrillo, E. (2007) Season of birth modifies puberty in female and male goats raised under subtropical condition. *Animal* 1, 858–864.

de Passillé, A.M. and Rushen, J. (2005) Can we measure human–animal interactions in on-farm animal welfare assessment? Some unresolved issues. *Applied Animal Behaviour Science* 92, 193–209.

Desserich, M., Ziswiler, V. and Fölsch, D.W. (1983) Die sensorische Versorgung des Hühnerschnabels. *Revue Suisse Zoologie* 90(4), 709–807.

Desserich, M., Fölsch, D.W. and Ziswiler, V. (1984) Das Schnabelkupieren bei Hühnern. *Tierärztliche Praxis* 12, 191–202.

Dewey, C.E. (1999) Diseases of the locomotor and nervous systems. In: Straw, B.E., D'Allaire, S., Mengeling, W.L. and Taylor, D.J. (eds) *Diseases of Swine*, 8th edn. Iowa State University Press, Ames, Iowa, pp. 861–882.

Dewey, C.E., Friendship, R.M. and Wilson, M.R. (1993) Clinical and postmortem examination of sows culled for lameness. *Canadian Veterinary Journal* 34, 555–556.

Dividich, J., Herpin, P., Geraert, P.A. and Vermorel, M. (1992) Cold stress. In: Phillips, C. and Piggins, D. (eds) *Farm Animals and the Environment*. CAB International, Wallingford, UK, pp. 3–25.

Donnelly, T.M. (1997) Basic anatomy, physiology and husbandry. In: Hillyer, E.W. and Quesenberry, K.E. (eds) *Ferrets, Rabbits and Rodents – Clinical Medicine and Surgery*. W.B. Saunders, London, pp. 147–159.

Donnelly, T.M. (2004) Basic anatomy, physiology, and husbandry. In: Quesenberry, K.E. and Carpenter, J.W. (eds) *Ferrets, Rabbits, and Rodents, Clinical Medicine and Surgery*, 2nd edn. Saunders, St Louis, Missouri, pp. 136–146.

Douglas-Hudson, C. and Waran, N. (1993) Do sheep find transport aversive (an assessment using direct and indirect measures)? *Applied Animal Behaviour Science* 38(1), 78–79.

Drenowatz, C., Sales, J.D., Sarasqueta, D.V. and Weilbrenner, A. (1995) In: Drenowatz, C. (ed.) *The Ratite Encyclopedia, Ostrich, Emu, Rhea*. Ratite Records, San Antonio, Texas, pp. 3–30.

Drescher, B. (1994) Housing systems for breeding rabbits with respect to animal welfare. *Applied Animal Behaviour Science* 40, 76–77.

Drescher, B. and Schlender-Bobbis, I. (1996) Pathological study of pododermatitis in breeding rabbits of a heavy broiler strain. *World Rabbit Science* 4, 143–148.

Drolet, R., D'Allaire, S. and Chagnon, M. (1992) Some observations on cardiac failure in sows. *Canadian Veterinary Journal* 33, 325–329.

Duke, G.E. (1977) Respiration in birds. In: Swenson, M.J. (ed.) *Dukes' Physiology of Domestic Animals*, 9th edn. Cornell University Press, Ithaca, New York, pp. 203–209.

Duncan, I.J.H. and Filshie, J.H. (1980) The use of radio telemetry device to measure temperature and heart rate in domestic fowl. In: Amlaner, C.J. and MacDonald, D.W. (eds) *A Handbook on Biotelemetry and Radio Tracking*. Pergamon Press, Oxford, UK, pp. 579–588.

Duncan, I.J.H. and Woodgush, D.G. (1971) Frustration and aggression in domestic fowl. *Animal Behaviour* 19, 500–504.

Duncan, I.J.H., Savory, C.J. and Wood-Gush, D.G.M. (1978) Observations on the reproductive behaviour of domestic fowl in the wild. *Applied Animal Ethology* 4, 29–42.

Duncan, I.J.H., Beatty, E.R., Hocking, P.M. and Duff, S.R.I. (1991) Assessment of pain associated with degenerative hip disorders in adult male turkeys. *Research in Veterinary Science* 50, 200–203.

Duncan, P. (1980) Time budgets of Camargue France horses, II. Time budgets of adult horses and weaned subadults. *Behaviour* 72, 26–49.

Duncan, P. (1985) Time budgets of Camargue horses, III. Environmental influences. *Behaviour* 92, 188–208.

Duncan, P., Harvey, P.H. and Wells, S.M. (1984) On lactation and associated behaviour in a natural herd of horses. *Animal Behaviour* 32, 255–263.

Dwyer, C. (2009) The behaviour of sheep and goats. In: Jensen, P. (ed.) *The Ethology of Domestic Animals – An Introductory Text*. CAB International, Wallingford, UK, pp. 161–176.

Ebert, C.S., Blanks, D.A., Patel, M.R., Coffey, C.S., Marshall, A.F. and Fitzpatrick, D.C. (2008) Behavioural sensitivity to interaural time differences in the rabbit. *Hearing Research* 235, 134–142.

Edwards, S.A. (1982) Factors affecting the time to first suckling in dairy calves. *Animal Production* 34, 330–346.

Egerton, J.R. (2000) Foot rot and other foot conditions. In: Martin, W.B. and Aitken, I.D. (eds) *Diseases of Sheep*. Blackwell Science, Ames, Iowa, pp. 243–249.

Ekesbo, I. (1963) Djuren i kölden. *Medlemsblad för Sveriges Veterinärförbund* 15, 21–23.

Ekesbo, I. (1966) Disease incidence in tied and loose housed dairy cattle and causes of this incidence variation with particular reference to the cowshed type. *Acta Agriculturae Scandinavica* Suppl. 15, 74 pp.

Ekesbo, I. (1981) Some aspects of sow health and housing. In: Sybesma, W. (ed.) *Current Topics in Veterinary Medicine and Animal Science, Vol. II. The Welfare of Pigs*. Martinas Nijhoff Publishers for the Commission of the European Communities, Zeist, The Netherlands, pp. 250–266.

Ekesbo, I. (1985) Hälsotillståndet hos kor som hålls på bete respektive bundna inomhus sommartid. *Svensk Veterinärtidning* 37, 425–429.

Ekesbo, I. (1991) *Kompendium i Husdjurshygien, del II*, Rapport 29. Institutionen för husdjurshygien, Swedish University for Agricultural Sciences (SLU), Skara, Sweden, 319 pp.

Ekesbo, I. (2003) *Kompendium i husdjurshygien*, 9th edn. Department of Animal Environment and Health, SLU, Skara, Sweden, ISBN 91-576-6289-4, 228 pp.

Ekesbo, I. (2006) The Swedish approach. In: *Animal Welfare. Council of Europe Ethical Eye Series*. Council of Europe Publishing, Strasbourg, France, pp. 185–197.

Ekesbo, I. (2009) Impact on and demands for health and welfare of range beef cattle in Scandinavian conditions. In: Aland, A. and Madec, F. (eds) *Sustainable Animal Production*. Academic Publishers, Wageningen, The Netherlands, pp. 173–188.

Ekesbo, I. and Högsved, O. (1976) Some results of health studies of calves in buildings with liquid and solid manure handling. *Proceedings 2nd Congress of the International Society for Animal Hygiene*. Department of Animal Hygiene, Croatian School of Veterinary Medicine, Zagreb, pp. 551–586.

Ekesbo, I., Jensen, P. and Lock, R. (1979) Några exempel på beteendeförändringar vid fixering av sinsuggor (summary in English). *Svensk Veterinärtidning* 31, 315–319.

Ekkel, E.D., Vandoorn, C.E.A., Hessing, M.J.C. and Tielen, J.M. (1995) The specific-stress-free housing system has positive effects on productivity, health, and welfare of pigs. *Journal of Animal Science* 73, 1544–1551.

Ekstrand, C. and Algers, B. (1997) Rearing conditions and foot-pad dermatitis in Swedish turkey poults. *Acta Veterinaria Scandinavica* 38, 167–174.

Erickson, H.H. and Detweiler, D.K. (2004) Regulation of the heart. In: Reece, W.O. (ed.) *Dukes' Physiology of Domestic Animals*, 12th edn. Cornell University Press, Ithaca, New York, pp. 261–274.

Evans, C.S. and Evans, L. (1999) Chicken food calls are functionally referential. *Animal Behaviour* 58, 307–319.

Evans, C.S., Evans, S. and Marler, P. (1993) On the meaning of alarm calls: functional reference in an avian vocal system. *Animal Behaviour* 46, 23–38.

Ewbank, R. (2010) Cattle. In: Hubrecht, R. and Kirkwood, J. (eds) *The Care and Management of Laboratory and Other Research Animals*. Wiley-Blackwell, Oxford, UK, pp. 495–509.

Ewbank, R. and Meese, G.B. (1971) Aggressive behaviour in groups of pigs on removal and return of individuals. *Animal Production* 13, 685–693.

FAWC (Farm Animals Welfare Council) (1995) *Report on the Welfare of Turkeys*. MAFF Publications, Tolworth, Surbiton, UK, 42 pp.

Fedde, M.R. (1993) Respiration in birds. In: Swenson, M.J. and Reece, W.O. (eds) *Dukes' Physiology of Domestic Animals*, 11th edn. Cornell University Press, Ithaca, New York, pp. 294–302.

Feh, C. (2005) Relationships and communication in socially natural horse herds. In: Mills, S.D. and McDonnell, S. (eds) *The Domestic Horse. The Evolution, Development and Management of its Behaviour*. Cambridge University Press, Cambridge, UK, pp. 83–93.

Feh, C. and de Mazières, J. (1993) Grooming at a preferred site reduces heart rate in horses. *Animal Behaviour* 46, 1191–1194.

Ferket, P.R. (2003) Growth of toms improves substantially. *WATT Poultry USA* 4, 38–48.

Fernie, K.J. and Reynolds, S.J. (2005) The effects of electromagnetic fields from power lines on avian reproductive biology and physiology. A review. *Journal of Toxicology and Environmental Health – Part B – Critical Reviews* 8, 127–140.

Fiedler, H.H. and König, K. (2006) Assessment of beak trimming in day-old turkey chicks by infrared irradiation in view of animal welfare. *Archiv für Geflügelkunde* 70, 241–249.

Fischer, G. (1975) The behaviour of chickens. In: Hafez, E.S.E. (ed.) *The Behaviour of Domestic Animals*, 3rd edn. Baillière Tindall, London, pp. 454–489.

Fitzpatrick, J., Scott, M. and Nolan, A. (2006) Assessment of pain and welfare in sheep. *Small Ruminant Research* 62, 55–61.

Fletcher, T.J. (1986) Reproduction. In: Alexander, T.L. (ed.) *Management and Diseases of Deer*. Veterinary Deer Society, BVA, London, pp. 173–177.

Fogarty, N.M., Ingham, V.M., Gilmour, A.R., Afolayan, R.A., Cummins, L.J., Edwards, J.E.H. and Gaunt, G.M. (2007) Genetic evaluation of crossbred lamb production. 5. Age of puberty and lambing performance of yearling crossbred ewes. *Australian Journal of Agricultural Research* 58, 928–934.

Fowler, M.E. (1996) Clinical anatomy of ratites. In: Tully, T.N. Jr and Shane, S.M. (eds) *Ratite Management, Medicine and Surgery*. Krieger Publishing Co., Malabar, Florida, pp. 1–10.

Frafjord, K. (2004) Vigilance in pre-nesting male geese: mate guarding or predator detection. *Ornis Norvegica* 27, 48–58.

Fraser, A.F. (1992) *The Behaviour of the Horse*. CAB International, Wallingford, UK, 288 pp.

Fraser, A.F. and Broom, D.M. (1990) *Farm Animal Behaviour and Welfare*, 3rd edn. CAB International, Wallingford, UK, 437 pp.

Fraser, D. (1973) The nursing and suckling behaviour of pigs, I. The importance of stimulation of the anterior teats. *British Veterinary Journal* 133, 126–133.

Fraser, D. (1980) A review of the behavioural mechanism of milk ejection of the domestic pig. *Applied Animal Ethology* 6, 247–255.

Fraser, M.A. and Girling, S.J. (2009) *Rabbit Medicine and Surgery for Veterinary Nurses*. Wiley-Blackwell, Oxford, UK, 231 pp.

Freeman, B.M. (1984) Transportation of poultry. *World's Poultry Science Journal* 40, 19–30.

Freire, R., Munro, U.H., Rogers, L.J., Wiltschko, R. and Wiltschko, W. (2005) Chickens orient using a magnetic compass. *Current Biology* 15(16), R620–R621.

Freitas, V.J.F., Lopes-Junior, E.S., Rondina, D., Salmito-Vanderley, C.S.B., Salles, H.O., Simplicio, A.A., Baril, G. and Saumande, J. (2004) Puberty in Anglo-Nubian and Saanen female kids raised in the semi-arid area of north-eastern Brazil. *Small Ruminant Research* 53, 167–172.

Fuller, J.M. (1928) Some physical and physiological activities of dairy cows under conditions of modern herd management. *New Hampshire Agricultural Experimental Station Technical Bulletin* 35, 2–30.

Funk, G.D., Steeves, J.I.D. and Milsom, W.K. (1992) Coordination of wingbeat and respiration in birds, II. 'Fictive' flight. *Journal of Applied Physiology* 73, 1025–1033.

Geigl, E.M. (2008) Palaeogenetics of cattle domestication: methodological challenges for the study of fossil bones preserved in the domestication centre in Southwest Asia. *Comptes Rendus Palevol* 7, 99–112.

Geist, V. (1971) *Mountain Sheep. A Study in Behaviour and Evolution*. University of Chicago Press, Chicago, Illinois, 383 pp.

Gentle, M.J. (1986) Neuroma formation following partial beak amputation in the chicken. *Research in Veterinary Science* 41, 383–385.

Gentle, M.J. and Hunter, L.N. (1991) Physiological and behavioural responses associated with feather removal in *Gallus gallus* var. *domesticus*. *Research in Veterinary Science* 50, 95–101.

Gentle, M. and Wilson, S. (2004) Pain and the laying hen. In: Perry, G.C. (ed.) *Welfare of the Laying Hen. Poultry Science Series* 27. CAB International, Wallingford, UK, pp. 165–175.

Gentle, M.J., Thorp, B.H. and Hughes, B.O. (1995) Anatomical consequences of partial beak amputation (beak trimming) in turkeys. *Research in Veterinary Science* 58, 158–162.

Getty, R. (1975) *Sissons and Grossman's The Anatomy of the Domestic Animals*, Volume 2, 5th edn. W.B. Saunders, Philadelphia, Pennsylvania, 2095 pp.

Geverink, N.A., Bradshaw, R.H., Lambooy, E. and Broom, D.M. (1996) Handling of slaughter pigs in lairage behavioural and physiological effects. In: Schütte, A. (ed.) *Proceedings 122, EU Seminar: New Information on Welfare and Meat Quality of Pigs as Related to Handling, Transport and Lairage Conditions*. Mariensee, 29–30 June 1995. Sonderheft 166, FAL, Völkenrode, pp. 207–212.

Geverink, N.A., Bradshaw, R.H., Lambooij, E., Wiegant, V.M. and Broom, D.M. (1998) Effects of simulated lairage conditions on the physiology and behaviour of pigs. *Veterinary Record* 143, 241–244.

Gibb, J.A. (1993) Sociality, time and space in a sparse population of rabbits (*Oryctolagus cuniculus*). *Journal of Zoology* 229, 581–607.

Gilbert, R.P. and Bailey, D.R. (1991) Hair coat characteristics and postweaning growth of Hereford and Angus cattle. *Journal of Animal Science* 69, 498–506.

Giuffra, E., Kijas, J.M.H., Amarger, V., Carlborg, Ö., Jeon, J.-T. and Andersson, L. (2000) The origin of the domestic pig: independent domestication and subsequent introgression. *Genetics* 154, 1785–1791.

Glatz, P.C. and Miao, Z.H. (2008) Husbandry of ratites and potential welfare issues: a review. *Australian Journal of Experimental Agriculture* 48, 1257–1265.

Goldschmidt-Rothschild, B. von and Tschanz, B (1978) Soziale Organisation und Verhalten einer Jungtierherde beim Camargue-Pferde. *Zeitschrift für Tierpsychologie* 46, 372–400.

Gonyou, H.W., Christopherson, R.J. and Young, B.A. (1979) Effects of cold temperature and winter conditions on some aspects of behaviour of feedlot cattle. *Applied Animal Ethology* 5, 113–124.

Gonzalez-Redondo, P. and Zamora-Lozano, M. (2008) Neonatal cannibalism in cage-bred wild rabbits (*Oryctolagus cuniculus*). *Archivos de Medicina Veterinaria* 40, 281–287.

Goossens, X., Sobry, L., Ödberg, F., Tuyttens, F., Maes, D., De Smet, S., Nevens, F., Opsomer, G., Lommelen, F. and Geers, R. (2008) A population-based on-farm evaluation protocol for comparing the welfare of pigs between farms. *Animal Welfare* 17, 35–41.

Grandin, T. (1991) Handling problems caused by excitable pigs. In: *Proceedings of the 37th International Congress of Meat Science and Technology*, Vol 1. Federal Centre for Meat Research, Kulmbach, Germany.

Grandin, T. (2000) Behavioural principles of handling cattle and other grazing animals under extensive conditions. In: Grandin, T. (ed.) *Livestock Handling and Transport*, 2nd edn. CAB International, Wallingford, UK, pp. 63–85.

Grandin, T., Curtis, S.E. and Taylor, I.A. (1987) Toys, mingling and driving reduce excitability in pigs. *Journal of Animal Science* 65 (Suppl. 1), 230–231.

Graves, H.B. (1984) Behaviour and ecology of wild and feral swine. *Journal of Animal Science* 38, 482–492.

Gundlach, H. (1968) Brutfürsorge, Brutpflege, Verhaltensontogenese und Tagesrhythmik beim Europäischen Wildschwein. *Zeitschrift für Tierpsychologie* 25, 955–995.

Gunnarsson, S. (2000) Laying hens in loose housing systems. Clinical, ethological and epidemiological aspects. Thesis. *Acta Universitatis Agriculturae Sueciae, Veterinaria* 73, 108 pp.

Gunnarsson, S., Keeling, L.J. and Svedberg, J. (1999) Effect of rearing factors on the prevalence of floor eggs, cloacal cannibalism and feather pecking in commercial flocks of loose housed laying hens. *British Poultry Science* 40, 12–18.

Gunnarsson, S., Yngvesson, J., Keeling, L.J. and Forkman, B. (2000) Rearing without early access to perches impairs the spatial skills of laying hens. *Applied Animal Behaviour Science* 67, 217–228.

Gustafsson, M., Jensen, P., de Jonge, F.H., Illmann, G. and Spinka, M. (1999) Maternal behaviour of domestic sows and crosses between domestic sows and wild boar. *Applied Animal Behaviour Science*, 65, 29–42.

Hafez, E.S.E. (1962) *The Behaviour of Domestic Animals*. Baillière Tindall and Cox, London, 619 pp.

Hafez, E.S.E. (ed.) (1975) *The Behaviour of Domestic Animals*. Baillière Tindall, London, 619 pp.

Hafez, E.S.E. and Bouissou, M.F. (1975) The behaviour of cattle. In: Hafez, E.S.E. (ed.) *The Behaviour of Domestic Animals*. Baillière Tindall, London, pp. 203–245.

Hagen, K. and Broom, D.M. (2003) Cattle discriminate between individual familiar members in a learning experiment. *Applied Animal Behaviour Science* 82, 13–28.

Hahn, G.L. (1999) Dynamic responses of cattle to thermal heat loads. *Journal of Animal Science* 77, 10–20.

Hale, E.B., Schleidt, W.M. and Schein, M.W. (1969) The behaviour of turkeys. In: Hafez, E.S.E. (ed.) *The Behaviour of Domestic Animals* 2nd edn. Baillière Tindall and Cox, London, pp. 554–592.

Hall, J.J. (2002) Behaviour of cattle. In: Jensen, P. (ed.) *The Ethology of Domestic Animals, An Introductory Text*. CAB International, Wallingford, UK, pp. 131–143.

Hall, S.J.G. (2005) The horse in humane society. In: Mills, S.D. and McDonnell, S. (eds) *The Domestic Horse. The Evolution, Development and Management of its Behaviour*. Cambridge University Press, Cambridge, UK, pp. 23–32.

Hall, S.J.G., Kirkpatrick, S.M. and Broom, D.M. (1998) Behavioural and physiological responses of sheep of different breeds to supplementary feeding, social mixing and taming, in the context of transport. *Animal Science* 67, 475–483.

Hallam, M.G. (1992) *The Topaz Introduction to Practical Ostrich Farming.* Superior Print and Packaging, Harare, 149 pp.

Hamada, T. (1971) Estimation of lower critical temperatures for dry and lactating dairy cows. *Journal of Dairy Science* 54, 1704–1705.

Hamilton, W.J. (1986) The background. In: Alexander, T.L. (ed.) *Management and Diseases of Deer.* Veterinary Deer Society, BVA, London, pp. 5–10.

Hammarberg, K. (2010) *Hälsovård och sjukdomar hos get.* Natur och Kultur, Stockholm.

Hancock, J. (1950) Grazing habits of dairy cows in New Zealand. *Empire Journal of Experimental Agriculture* 18, 249–263.

Hancock, J. (1953) Grazing behaviour of cattle. *Animal Breeding Abstracts* 21, 1–13.

Hancock, J. (1954) Uniformity trials: grazing behaviour. In: *Studies of Monozygotic Cattle Twins.* Publication No. 63. New Zealand Department of Agriculture Animal Research Division, Hamilton, New Zealand, pp. 85–122.

Hansen, L.T. and Berthelsen, H. (2000) The effect of environmental enrichment on the behaviour of caged rabbits (*Oryctolagus cuniculus*). *Applied Animal Behaviour Science* 68, 163–178.

Hansen, R.S. (1970) Hysteria in mature hens in cages. *Poultry Science* 49, 1392–1393.

Harcourt-Brown, F. (2002) *Textbook of Rabbit Medicine.* Butterworth-Heineman, Oxford, 410 pp.

Hare, E., Norman, H.D. and Wright, J.R. (2006) Trends in calving ages and calving intervals for dairy cattle breeds in the United States. *Journal of Dairy Science* 89, 365–370.

Hart, B.L. and Pryor, P.A. (2004) Developmental and hair-coat determinants of grooming behaviour in goats and sheep. *Animal Behaviour* 67, 11–19.

Hartigan, P.J. (2004) Stress and the pathogenesis of diseases. In: Andrews, A.H., Blowey, R.W., Boyd, H. and Eddy, R.G. (eds) *Bovine Medicine: Diseases and Husbandry of Cattle,* 2nd edn. Wiley-Blackwell, Oxford, UK, pp. 1133–1148.

Hartsock, T.G. and Graves, H.B. (1976) Neonatal behaviour and nutrition-related mortality in domestic swine. *Journal of Animal Science* 42, 235–241.

Harun, M.A.S., Veeneklaas, R.J., Kampen, M. van and Mabasso, M. (1998) Breeding biology of Muscovy duck *Cairina moschata* in natural incubation. The effect of nesting behaviour on hatchability. *Poultry Science* 77, 1280–1286.

Harwood, D. (2006) *Goat Health and Welfare. A Veterinary Guide.* Crowood Press, Marlborough, UK, 175 pp.

Hausberger, M., Gautier, E., Muller, C. and Jego, P. (2007) Lower learning abilities in stereotypic horses. *Applied Animal Behaviour Science* 107, 299–306.

Hausberger, M., Roche, H., Henry, S. and Visser, E.K. (2008) A review of the human–horse relationship. *Applied Animal Behaviour Science* 109, 1–24.

Hay, M., Vulin, A., Génin, S., Sales, P. and Prunier, A. (2003) Assessment of pain induced by castration in piglets; behavioural and physiological responses over the subsequent 5 days. *Applied Animal Behavioural Science* 82, 210–218.

Hay, M., Rue, J., Sansac, C., Brunel, G. and Prunier, A. (2004) Long-term detrimental effects of tooth clipping or grinding in piglets; a histological approach. *Animal Welfare* 13, 27–32.

Heffner, R.S. and Heffner, H.E. (1983) Hearing in large mammals: horses (*Equus caballus*) and cattle (*Bos taurus*). *Behavioural Neuroscience* 97, 299–309.

Heffner, R.S. and Heffner, H.E. (1990) Hearing in domestic pigs (*Sus scrofa*) and goats (*Capra hircus*). *Hearing Research* 48, 231–240.

Heffner, R.S. and Heffner, H.E. (1992a) Hearing in large mammals: sound localization acuity in cattle (*Bos taurus*) and goats (*Capra hircus*). *Journal of Comparative Psychology* 106, 107–113.

Heffner, R.S. and Heffner, H.E. (1992b) Auditory perception. In: Phillips, C.J.C. and Piggins, D. (eds) *Farm Animals and the Environment.* CAB International, Wallingford, UK, pp. 159–184.

Heffner, R.S. and Masterton, R.B. (1990) Sound localisation in mammals: brain stem mechanisms. In: Berkely, M.A. and Stebbins, W.C. (eds) *Comparative Perception. Vol 1, Basic Mechanism.* John Wiley and Sons, Chichester, UK, pp. 285–314.

Held, S., Baumgartner, J., KilBride, A., Byrne, R.W. and Mendl, M. (2005) Foraging behaviour in domestic pigs (*Sus scrofa*): remembering and prioritizing food sites of different value. *Animal Cognition* 8, 114–121.

Hemsworth, P.H. (2000) Behavioural principles of pig handling. In: Grandin, T. (ed.) *Livestock Handling and Transport*, 2nd edn. CAB International, Wallingford, UK, pp. 255–274.

Hemsworth, P.H. (2003) Human–animal interactions in livestock production. *Applied Animal Behaviour Science* 81, 185–198.

Hemsworth, P.H. (2004) Human–animal interactions. *Welfare of the Laying Hen* 27, 329–343.

Hemsworth, P.H. (2007) Behavioural principles of pig handling. In: Grandin, T. (ed.) *Livestock Handling and Transport*, 3rd edn. CAB International, Wallingford, UK, pp. 214–227.

Hemsworth, P.H. and Barnett, J.L. (1987) Human–animal interactions. *Veterinary Clinics of North America – Food Animal Practice* 3, 339–356.

Hemsworth, P.H. and Coleman, G.J. (1998) *Human–Livestock Interactions: The Stockperson and the Productivity and Welfare of Intensively Farmed Animals*. CAB International, Wallingford, UK, 152 pp.

Hemsworth, P.H., Brand, A. and Willems, P. (1981) The behavioural-response of sows to the presence of human beings and its relation to productivity. *Livestock Production Science* 8, 67–74.

Hemsworth, P.H., Barnett, J.L., Hansen, C. and Gonyou, H.W. (1986a) The influence of early contact with humans on subsequent behavioural-response of pigs to humans. *Applied Animal Behaviour Science* 15, 55–63.

Hemsworth, P.H., Barnett, J.L. and Hansen, C. (1986b) The influence of handling by humans on the behaviour, reproduction and corticosteroids of male and female pigs. *Applied Animal Behaviour Science* 15, 303–314.

Hemsworth, P.H., Barnett, J.L., Beviridge, L. and Matthews, L.R. (1995) The welfare of extensively managed dairy cattle: a review. *Applied Animal Behaviour Science* 42, 161–182.

Hemsworth, P.H., Coleman, G.J., Barnett, J.L. and Borg, S. (2000) Relationships between human–animal interactions and productivity of commercial dairy cows. *Journal of Animal Science* 78, 2821–2831.

Herbold, H., Suchentrunk, F., Wagner, S. and Willing, R. (1992) The influence of anthropogenic disturbances on the heart frequency of red deer (*Cervus elaphus*) and roe deer (*Capreolus capreolus*). *Zeitschrift für Jagdwissenschaft* 38, 145–159.

Hicks-Alldredge, K.D. (1996) Reproduction. In: Tully, T.N. Jr and Shane, S.M. (eds) *Ratite Management, Medicine and Surgery*. Krieger Publishing Co., Malabar, Florida, pp. 47–57.

Hillman, P.E., Gebremedhin, K.G. and Warner, R.G. (1989) Heat transfer through wetted fur of neonatal calves. *American Society Agriculture Engineering* No. 89-4515, 8 pp.

Hinde, R.A. (1970) *Behaviour: A Synthesis of Ethology and Comparative Psychology*. McGraw-Hill Kogakusha, Japan. Cited by Hemsworth in Behavioural principles of pig handling. In: Grandin, T. (ed.) *Livestock Handling and Transport* (2000). CAB International, Wallingford, UK, 449 pp.

Hjortberg, G.F. (1776) *Swenska Boskaps-Afwelen till sin rätta Wård och Skötsel uti Helsos och Sjukdoms Tid igenom Pröfwade Medel och Nyttige Råd till Landtmanna tjenst ihopsamlade af G.F.H.* Immanuel Smitt Tryckeri, Göteborg, 575 pp.

Hocking, P.M. (1999) Assessment of pain during locomotion and the welfare of adult male turkeys with destructive cartilage loss of the hip joint. *British Poultry Science* 40, 30–34.

Hoffmeister, H. (1979) Untersuchungen über die Reaktion von Reh- und Damwild auf verschiedene Umwelteinflüsse unter Einsatz der telemetrischen Erfassung von Herzfrequenz und EKG. Thesis. Tierärztliche Hochschule, Hanover, Germany, 84 pp.

Holmgren, N. and Lundeheim, N. (2010) Bogsår hos suggor i moderna grisningsboxar [Shoulder lesions in sows in modern farrowing pens]. *Svensk Veterinärtidning* 62, 19–23.

Horrell, R.I., A'Ness, P.J., Edwards, S.A. and Eddison, J.C. (2001) The use of nose-rings in pigs: consequences for rooting, other functional activities, and welfare. *Animal Welfare* 10, 3–22.

Houpt, K.A. (1982) Misbehaviour of horses: trailer problems. *Equine Practice* 4, 12–16.

Houpt, K.A. (1998) *Domestic Animal Behaviour for Veterinarians and Animal Scientists*, 3rd edn. Iowa State University Press, Ames, Iowa, 495 pp.

Houpt, K.A. (2004) Behavioural physiology. In: Reece, W.O. (ed.) *Dukes' Physiology of Domestic Animals*, 12th edn. Cornell University Press, Ithaca, New York, pp. 952–961.

Houpt, K.A. (2005) Maintenance behaviour. In: Mills, S.D. and McDonnell, S. (eds) *The Domestic Horse. The Evolution, Development and Management of its Behaviour*. Cambridge University Press, Cambridge, UK, pp. 94–109.

Houpt, K.A., Law, K. and Martinisi, V. (1978) Dominance hierarchies in horses. *Applied Animal Ethology* 4, 273–283.

Hoy, S., Ruis, M. and Szendro, Z. (2006) Housing of rabbits – results of a European research network. *Archiv für Geflügelkunde* 70, 223–227.

Huber, H.U., Fölsch, D.W. and Stähli, U. (1985) Influence of various nesting materials on nest site selection of the domestic hen. *British Poultry Science* 26, 367–373.

Huber, H.-U.E. (1987) Untersuchungen zum Einfluss von Tages- und Kunstlicht auf das Verhalten von Hühnern. Dissertation. Eidgenössischen Technischen Hochschule, Zurich, Switzerland, 142 pp.

Huchzermeyer, F.W. (1994) *Ostrich Diseases*. ARC, Ondersteport Vet. Inst., Gutenberg Boo Printers, Pretoria, South Africa, 121 pp.

Huerkamp, M.J. (2003) The rabbit. In: Ballard, B. and Cheek, R. (eds) *Exotic Animal Medicine for the Veterinary Technician*. Iowa State Press, Ames, Iowa, pp. 191–226.

Hughes, B.O. and Black, A.J. (1974) The effect of environmental factors on activity, selected behaviour patterns and 'fear' of fowls in cages and pens. *British Poultry Science* 15, 375–380.

Hughes, B.O. and Elson, H.A. (1977) Use of perches by broilers in floor pens. *British Poultry Science* 18, 715–722.

Hultgren, J. (1990a) Small electric currents affecting farm animals and man: a review with special reference to stray voltage, I. Electric properties of the body and the problem of stray voltage. *Veterinary Research Communications* 14, 287–298.

Hultgren, J. (1990b) Small electric currents affecting farm animals and man: a review with special reference to stray voltage, II. Physiological effects and the concept of stress. *Veterinary Research Communications* 14, 299–308.

Hutson, G.D. (1993) Behavioural principles of sheep handling. In: Grandin, T. (ed.) *Livestock Handling and Transport*. CAB International, Wallingford, UK, pp. 127–146.

Hutson, G.D. (2007) Behavioural principles of sheep handling. In: Grandin, T. (ed.) *Livestock Handling and Transport*, 3rd edn. CAB International, Wallingford, UK, pp. 155–174.

Iggo, A. (1984) cited by Phillips, C.J.C. (2002) *Cattle Behaviour and Welfare*, 2nd edn. Blackwell, Oxford, UK, 264 pp.

Ingram, D.L. (1965) Evaporative cooling in the pig. *Nature* 207, 415–416.

Ingram, D.L. (1972) Meteorological effects on pigs. In: Tromp, S.W. (ed.) *Progress in Human Meteorology*. Swets and Zeitlinger, Amsterdam, pp. 22–29.

Ingram, D.L. and Legge, K.F. (1970) The thermoregulatory behaviour of young pigs in a natural environment. *Physiology and Behaviour* 5, 981–990.

Ingram, D.L. and Legge, K.F. (1972) The influence of deep body temperatures and skin temperatures on respiratory frequency in pigs. *Journal of Physiology* 220, 283–286.

Innes, L. and McBride, S. (2008) Negative versus positive reinforcement: an evaluation of training strategies for rehabilitated horses. *Applied Animal Behaviour Science* 112, 357–368.

Ivarsson, E., Mattson, B., Lundeheim, N. and Holmgren, N. (2009) Bogsår – förekomst och riskfaktorer. *Pig Rapport Nr. 42.* Svenska Pig, Skara, Sweden, pp. 1–8.

Jackson, K.M.A. and Hackett, D. (2007) A note: the effects of human handling on heart girth, behaviour and milk quality in dairy goats. *Applied Animal Behaviour Science* 108, 332–336.

Jacobs, G.H., Deegen, J.F. and Neitz, J. (1998) Photopigment basis for dichromatic colour vision in cows, goats and sheep. *Visual Neuroscience* 15, 581–584.

Jansen, T., Forster, P., Levine, M.A., Oelke, H., Hurles, M., Renfrew, C., Weber, J. and Olek, K. (2002) Mitochondrial DNA and the origins of the domestic horse. *Proceedings of the National Academy of Sciences of the United States of America* (PNAS) 99, 10905–10910.

Jayakody, S., Sibbald, A.M., Gordon, I.J. and Lambin, X. (2008) Red deer (*Cervus elaphus*) vigilance behaviour differs with habitat and type of human disturbance. *Wildlife Biology* 14, 81–91.

Jensen, M.B., Pedersen, L.J. and Munksgaard, L. (2005) The effect of reward duration on demand functions for rest in dairy heifers and lying requirements as measured by demand functions. *Applied Animal Behaviour Science* 90, 207–217.

Jensen, P. (1984) Effects of confinement on social interaction patterns in dry sows. *Applied Animal Behaviour Science* 12, 93–101.

Jensen, P. (1986) Observations on the maternal behaviour of free-ranging domestic pigs. *Applied Animal Behaviour Science* 16, 131–142.

Jensen, P. (1988a) Maternal behaviour and mother–young interactions during lactation in

---

free-ranging domestic pigs. *Applied Animal Behaviour Science* 20, 297–308.

Jensen, P. (1988b) Diurnal rhythm of bar-biting in relation to other behaviour in pregnant sows. *Applied Animal Behaviour Science* 21, 337–346.

Jensen, P. (1988c) *Maternal Behaviour of Free-ranging Domestic Pigs, I. Results of a Three-year Study*. Report 22, Swedish University of Agriculture Science, Department of Animal Hygiene, Skara, Sweden, 56 pp.

Jensen, P. (1989) Nest site choice and nest building of free-ranging Domestic Pigs due to farrow. *Applied Animal Behaviour Science* 22, 13–21.

Jensen, P. (1993) Nest building in domestic sows: the role of external stimuli. *Animal Behaviour* 45, 351–358.

Jensen, P. (2002) Behaviour of pigs. In: Jensen, P. (ed.) *The Ethology of Domestic Animals, An Introductory Text*. CAB International, Wallingford, UK, pp. 159–172.

Jensen, P. and Algers, B. (1982) An ethogram of piglet vocalizations during suckling. *Applied Animal Ethology* 11, 237–248.

Jensen, P. and Ekesbo, I. (1986) Some results from a study of maternal behaviour in free-ranging sows. *Proceedings, 37th Annual Meeting of the European Association for Animal Production*, Volume I, Budapest, pp. 576–577.

Jensen, P. and Recén, B. (1989) When to wean – observations from free-ranging domestic pigs. *Applied Animal Behaviour Science* 23, 49–60.

Jensen, P. and Redbo, I. (1987) Behaviour during nest leaving in free-ranging domestic pig. *Applied Animal Behaviour Science* 18, 355–362.

Jensen, P. and Wood-Gush, D.G.M. (1984) Social interactions in a group of free-ranging sows. *Applied Animal Behaviour Science* 12, 327–337.

Jensen, P., Recén, B. and Ekesbo, I. (1988) Preference of loose housed dairy cows for two different cubicle floor coverings. *Swedish Journal of Agricultural Research* 18, 141–146.

Jensen, P., Stangel, G. and Algers, B. (1991) Nursing and suckling behaviour of semi-naturally kept pigs during the first 10 days postpartum. *Applied Animal Behaviour Science* 31, 195–209.

Jerina, K., Dajcman, M. and Adamic, M. (2008) Red deer (*Cervus elaphus*) bark stripping on spruce with regard to spatial distribution of supplemental feeding places. *Zbornik Gozdarstva in Lesarstva* 86, 33–43.

Jiang, M., Gebremedhin, K.G. and Albright, L.D. (2005) Simulation of skin temperature and sensible and latent heat losses through fur layers. *American Society Agricultural Engineering* 48, 767–775.

Johansson, E. (1989) *Hjortvilt i hägn*. Norra Skåne Offset, Tyringe, Sweden, 123 pp.

Johansson, G. and Jönsson, L. (1977) Myocardial cell damage in the porcine stress syndrome. *Journal of Comparative Pathology* 87, 67–74.

Johansson, G. and Jönsson, L. (1979) Porcine stress syndrome. *Acta Agriculturae Scandinavica* Suppl. 21, 327–329.

Johnson, S.B., Hamm, R.J. and Leahey, T.H. (1986) Observational learning in *Gallus gallus domesticus* with and without a conspecific model. *Bulletin of the Psychonomic Society* 24, 237–239.

Jones, E.K.M., Wathes, C.A. and Webster, A.J.F. (2005) Avoidance of atmospheric ammonia by domestic fowl and the effect of early experience. *Applied Animal Behaviour Science* 90, 293–308.

Jones, R.B. (1986) The tonic immobility reaction of the domestic fowl: a review. *World's Poultry Science Journal* 42, 82–96.

Jones, R.B. and Faure, J.M. (1982) Open field behaviour of male and female domestic chicks as a function of housing conditions, test situations and novelty. *Biology of Behaviour* 7, 17–25.

Jones, R.B. and Gentle, M.J. (1985) Olfaction and behavioural modification in domestic chicks (*Gallus domesticus*). *Physiology and Behaviour* 34, 917–924.

Jones, R.B. and Roper, T.J. (1997) Olfaction in the domestic fowl; a critical review. *Physiology and Behaviour* 62, 1009–1018.

Jones, T.A., Waitt, C.D. and Dawkins, M.S. (2009) Water off a duck's back: showers and troughs match ponds for improving duck welfare. *Applied Animal Behaviour Science* 116, 52–57.

Jönsson, L. and Johansson, G. (1979) Cardiac muscle cell damage of the porcine stress syndrome and experimental restraint stress. *Acta Agriculturae Scandinavica* Suppl. 21, 330–338.

Jordbruksverket (The Swedish Board of Agriculture) (2004) *Jordbruksstatistisk årsbok [Statistical Year-book of Agriculture]*. Jordbruksverket, Jönköping, Sweden.

Jordbruksverket (The Swedish Board of Agriculture) (2007) *Jordbruksstatistisk årsbok [Statistical Year-book of Agriculture]*. Jordbruksverket, Jönköping, Sweden.

Jørgensen, G.H.M., Andersen, I.L. and Bøe, K.E. (2007) Feed intake and social interactions in dairy goats – the effects of feeding, space and type of roughage. *Applied Animal Behaviour Science* 107, 239–251.

Kadzere, C.T., Murphy, M.R., Silanikove, N. and Maltz, E. (2002) Heat stress in lactating dairy cows: a review. *Livestock Production Science* 77, 59–91.

Kalmbach, E. (2006) Why do goose parents adopt unrelated goslings? A review of hypotheses and empirical evidence, and new research questions. *Ibis The International Journal of Avian Science* 148, 66–78.

Kalmbach, E., Aa van der, P. and Komdeur, J. (2005) Experimental evidence for gosling choice and age-dependency of adoption in greylag geese. *Behaviour* 142, 1515–1533.

Kaminski, G., Brandt, S., Baubet, E. and Baudoin, C. (2005a) Life-history patterns in female wild boars (*Sus scrofa*): mother–daughter postweaning associations. *Canadian Journal of Zoology* 83, 474–480.

Kaminski, J., Riedel, J., Call, J. and Tomasello, M. (2005b) Domestic goats, *Capra hircus*, follow gaze direction and use social cues in an object choice task. *Animal Behaviour* 69, 11–18.

Kare, M.R., Pond, W.C. and Campbell, J. (1965) Observations on the taste reactions in pigs. *Animal Behaviour* 13, 265–269.

Kassim, H. and Sykes, A. (1982) The respiratory responses of the fowl to hot climates. *Journal of Experimental Biology* 97, 301–309.

Keeling, L. (2002) Behaviour of fowl and other domesticated birds. In: Jensen, P. (ed.) *The Ethology of Domestic Animals, An Introductory Text*. CAB International, Wallingford, UK, pp. 101–117.

Kelley, K.W. (1985) Immunological consequences of changing environmental stimuli. In: Moberg, G.P. (ed.) *Animal Stress*. American Physiology Society, Bethesda, Maryland, pp. 193–223.

Kendrick, K.M., Levy, F. and Keverne, E.B. (1992) Changes in the sensory processing of olfactory signals induced by birth in sheep. *Science* 256, 833–836.

Kennedy, A.D., Bergen, R.D., Christopherson, R.J., Glover, N.D. and Small, J.A. (2005) Effect of once daily 5-h or 10-h cold-exposures on body temperature and resting heat production of beef cattle. *Canadian Journal of Animal Science* 85, 177–183.

Kent, J.P. (1992) Maternal aggression and inter-individual distance in the broody hen. *Behavioural Processes* 27, 37–44.

Kent, J.P. and Murphy, K.J. (2003) Synchronized egg laying in flocks of domestic geese (*Anser anser*). *Applied Animal Behaviour Science* 82, 219–228.

Kersten, A.M.P. (1995) Nesting behaviour and reproduction of individually caged and group housed rabbits. In: Rutter, S.M., Rushen, J., Randle, H.D. and Eddison, J.C. (eds) *Proceedings 29th International ISAE Congress*. Universities Federation for Animal Welfare, Exeter, pp. 189–190.

Kersten, A.M., Meijsser, F.M. and Metz, J.H. (1989) Effects of early handling on later open-field behaviour in rabbits. *Applied Animal Behaviour Sciences* 24, 157–167.

Keys, J.E., Smith, L.W. and Weinland, B.T. (1976) Response of dairy cattle given a free choice of free stall location and three bedding materials. *Journal Dairy Science* 59, 1157–1162.

Kiley-Worthington, M. (1987) *The Behaviour of Horses in Relation to Management and Training*. A. Müller, Ruschlikon, Zurich, Switzerland, 216 pp.

Klemm, W.R. (1984) Neurophysiology of consciousness and behavioral physiology. In: Swenson, M.J. (ed.) *Dukes' Physiology of Domestic Animals*, 10th edn. Cornell University Press, Ithaca, New York, pp. 671–706.

Klingel, H. (1974) A comparison of the social behaviour in the Equidae. In: Geist, V. and Walther, F. (eds) *The Behaviour of Ungulates and its Relation to Management*. International Union for the Conservation of Nature (IUCN) Publications, Alberta, Canada, 24, 1, pp. 124–133.

Knowles, T.G. and Broom, D.M. (1990) The handling and transport of broilers and spent hens. *Applied Animal Behaviour Science* 28, 75–91.

Kowalski, K. (1967) The Pleistocene extinction of mammals in Europe. In: Martin, P.S. and Wright, H.E. (eds) *Pleistocene Extinctions*. Yale University Press, London, pp. 349–365.

Kraft, R. (1979) Vergleichende Verhaltensstudien an Wild- und Hauskaninchen. *Zeitschrift für Tierzüchtungsbiologie* 95, 165–179.

Kreibich, A. and Sommer, M. (1995) *Ostrich Farm Management*. Landwirtschaftsverlag GmbH, Münster-Hiltrup, Germany, 223 pp.

Kridli, R.T., Abdullah, A.Y., Shaker, M.M. and Masa'deh, M.K. (2007) Sexual activity and puberty in pure Awassi and crosses and back-crosses with Charollais and Romanov sheep breed. *New Zealand Journal of Agricultural Research* 50, 429–436.

Krohn, C.C. and Munksgaard, L. (1993) Behaviour of dairy cows kept in extensive (loose housing/pasture) or intensive (tie stall) environments, II. Lying and lying down behaviour. *Applied Animal Behaviour Science* 37, 1–16.

Kumar, D., Singh, U., Bhatt, R.S. and Risam, K.S. (2005) Reproductive efficiency of female German Angora rabbits under Indian sub-temperate climatic conditions. *World Rabbit Science* 13, 113–122.

Labaque, M.C., Navarro, J.L. and Martella, M.B. (1999) A note on chick adoption: a complementary strategy for rearing rheas. *Applied Animal Behaviour Science* 63, 165–170.

Lammers, G.J. and de Lange, A. (1986) Pre-farrowing and post-farrowing behaviour in primiparous domesticated pigs. *Applied Animal Behaviour Science* 15, 31–43.

Larson, G., Dobney, K., Albarella, U., Fang, M.Y., Matisoo-Smith, E., Robins, J., Lowden, S., Finlayson, H., Brand, T., Willerslev, E., Rowley-Conwy, P., Andersson, L. and Cooper, A. (2005) Worldwide phylogeography of wild boar reveals multiple centres of pig domestication. *Science* 307, 1618–1621.

Laut, J.E., Houpt, K.A., Hintz, H.F. and Houpt, T.R. (1985) The effects of caloric dilution on meal patterns and food intake of ponies. *Physiology and Behaviour* 35, 549–554.

Lawrence, A.B., Petherick, J.C., McLean, K., Gilbert, C.L., Chapman, C. and Russell, J.A. (1992) Naloxone prevents interruption of parturition and increases plasma oxytocin following environmental disturbance in parturient sows. *Physiology and Behaviour* 52, 917–923.

Lee, J.C., Taylor, J.F.N. and Downing, S.E. (1975) A comparison of ventricular weights and geometry in newborn, young and adult mammals. *Journal of Applied Physiology* 38, 147–150.

Lehmann, M. (1991) Social behaviour in young domestic rabbits under semi-natural conditions. *Applied Animal Behaviour Science* 32, 269–292.

Leonard, M.L. and Horn, A.G. (1995) Crowing in relation to status in roosters. *Animal Behaviour* 49, 1283–1290.

Leonard, M.L., Horn, A.G. and Fairfull, R.W. (1995) Correlates and consequences of allopecking in White Leghorn chickens. *Applied Animal Behaviour Science* 43(1), 17–26.

Lewis, N.J. and Hurnik, J.F. (1986) An approach response of piglets to the sow's nursing vocalizations. *Canadian Journal of Animal Science* 66, 537–539.

Lickliter, R.E. (1984) Behaviour associated with parturition in the domestic goat. *Applied Animal Behaviour Science* 13, 335–345.

Lidfors, L. (1994) Mother–young behaviour in cattle, parturition, development of cow–calf attachment, suckling and effects of separation. Report 33, Thesis. Department of Animal Hygiene, SLU, Sweden, 56 pp.

Lidfors, L. (1997) Behavioural effects of environmental enrichment for individually caged rabbits. *Applied Animal Behaviour Science* 52, 157–169.

Lidfors, L. and Isberg, L. (2003) Intersucking in dairy cattle – review and questionnaire. *Applied Animal Behaviour Science* 80, 207–231.

Lindgren, N.O. (1964) On the aetiology of salpingitis and salpingoperitonitis of the domestic fowl. A statistical and experimental investigation. Thesis. Stockholm, 95 pp.

Lindhé, B. (1968) Cross breeding for beef with Swedish red and white cattle. Part 1. Performance under varying field conditions. Thesis. *Annals of the Agriculture College of Sweden* 34(5), 465.

Lister, A.M. (2001) Tales from the DNA of domestic horses. *Science* 292, 218.

Lister, A.M., Kadwell, M., Kaagan, L.M., Jordan, W.C., Richards, M.B. and Stanley, H. (1998) Ancient and modern DNA in a study of horse domestication. *Ancient Biomolecules* 2, 267–280.

Livingstone, A., Ley, S. and Warerman, A. (1992) Tactile and pain perception. In: Phillips, C.J.C. (ed.) *Farm Animals and the Environment*. CAB International, Wallingford, UK, pp. 201–208.

Lloyd, A.S., Martin, J.E., Bornett-Gauci, H.L.I. and Wilkinson, R.G. (2008) Horse personality: variation between breeds. *Applied Animal Behaviour Science* 112, 369–383.

Losinger, W.C. and Heinrichs, A.J. (1997) Management practices associated with high mortality among preweaned dairy heifers. *Journal of Dairy Research* 64, 1–11.

Ludt, C.J., Schroeder, W., Rottmann, O. and Kuehn, R. (2004) Mitochondrial DNA phylogeography of red deer (*Cervus elaphus*). *Molecular Phylogenetics and Evolution* 31, 1064–1083.

Lundberg, A., Keeling, L. and Petersson, L. (2006) Training and timing – how to facilitate the daily inspections of extensively kept cattle. In: Mendl, M., Bradshaw, J.W.S., Burman, O.H.P., Butterworth, A., Harris, M.J., Held, S.D.E., Jones, S.M., Littin, K.E., Main, D.C.J., Nicol, C.J., Parker, R.M.A., Paul, E.S., Richards, G., Sherwin, C.M., Statham, P.T.E., Toscano, M.J. and Warriss, P.D. (eds) *Proceedings 40th International ISAE Congress*. International Society for Applied Ethology (ISAE) Scientific Committee, Cranfield University Press, Bedfordshire, UK, 240 pp.

Lynch, J.J., Hinch, G.N. and Adams, D.B. (1992) *The Behaviour of Sheep*. CAB International, Wallingford, UK, 248 pp.

Lyons, D.M. and Price, E.O. (1987) Relationships between heart rates and behaviour of goats in encounters with people. *Applied Animal Behaviour Science* 18, 363–369.

McBride, G. (1970) The social control of behaviour in fowls. In: Freeman, B.M. and Gordon, R.F. (eds) *Aspects of Poultry Behaviour*. British Poultry Science, Edinburgh, pp. 3–13.

McBride, G.E., Christopherson, R.J. and Sauer, W. (1985) Metabolic rate and plasma thyroid hormone concentrations of mature horses in response to changes in ambient temperature. *Canadian Journal of Animal Science* 65, 375–382.

McConaghy, F. (1994) Thermoregulation. In: Hodgson, D.R. and Rose, R.J. (eds) *The Athletic Horse*. W.B. Saunders, Philadelphia, Pennsylvania, pp. 181–202.

McDonnel, S.M. (2002) Behaviour of horses. In: Jensen, P (ed.) *The Ethology of Domestic Animals, An Introductory Text*. CAB International, Wallingford, UK, pp. 119–129.

McGreevy, P. and Nicol, C. (1998) Prevention of crib-biting: a review. *Equine Veterinary Journal* 27, 35–39.

McGreevy, P.D., Webster, A.J.F. and Nicol, C.J. (2001) Study of the behaviour, digestive efficiency and gut transit times of crib-biting horses. *Veterinary Record* 148, 592–596.

McKeegan, D.E.F., (2004) Sensory perception: chemoreception. *Welfare of the Laying Hen* 27, 139–153.

McKinney, F. (1969) The behaviour of ducks. In: Hafez, E.S.E (ed.) *The Behaviour of Domestic Animals*, 2nd edn. Baillière Tindall and Cassell, London, pp. 593–626.

McKinney, F. (1975) The behaviour of ducks. In: Hafez, E.S.E. (ed.) *The Behaviour of Domestic Animals*, 3rd edn. Baillière Tindall and Cassell, London, pp. 490–519.

McLean, A. (2004) Short-term spatial memory in the domestic horse. *Applied Animal Behaviour Science* 85, 93–105.

McLeman, M.A., Mendl, M.T., Jones, R.B. and Wathes, C.M. (2008) Social discrimination of familiar conspecifics by juvenile pigs, *Sus scrofa*: development of a non-invasive method to study the transmission of unimodal and bimodal cues between live stimuli. *Applied Animal Behaviour Science* 115, 123–137.

Mader, D.R. (1997) Basic approach to veterinary care. In: Hillyer, E.W. and Quesenberry, K.E. (eds) *Ferrets, Rabbits and Rodents – Clinical Medicine and Surgery*. W.B. Saunders, London, pp. 160–168.

Mader, T.L. (2003) Environmental stress in confined beef cattle. *Journal of Animal Science* 81, 110–119.

Main, D.C.J., Clegg, J., Spatz, A. and Green, L.E. (2000) Repeatability of a lameness scoring system for finishing pigs. *Veterinary Record* 147, 574–576.

Mal, M.E. and McCall, C.A. (1996) The influence of handling during different ages on a halter training test in foals. *Applied Animal Behaviour Science* 50, 115–120.

Maloney, S.K. (2008) Thermoregulation in ratites: a review. *Australian Journal of Experimental Agriculture* 48, 1293–1301.

Manser, C. (1996) Effects of lighting on the welfare of domestic poultry: a review. *Animal Welfare* 5, 341–360.

Manteuffel, G., Puppe, B. and Schon, P.C. (2004) Vocalization of farm animals as a measure of welfare. *Applied Animal Behaviour Science* 88, 163–182.

Marai, F.M. and Rashwan, A.A. (2003) Rabbits' behaviour under modern commercial production conditions – a review. *Archiv für Tierzucht – Archives of Animal Breeding* 46, 357–376.

Marai, I.F.M. and Rashwan, A.A. (2004) Rabbits behavioural response to climatic and managerial conditions – a review. *Archiv für Tierzucht – Archives of Animal Breeding* 47, 469–482.

Marai, I.F.M., Habeeb, A.A.M. and Gad, A.E. (2004) Reproductive traits of female rabbits as affected by heat stress and lighting regime under subtropical conditions of Egypt. *Animal Science* 78, 119–127.

Marinier, S.L., Alexander, A.J. and Waring, G.H. (1988) Flehmen behaviour in the domestic horse. Discrimination of conspecific odours. *Applied Animal Behaviour Science* 19, 227–237.

Marquez-Olivas, M., Garcia-Moya, E., Gonzalez-Rebeles Islas, C. and Vaquera-Huerta, H. (2007) Roost sites characteristics of wild turkey (*Meleagris gallopavo mexicana*) in Sierra Fria, Aguascalientes, Mexico. *Revista Mexicana de Biodiversidad* 78, 163–173.

Martrenchar, A. (1999) Animal welfare and intensive production of turkey broilers. *World's Poultry Science Journal* 55, 143–152.

Martrenchar, A., Huonnic, D., Cotte, J.P., Boilletot, E. and Morisse, J.P. (1999) Influence of stocking density on behavioural, health and productivity traits of turkeys in large flocks. *British Poultry Science* 40, 323–331.

Marx, D. and Buchholz, M. (1991) Ethological free-choice experiments with early weaned piglets in pens with different application of straw. 2. Effects of different applications of straw and different area-dimensions. *Deutsche Tierärztliche Wochenschrift* 98, 50–56.

Mason, I.L. (1984) *Evolution of Domesticated Animals*. Longman, London, 452 pp.

Mauget, R. (1982) Seasonality of reproduction in the wild boar. In: Cole, D.J.A. and Foxcroft, G.R. (eds) *Control of Pig Reproduction*. Butterworth Scientific, London, pp. 509–526.

Mayne, R.K., Else, R.W. and Hocking, P.M. (2007) High litter moisture alone is sufficient to cause footpad dermatitis in growing turkey. *British Poultry Science* 48, 538–545.

Meese, G.B. and Ewbank, R. (1973a) Establishment and nature of the dominance hierarchy in the domesticated pig. *Animal Behaviour* 21, 326–334.

Meese, G.B. and Ewbank, R. (1973b) Exploratory behaviour and leadership in the domesticated pig. *British Veterinary Journal* 129, 251–259.

Mejdell, C.M. and Bøe, K.E. (2005) Responses to climatic variables of horses housed outdoors under Nordic winter conditions. *Canadian Journal of Animal Science* 85, 301–308

Mench, J.J. and Keeling, L.J. (2001) The social behaviour of domestic birds. In: Keeling, L.J. and Gonyou, H. (eds) *Social Behaviour in Farm Animals*. CAB International, Wallingford, UK, pp. 177–209.

Meredith, A. (2000) General biology and husbandry. In: Flecknell, P. (ed.) *Manual of Rabbit Medicine and Surgery*. British Small Animal Veterinary Association, London, pp. 13–23.

Meredith, A. (2006) General biology and husbandry. In: Meredith, A. and Flecknell, P. (eds) *BSAVA Manual of Rabbit Medicine and Surgery*, 2nd edn. British Small Animal Veterinary Association, Gloucester, UK, pp. 1–17.

Michanek, P. and Ventorp, M. (1996) Time spent in shelter in relation to weather by two free-ranging thoroughbred yearlings during winter. *Applied Animal Behaviour Science* 49, 104.

Michel, V. and Boilletot, E. (2003) Influence of environmental enrichment on the behaviour, body state and zootechnical performances of turkey broilers. *Sciences et Techniques Avicoles* 44, 18–27.

Mills, D.S. and Davenport, K. (2002) The effect of a neighbouring conspecific versus the use of a mirror for the control of stereotypic weaving behaviour in the stabled horse. *Animal Science* 74, 95–101.

Mills, D.S. and Taylor, K. (2003) Field study of the efficacy of three types of nose net for the treatment of headshaking in horses. *Veterinary Record* 152, 41–44.

Minero, M., Canali, E., Ferrante, V., Verga, M. and Ödberg, F.O. (1999) Heart rate and behavioural responses of crib-biting horses to two acute stressors. *Veterinary Record* 145, 430–433.

Mitchell, M.A. (1999) Welfare. In: Deeming, D.C. (ed.) *The Ostrich, Biology, Production and Health*. CAB International, Wallingford, UK, pp. 217–230.

Mohr, E. (1960) *Wilde Schweine. Die neue Brehm-Bucherei*. Verlag A. Ziemsen, Wittenberg, Germany, 247 pp.

Moinard, C. and Sherwin, C.M. (1999) Turkeys prefer fluorescent light with supplementary ultraviolet radiation. *Applied Animal Behaviour Science* 64, 261–267.

Möller, A.P., Sanotra, G.S. and Vestergaard, K.S. (1999) Developmental instability and light regime in chickens. *Applied Animal Behaviour Science* 62, 57–71.

Molnar, M., Bogenfurst, F., Molnar, T. and Almasi, A. (2002) The effects of the domestication on the behaviour of goose under intensive conditions. *Acta Agraria Kaposvariensis* 6, 287.

Moncomble, A.S., Coureaud, G., Quennedey, B., Langlois, D., Perrier, G. and Schaal, B. (2005) The mammary pheromone of the rabbit: from where does it come? *Animal Behaviour* 69, 29–38.

Morales, A., Garza, A. and Sotomayor, J.C. (1997) Wild turkey's diet in Durango, Mexico. *Revista Chilena de Historia Natural* 70, 403–414.

Morgan, K., Ehrlemark, A. and Sällvik, K. (1997) Dissipation of heat from standing horses exposed to ambient temperatures between –3°C and 37°C. *Journal of Thermal Biology* 22, 177–186.

Mount, L.E., Holmes, C.W., Close, W.H., Morrison, S.R. and Start, I.B. (1971) A note on the consumption of water by the growing pig at several environmental temperatures and levels of feeding. *Animal Production* 13, 561–563.

Mount, N.C. and Seabrook, M.F. (1993) A study of aggression when group housed sows are mixed. *Applied Animal Behaviour Science* 36, 377–383.

Mouser, D. (1996) Restraint and handling of the emu. In: Tully, T.N. Jr and Shane, S.M. (eds) *Ratite Management, Medicine and Surgery*. Krieger Publishing Co., Malabar, Florida, pp. 41–45.

Muggenburg, B.A., Kowalczyk, T., Hoekstra, W.G. and Gummer, R.H. (1967) Effect of certain management variables on the incidence and severity of gastric lesions in swine. *Veterinary Medicine and Small Animal Clinician* 62, 1090–1094.

Mulley, R.C. and English, A.W. (1991) Velvet antler harvesting from fallow deer (*Dama dama*). *Australian Veterinary Journal* 68, 309–311.

Mulley, R.C., English, A.W. and Kirby, A.C. (1990) Rearing performance of farmed fallow deer (*Dama dama*). *Australian Veterinary Journal* 67, 454–456.

Munksgaard, L. and Simonsen, H.B. (1996) Behavioural and pituitary–adrenal axis responses of dairy cows to social isolation and deprivation of lying down. *Journal of Animal Science* 74, 769–777.

Munksgaard, L., Ingvartsen, K.L., Pedersen, L.J. and Nielsen, V.K.M. (1999) Deprivation of lying down affects behaviour and pituitary–adrenal axis responses in young bulls. *Acta Agriculturae Scandinavica Section A – Animal Science* 49, 172–178.

Murphy, M.R. (1992) Water metabolism of dairy cattle. *Journal of Dairy Science* 75, 326–333.

Myers, L.J. and Coulter, D.B. (2004) Vision. In: Reece, W.O. (ed.) *Dukes' Physiology of Domestic Animals*, 12th edn. Cornell University Press, Ithaca, New York, pp. 843–851.

Mykytowycz, R. (1968) Territorial marking by rabbits. *Scientific American* 218, 116–126.

Nagy, K., Schrott, A. and Kabai, P. (2008) Possible influence of neighbours on stereotypic behaviour in horses. *Applied Animal Behaviour Science* 111, 321–328.

Navarro, J.L. and Martella, M.B. (2002) Reproductivity and raising of Greater Rhea (*Rhea Americana*) and Lesser Rhea (*Pterocnemia pennata*) – a review. *Archiv für Geflügelkunde* 66, 124–132.

Nicol, C. (1999) Understanding equine stereotypies. In: Harris, P., Goodwin, D. and Green, R. (eds) *The Role of the Horse in Europe, Proceedings of the Waltham Symposium. Equine Veterinary Journal* Suppl. 28, 20–25.

Nicol, C. (2006) How animals learn from each other. *Applied Animal Behaviour Science* 100, 58–63.

Nicol, C.J., Davidson, H.P.D., Harris, P.A., Waters, A.J. and Wilson, A.D. (2002) Study of crib-biting and gastric inflammation and ulceration in young horses. *Veterinary Record* 151, 658–662.

Nolet, B.A., Butler, P.J. and Woakes, A.J. (1992) Estimation of daily energy expenditure from heart rate and doubly labelled water in exercising geese. *Physiological Zoology* 65, 1188–1216.

Nuboer, K.F.W. (1993) Visual ecology in poultry houses. In: Savory, C.J. and Hughes, B.O. (eds) *Fourth European Symposium on Poultry Welfare*. University Federation of Animal Welfare, Potters Bar, UK, pp. 39–44.

Nyström, B. and Hagbarth, K.-E. (1981) Microelectrode recordings from transected nerves in amputees with phantom limb pain. *Neuroscience Letters* 27, 211–216.

Obel, A.-L. (1971) Frakturer och burtrötthet resultat av burhönsuppfödning. *Svensk Veterinärtidning* 23, 54–55.

Odén, K., Vestergaard, K.S. and Algers, B. (1999) Agonistic behaviour and feather pecking in single-sexed and mixed groups of laying hens. *Applied Animal Behaviour Science* 62, 219–231.

Odén, K., Vestergaard, K.S. and Algers, B. (2000) Space use and agonistic behaviour in relation to sex composition in large flocks of laying hens. *Applied Animal Behaviour Science* 67, 307–320.

Odén, K., Keeling, L.J. and Algers, B. (2002) Behaviour of layers in two types of aviary system on 25 commercial farms in Sweden. *British Poultry Science* 43, 169–181.

Odén, K., Berg, C., Gunnarsson, S. and Algers, B. (2004) Male rank order, space use and female attachment in large flocks of laying hens. *Applied Animal Behaviour Science* 87, 83–94.

Olanrewaju, H.A., Miller, W.W., Maslin, W.R., Thaxton, J.P., Dozier, W.A. III, Purswell, J. and

Branton, S.L. (2007) Interactive effects of ammonia and light intensity on ocular, fear and leg health in broiler chickens. *International Journal of Poultry Science* 6, 762–769.

Oliviero, C., Heinonen, M., Valros, A. and Peltoniemi, O. (2010) Environmental and sow-related factors affecting the duration of farrowing. *Animal Reproduction Science* 119, 85–91.

Olson, D.P., Papasian, C.J. and Ritter, R.C. (1980) The effects of cold stress on neonatal calves. II. Absorption of colostral immunoglobulin. *Canadian Journal of Comparative Medicine* 44, 19–23.

Olson, D.P., Bull, R.C., Kelley, K.W., Ritter, R.C., Woodward, L.F. and Everson, D.O. (1981) Effects of maternal nutritional restriction and cold stress on young calves: absorption of colostral immunoglobulin. *American Journal of Veterinary Research* 42, 876–880.

Olsson, I.A.S. and Keeling, L.J. (2000) Night-time roosting in laying hens and the effect of thwarting access to perches. *Applied Animal Behaviour Science* 68, 243–256.

Olsson, I.A.S. and Keeling, L.J. (2005) Why in earth? Dustbathing behaviour in jungle and domestic fowl reviewed from a Tinbergian and animal welfare perspective. *Applied Animal Behaviour Science* 93, 259–282.

Olsson, I.A.S., Keeling, L.J. and McAdie, T.M. (2002) The push-door for measuring motivation in hens: an adaptation and a critical discussion of the method. *Animal Welfare* 11, 1–10.

O'Malley, B. (2005) *Clinical Anatomy and Physiology of Exotic Species*. Elsevier Saunders, Edinburgh, UK, 269 pp.

O'Reilly, K.M., Harris, M.J., Mendl, M., Held, S., Moinard, C., Statham, R., Marchant-Forde, J. and Green, L.E. (2006) Factors associated with preweaning mortality on commercial pig farms in England and Wales. *Veterinary Record* 159, 193–196.

Ousey, J.C., McArthur, A.J., Murgatroyd, P.R., Stewart, J.H. and Rossdale, P.D. (1992) Thermoregulation and total body insulation in the neonatal foal. *Journal of Thermal Biology* 17, 1–10.

Outram, A.K., Stear, N.A., Bendrey, R., Olsen, S., Kasparov, A., Zaibert, V., Thorpe, N. and Evershed, R.P. (2009) The earliest horse harnessing and milking. *Science* 323, 1332–1335.

Papachristoforou, C., Koumas, A. and Photiou, C. (2000) Seasonal effects on puberty and reproductive characteristics of female Chios sheep and Damascus goats born in autumn or in February. *Small Ruminant Research* 38, 9–15.

Parer, I. (1982) Dispersal of the wild rabbit, *Oryctolagus cuniculus*, at Urana in New South Wales. *Australian Wildlife Research* 9, 427–441.

Patris, B., Perrier, G., Schaal, B. and Coureaud, G. (2008) Early development of filial preferences in the rabbit: implications of nursing- and pheromone-induced odour learning? *Animal Behaviour* 76, 305–314.

Pedersen, V., Barnett, J.L., Hemsworth, P.H., Newman, E.A. and Schirmer, B. (1998) The effects of handling on behavioural and physiological responses to housing in tether-stalls among pregnant pigs. *Animal Welfare* 7, 137–150.

Pepin, D., Adrados, C. and Angibault, J.M. (2005) Forests and deer: monitoring of red deer in the Cevennes using GPS devices. *Rendez-Vous Techniques* 7, 51–56.

Perelman, B. (1999) Health management and veterinary procedures. In: Deeming, D.C. (ed.) *The Ostrich, Biology, Production and Health*. CAB International, Wallingford, UK, pp. 321–346.

Perry, G.C. (2004) Lighting. In: Perry, G.C. (ed.) *Welfare of the Laying Hen*, Poultry Science Series 27. CAB International, Wallingford, UK, pp. 299–311.

Petherick, J.C. (1983) A note on the space use for excretory behaviour of suckling piglet. *Applied Animal Ethology* 9, 367–371.

Pfister, J.A., Stegelmeier, B.L., Cheney, C.D., Ralphs, M.H. and Gardner, D.R. (2002) Conditioning taste aversions to locoweed (*Oxytropis sericea*) in horses. 1. *Journal of Animal Science* 80, 79–83.

Phillips, C.J.C. (1993) *Cattle Behaviour*. Farming Press Books, Ipswich, UK, 212 pp.

Phillips, C.J.C. (2002) *Cattle Behaviour and Welfare*, 2nd edn. Blackwell, Oxford, UK, 264 pp.

Phillips, C.J.C. and Lomas, C.A. (2001) The perception of colour by cattle and its influence on behaviour. *Journal of Dairy Science* 84, 801–813.

Phillips, C. and Piggins, D. (1992) *Farm Animals and the Environment*. CAB International, Wallingford, UK, 430 pp.

Pick, D.F., Lovell, G., Brown, S. and Dail, D. (1994) Equine colour-perception revisited. *Applied Animal Behaviour Science* 42, 61–65.

Pollard, J.C. and Littlejohn, R.P. (1995) Effects of lighting on heart rate and positional preferences during confinement in farmed red deer. *Animal Welfare* 4, 329–337.

Pollard, J.C. and Littlejohn, R.P. (1999) Activities and social relationships of red deer at pasture. *New Zealand Veterinary Journal* 47, 83–87.

Pollock, M.L., Lee, W.G., Walker, S. and Forres, G. (2007) Ratite and ungulate preferences for woody New Zealand plants – influence of chemical and physical traits. *New Zealand Journal of Ecology* 31, 68–78.

Porter, R.H., Roelofsen, R., Picard, M. and Arnould, C. (2005) The temporal development and sensory mediation of social discrimination in domestic chicks. *Animal Behaviour* 70, 359–364.

Prescott, N.B., Wathes, C.M. and Jarvis, J.R. (2003) Light, vision and the welfare of poultry. *Animal Welfare* 12, 269–288.

Prescott, N.B., Jarvis, J.R. and Wathes, C.M. (2004) Vision in the laying hen. *Welfare of the Laying Hen* 27, 155–164.

Price, E.O. and Thos, J. (1980) Behavioural responses to short-term social isolation in sheep and goats. *Applied Animal Ethology* 6, 331–339.

Price, S., Sibly, R.M. and Davies, M.H. (1993) Effects of behaviour and handling on heart rate in farmed red deer. *Applied Animal Behaviour Science* 37, 111–123.

Prime, R.W., Fahy, V.A., Ray, W., Cutler, R.S. and Spicer, E.M. (1999) On-farm validation of research: lowering preweaning mortality in pigs. Cited by Cutler, R.S., Fahy, V.A., Spicer, E.M. and Cronin, G.M. In: Straw, B.E., D'Allaire, S., Mengeling, W.L. and Taylor, D.J. (eds) (1999) *Diseases of Swine*. Iowa State University Press, Ames, Iowa, p. 988.

Prince, J.H. (1977) The eye and vision. In: Swenson, M.J. (ed.) *Dukes' Physiology of Domestic Animals*, 9th edn. Cornell University Press, Ithaca, New York, pp. 696–712.

Radostits, O.M., Gay, C.C., Blood, D.C. and Hincliff, K.W. (1999) *Veterinary Medicine, A Textbook of the Diseases of Cattle, Sheep, Pigs, Goats and Horses*, 9th edn. Saunders, London, 1881 pp.

Raines, A.M. (1998) Restraint and housing of ratites. *The Veterinary Clinics of North America, Food Animal Practice* 14, 387–399.

Randler, C. (2004) Aggressive interactions in Swan Geese *Anser cygnoides* and their hybrids. *Acta Ornithologica* 39, 147–153.

Randolph, J.H., Cromwell, G.L., Stahly, T.S. and Kratzer, D.D. (1981) Effects of group size and space allowance on performance and behaviour of swine. *Journal of Animal Science* 53, 922–927.

Rauch, H.W., Pingel, H. and Bilsing, A. (1993) Welfare of waterfowl. In: Savory, C.J. and Hughes, B.O. (eds) *Fourth European Symposium on Poultry Welfare*. Universities Federation for Animal Welfare, Potters Bar, UK, pp. 139–147.

Rawson, R.E., Bates, D.W., Dziuk, H.E., Ruth, G.R., Good, A.L., Serfass, R.C. and Anderson, J.F. (1988) Health and physiology of newborn calves housed in severe cold. In: *Proceedings of III International Livestock Environment Symposium*, 25–27 April, Toronto, Canada. ASAE Publication 1-88, Michigan, pp. 365–368.

Rawson, R.E., Dziuk, H.E., Good, A.L., Anderson, J.F., Bates, D.W., Ruth, G.R. and Serfass, R.C. (1989a) Health and metabolic responses of young calves housed at –30°C to –8°C. *Canadian Journal of Veterinary Research* 53, 268–274.

Rawson, R.E., Dziuk, H.E., Good, A.L., Anderson, J.F., Bates, D.W. and Ruth, G.R. (1989b) Thermal insulation of young calves exposed to cold. *Canadian Journal of Veterinary Research* 53, 275–278.

Recuerda, P., Arias de Reyna, L., Redondo, T. and Trujillo, J. (1986) Analyzing stereotypy in red deer alarm postures by means of informational redundancy. *Behavioural Processes* 14, 71–87.

Reece, W.O. (2004) Respiration in mammals. In: Reece, W.O. (ed.) *Dukes' Physiology of Domestic Animals*, 12th edn. Cornell University Press, Ithaca, New York, pp. 114–149.

Reiland, S. (1975) Osteochondrosis in the pig. A morphologic and experimental investigation with special reference to the leg weakness syndrome. Thesis, Stockholm, 118 pp.

Reinhardt, V. and Reinhardt, A. (1981) Cohesive relationships in a cattle herd (*Bos indicus*). *Behaviour* 77, 121–151.

Reinken, G., Hartfiel, W. and Körner, E. (1990) *Deer Farming*. Farming Press Books, London, 289 pp.

Reiter, K. (1993) Analysis of short-term rhythms of feeding and drinking behaviour in Muscovy ducks. In: Savory, C.J. and Hughes, B.O. (eds) *Fourth European Symposium on Poultry Welfare*. Universities Federation for Animal Welfare, Potters Bar, UK, p. 308.

Reschmagras, C., Cherel, Y., Wyers, M. and Abourachid, A. (1993) Locomotion analysis of male commercial turkey – a comparative study of healthy and lame turkeys. *Veterinary Research* 24, 5–20.

Riber, A.B., Nielsen, B.L., Ritz, C. and Forkman, B. (2007) Diurnal activity cycles and synchrony in layer hen chicks. *Applied Animal Behaviour Science* 108, 276–287.

Robert, S., Dancosse, J. and Dallaire, A. (1987) Some observations on the role of environment and genetics in behaviour of wild and domestic forms of *Sus scrofa* (European wild boars and domestic pigs). *Applied Animal Behaviour Science* 17, 253–262.

Roberts, S.M. (1992) Equine vision and optics. *The Veterinary Clinics of North America: Equine Practice* 8, 451–457.

Robertshaw, D. (2004) Temperature regulation and the thermal environment. In: Reece, W.O. (ed.) *Dukes' Physiology of Domestic Animals*, 12th edn. Cornell University Press, Ithaca, New York, pp. 962–973.

Rodenburg, T.B. and Koene, P. (2004) Feather pecking and feather loss. In: Perry, G.C. (ed.) *Welfare of the Laying Hen, Poultry Science Series 27.* CAB International, Wallingford, UK, pp. 227–238.

Rodenburg, T.B., Bracke, M.B.M., Berk, J., Cooper, J., Faure, J.M., Guémené, D., Guy, G., Harlander, A., Jones, T., Knierim, U., Kuhnt, K., Pingel, H., Reiter, K., Servière, J. and Ruis, M.A.W. (2005) Welfare of ducks in European duck husbandry systems. *World's Poultry Science Journal* 61, 633–646.

Rodriguez-Estevez, V., Garcia, A., Pena, F. and Gomez, A.G. (2009) Foraging of Iberian fattening pigs grazing natural pasture in the Dehesa. *Livestock Science* 20, 135–143.

Rolandsdotter, E., Westin, R. and Algers, B. (2009) Maximum lying bout duration affects the occurrence of shoulder lesions in sows. *Acta Veterinaria Scandinavica* 51, 44, doi: 10.1186/1751-0147-51-44.

Rommers, J.M., Boiti, C., De Jong, I. and Brecchia, G. (2006) Performance and behaviour of rabbit does in a group-housing system with natural mating or artificial insemination. *Reproduction Nutrition Development* 46, 677–687.

Rosell, J.M. and de la Fuente, L.F. (2009) Culling and mortality in breeding rabbits. *Preventive Veterinary Medicine* 88, 120–127.

Rossdale, P.D. and Ricketts, S.W. (1980) *Equine Stud Farm Medicine*, 2nd edn. Baillière Tindall, London.

Ruckebusch, Y. (1972) The relevance of drowsiness in the circadian cycle of farm animals. *Animal Behaviour* 20, 637–643.

Ruckebusch, Y. (1974) Sleep deprivation in cattle. *Brain Research* 78, 495–499.

Ruckebusch, Y. (1975) Feeding and sleep patterns of cows prior to and post parturition. *Applied Animal Ethology* 1, 283–292

Rushen, J., de Passillé, A.M. and Munksgaard, L. (1999) Fear of people by cows and effects on milk yield, behaviour, and heart rate at milking. *Journal of Dairy Science* 82, 720–727.

Rushen, J., de Passillé, A.M., Munksgaard, L. and Tanida, H. (2001) People as social actors in the world of farm animals. In: Keeling, L.J. and Gonyou, H.W. (eds) *Social Behaviours in Farm Animals.* CAB International, New York, pp. 353–372.

Rushen, J., Passillé, A., Keyserlingk, M. and Weary, D. (2008) *The Welfare of Cattle.* Springer, Dordrecht, The Netherlands, 310 pp.

Rutter, S.M. (2002) Behaviour of sheep and goats. In: Jensen, P. (ed.) *The Ethology of Domestic Animals, An Introductory Text.* CAB International, Wallingford, UK, 218 pp.

Rydhmer, L., Zamaratskaia, G., Andersson, H.K., Algers, B., Guillemet, R. and Lundström, K. (2006) Aggressive and sexual behaviour of growing and finishing pigs reared in groups, without castration. *Acta Agriculturae Scandinavica Section A – Animal Science* 56, 109–119.

Sainsbury, D.W.B. (1992) *Poultry Health and Management.* Blackwell, London, 214 pp.

Sales, J. (2006a) Digestive physiology and nutrition of ratites. *Avian and Poultry Biology Reviews* 17, 41–55.

Sales, J. (2006b) The rhea, a ratite native to South America. *Avian and Poultry Biology Reviews* 17, 105–124.

Sales, J. (2007) The emu (*Dromaius novaehollandiae*): a review of its biology and commercial products. *Avian and Poultry Biology Reviews* 18, 1–20.

Sällvik, K. and Wejfeldt, B. (1993) Lower critical temperature for fattening pigs on deep straw bedding. *Proceedings American Society Agriculture Engineering, 4th International Livestock Environment Symposium*, St Joseph, Michigan, pp. 904–914.

Sanchez-Andrade, G. and Kendrick, K.M. (2009) The main olfactory system and social learning in mammals. *Behavioural Brain Research* 200, 323–335.

SCAHAW (EU Scientific Committee on Animal Health and Welfare) (2002) *The Welfare of Animals during Transport (Details for Horses, Pigs, Sheep and Cattle)*. European Commission, Brussels, 130 pp.

Schaal, B., Coureaud, G., Doucet, S., Allam, M.D.E., Moncomble, A.S., Montigny, D., Patris, B. and Holley, A. (2009) Mammary olfactory signalisation in females and odour processing in neonates: ways evolved by rabbits and humans. *Behavioural Brain Research* 200, Special Issue, 346–358.

Schaefer, T. (1968) Some methodological implications of the research on early handling in the rat. In: Newton, G. and Levine, S. (eds) *Psychobiology of Development*. C. Thomas, Springfield, Illinois, pp. 102–141.

Schake, L.M. and Riggs, J.K. (1966) Diurnal and nocturnal activities of lactating beef cows in confinement. *Journal of Animal Science* 25, 254.

Scheiber, I.B.R., Kotrschal, K. and Weiss, B.M. (2009) Serial agonistic attacks by greylag goose families, *Anser anser*, against the same opponent. *Animal Behaviour* 77, 1211–1216.

Schloeth, R. (1958) Über die Mutter-Kind-Beziehungen beim halbwilder Camarque-Rind. *Saugetierkundliche Mitteilungen* 6, 145–150.

Schmidt, G.H. (1971) *Biology of Lactation*. Freeman and Co., San Francisco, 317 pp.

Schrader, L. and Rohn, C. (1997) Lautäußerungen von Hausschweinen als Indikator für Stressreaktionen. *Landbauforschung Völkenrode* 47, 89–95.

Schumacher, J., Citino, S.B., Hernandez, K., Hutt, J. and Dixon, B. (1997) Cardiopulmonary and anaesthetic effects of propofol in wild turkeys. *American Journal of Veterinary Research* 58, 1014–1017.

Schütz, K. (2002) Trade-off in resource allocation between behaviour and production in fowl. Thesis. *Acta Universitatis Agriculturae Sueciae, Veterinaria*, 115 pp.

Seabrook, M.F. (1995) Behavioural interactions. *Pig Journal* 34, 31–40.

Seabrook, M.F. and Bartle, N.C. (1992) Human factors. In: Phillips, C. and Piggins, D. (eds) *Farm Animals and the Environment*. CAB International, Wallingford, UK, pp. 111–125.

Seabrook, M.F. and Mount, N.C. (1995a) Individuality in the reactions of pigs to humans. *Pig Journal* 34, 41–48.

Seabrook, M.F. and Mount, N.C. (1995b) Good stockmanship – good for animals, good for profit. *Journal of Royal Agriculture Society of Scotland* 154, 104–115.

Sherwin, C.M., Lewis, P.D. and Perry, G.C. (1999) The effects of environmental enrichment and intermittent lighting on the behaviour and welfare of male domestic turkeys. *Applied Animal Behaviour Science* 62, 319–333.

Shutt, D.A., Fell, L.R., Connell, R., Bell, A.K., Wallace, C.A. and Smith, A.I. (1987) Stress-induced changes in plasma concentrations of immunoreactive beta-endorphin and cortisol in response to routine surgical procedures in lambs. *Australian Journal of Biological Sciences* 40, 97–103.

Siegel, P.B., Haberfeld, A., Mukherjee, T.K., Stallard, L.C., Marks, H.L., Anthony, N.B. and Dunnington, E.A. (1992) Jungle fowl–domestic fowl relationships: a use of DNA fingerprinting. *World's Poultry Science Journal* 48, 147–155.

Silerova, J., Spinka, M., Sarova, J. and Algers, B. (2010) Playing and fighting by piglets around weaning on farms employing individual or group housing of lactating sows. *Applied Animal Behaviour Science* 124, 83–89.

Simonsen, H.B. (1983) Ingestive behaviour and wing-flapping in assessing welfare of laying hens. *Current Topics in Veterinary Medicine and Animal Science* 23, 89–95.

Sinding-Larsen, T. (1979) *Kronvilt (Red Deer)*. Signum, Lund, Sweden, 144 pp.

Siopes, T.D. (1995) Incidence of prelay squatting behaviour is not related to subsequent egg-laying in turkey breeder hens. *Poultry Science* 74, 1039–1043.

Sisson, S. (1975) *Sisson and Grossman's The Anatomy of the Domestic Animals*, 5th edn. Saunders, Philadelphia, Pennsylvania, pp. 1217–2095.

Sisson, S. and Grossman, J.D. (1948/1938) *The Anatomy of the Domestic Animals*, 3rd edn. Saunders, London, 972 pp.

Skadhauge, E. and Dawson, A. (1999) Physiology. In: Deeming, D.C. (ed.) *The Ostrich, Biology, Production and Health*. CAB International, Wallingford, UK, pp. 51–81.

Smith, M.C. and Sherman, D. (2009) *Goat Medicine*, 2nd edn. Wiley-Blackwell, Ames, Iowa, 871 pp.

Soley, J.T. and Groenewald, H. (1999) Reproduction. In: Deeming, D.C. (ed.) *The Ostrich, Biol-*

ogy, *Production and Health*. CAB International, Wallingford, UK, pp. 129–158.

Spears, B.L., Wallace, M.C., Ballard, W.B., Phillips, R.S., Holdstock, D.P., Brunjes, J.H., Applegate, R., Miller, M.S. and Gipson, P.S. (2007) Habitat use and survival of preflight wild turkey broods. *Journal of Wildlife Management* 71, 69–81.

Spicer, E.M., Driesen, S.J., Fahy, V.A., Horton, B.J., Sims, L.D., Jones, R.T., Cutler, R.S. and Prime, R.W. (1986) Causes of preweaning mortality on a large intensive piggery. *Australian Veterinary Journal* 63, 71–75.

Spinka, M. (2006) How important is natural behaviour in animal farming systems? *Applied Animal Behaviour Science* 100, 117–128.

Spinka, M. (2009) Behaviour in pigs. In: Jensen, P. (ed.) *The Ethology of Domestic Animals – An Introductory Text*. CAB International, Wallingford, UK, pp. 177–191.

Spinka, M., Newberry, R.C. and Bekoff, M. (2001) Mammalian play: training for the unexpected. *Quarterly Review of Biology* 76, 141–168.

Spitz, F. (1986) Current state of knowledge of the wild boar. *Pig News and Information* 7, 171–175.

Stamatopoulis, K., English, P.R., Macpherson, O., Roden, J.A., Davidson, F.M. and Williams, J. (1993) Evaluation of the usefulness of the provision of shredded paper as a bedding in slatted-floored farrowing pen. *Animal Production* 56, 475.

Stolba, A. and Wood-Gush, D.G.M. (1984) The identification of behavioural key features and their incorporation into a housing design for pigs. *Annales de Recherches Vétérinaires* 15, 287–298.

Stolba, A. and Wood-Gush, D.G.M. (1989) The behaviour of pigs in a semi-natural environment. *Animal Production* 48, 419–425.

Stookey, J.M. and Gonyou, H.W. (1994) Production parameters in finishing swine. *Journal of Animal Science* 72, 2804–2811.

Strain, G.M. and Myers, L.J. (2004) Hearing and equilibrium. In: Reece, W.O. (ed.) *Dukes' Physiology of Domestic Animals*, 12th edn. Cornell University Press, Ithaca, New York, pp. 852–864.

Street, B.R. and Gonyou, H.W. (2008) Effects of housing finishing pigs in two group sizes and at two floor space allocations on production, health, behaviour, and physiological variables. *Journal of Animal Science* 86, 982–991.

Studnitz, M., Bak Jensen, M. and Juul Pedersen, L. (2007) Why do pigs root and in what will they root? A review on the exploratory behaviour of pigs in relation to environmental enrichment. *Applied Animal Behaviour Science* 107, 183–197.

Stünzi, H., Teuscher, E. and Glaus, A. (1959) Systematische Untersuchungen am Herzen von Haustieren, 2 Mitteilung, Untersuchungen am Herzen des Schweines. *Zentralblatt, Veterinary Medicine* 6, 640–654.

Surridge, A.K., Bell, D.J. and Hewitt, G.M. (1999) From population structure to individual behaviour: genetic analysis of social structure in the European wild rabbit (*Oryctolagus cuniculus*). *Biological Journal of the Linnean Society* 68, 57–71.

Swedish Dairy Association (2009) http://www.sweebv.info/

Tanida, H., Miura, A., Tanaka, T. and Yoshimoto, T. (1994) The role of handling in communication between humans and weanling pigs. *Applied Animal Behaviour Science* 40, 219–228.

Taylor, C.R. (1966) The vascularity and possible thermoregulatory function of the horns in goats. *Physiological Zoology* 39, 127–139.

Taylor, C.R., Dmi'el, R., Fedak, M. and Schmidt-Nielsen, K. (1971) Energetic cost of running and heat balance in a large bird, the rhea. *American Journal of Physiology* 221, 597–601.

Taylor, E.L., Blache, D., Groth, D., Wetherall, J.D. and Martin, G.B. (2000) Genetic evidence for mixed parentage in nests of the emu (*Dromaius novaehollandiae*). *Behavioural Ecology and Sociobiology* 47, 359–364.

Temple, W., Foster, P.M. and O'Donnel, C.S. (1984) Behavioural estimates of auditory thresholds in hens. *British Poultry Science* 25, 487–493.

Terrazas, A., Nowak, R., Serafin, N., Ferreira, G., Levy, F. and Poindron, P. (2002) Twenty-four-hour-old lambs rely more on maternal behaviour than on the learning of individual characteristics to discriminate between their own and an alien mother. *Developmental Psychobiology* 40, 408–418.

Thompson, G.E. and Clough, D.P. (1970) Temperature regulation in the newborn ox. *Biology of the Neonate* 15, 19–25.

Tolu, C. and Savas, T. (2007) A brief report on intra-species aggressive biting in a goat herd. *Applied Animal Behaviour Science* 102, 124–129.

Troy, C.S., MacHugh, D.E., Bailey, J.F., Magee, D.A., Loftus, R.T., Cunningham, P., Chamberlain, A.T., Sykes, B.C. and Bradley, D.G. (2001) Genetic evidence for Near-Eastern origins of European cattle. *Nature* 410, 1088–1091.

Tucker, C.B., Rogers, A.R., Verkerk, G.A., Kendall, P.E., Webster, J.R. and Matthews, L.R. (2007) Effects of shelter and body condition on the behaviour and physiology of dairy cattle in the winter. *Applied Animal Behaviour Science* 105, 1–13.

Tucker, C.B., Rogers, A.R. and Schutz, K.E. (2008) Effect of solar radiation on dairy cattle behaviour, use of shade and body temperature in a pasture-based system. *Applied Animal Behaviour Science* 109, 141–154.

Tully, T.N. Jr and Shane, S.M. (eds) (1996) *Ratite Management, Medicine and Surgery.* Krieger Publishing Co., Malabar, Florida, 188 pp.

USDA (United States Department of Agriculture) (2005) *Swine 2000 Part IV: Changes in the US Pork Industry 1990–2000.* National Animal Health Monitoring System (NAHMS), USDA, Fort Collins, Colorado, 50 pp.

USDA (United States Department of Agriculture) (2007a) *Swine 2006, Part I: Reference of Swine Health and Management Practices in the United States 2006.* National Animal Health Monitoring System (NAHMS), USDA, Fort Collins, Colorado, 87 pp.

USDA (United States Department of Agriculture) (2007b) *Swine 2006, Part II: Reference of Swine Health and Management Practices in the United States 2006.* National Animal Health Monitoring System (NAHMS), USDA, Fort Collins, Colorado, 79 pp.

USDA (United States Department of Agriculture) (2007c) *Swine 2006, Part III: Reference of Swine Health and Management Practices in the United States 2006.* National Animal Health Monitoring System (NAHMS), USDA, Fort Collins, Colorado, 52 pp.

USDA (United States Department of Agriculture) (2008) *Dairy 2007 Part III.* National Animal Health Monitoring System (NAHMS), USDA, Fort Collins, Colorado, 154 pp.

USDA (United States Department of Agriculture) (2009) *Small-Enterprise Swine 2007: Reference of Management Practices on Small-Enterprise Swine Operations in the United States, 2007.* National Animal Health Monitoring System (NAHMS), USDA, Fort Collins, Colorado, 92 pp.

Val-Laillet, D. and Nowak, R. (2008) Early discrimination of the mother by rabbit pups. *Applied Animal Behaviour Science* 111, 173–182.

Vannoni, E. and McElligott, A.G. (2009) Fallow bucks get hoarse: vocal fatigue as a possible signal to conspecifics. *Animal Behaviour* 78, 3–10.

van Liere, D.W. (1992) The significance of fowls' bathing in dust. *Animal Welfare* 1, 187–202.

van Liere, D.W. and Bokma, S. (1987) Short-term feather maintenance as a function of dustbathing in laying hens. *Applied Animal Behaviour Science* 18, 197–204.

van Putten, G. and Elshof, W.J. (1978) Observations on the effect of transport on the wellbeing and lean quality of slaughter pigs. *Animal Regulation Studies* 1, 247–271.

van Putten, G. and Elshof, W.J. (1983) Der Einfluß von drei Lichtniveaus auf das Verhalten von Mastschweinen. *Aktuelle Arbeiten zur artgemäßen Tierhaltung, KTBL Schrift* 299, Darmstadt-Kranichstein, pp. 197–216.

Van Tien, D., Lynch, J., Hinch, G.N. and Nolan, J.V. (1999) Grass odour and flavour overcome feed neophobia in sheep. *Small Ruminant Research* 32, 223–229.

Verheyden, H., Ballon, P., Bernard, V. and Saint-Andrieux, C. (2006) Variations in bark-stripping by red deer *Cervus elaphus* across Europe. *Mammal Review* 36, 217–234.

Verstege, M.W. and Vanderhe, W. (1974) The effects of temperature and type of floor on metabolic rate and effective critical temperature in groups of growing pigs. *Animal Production* 18, 1–11.

Verwoerd, D.J., Deeming, D.C., Angel, C.R. and Perelman, B. (1999) Rearing environments around the world. In: Deeming, D.C. (ed.) *The Ostrich, Biology, Production and Health.* CAB International, Wallingford, UK, pp. 191–216.

Vestergaard, K. (1981) The well-being of the caged hen – an analysis based on the normal behaviour of fowls. *Tierhaltung* 12, 145–167.

Vestergaard, K. (1982) Dust-bathing in the domestic fowl – diurnal rhythm and dust deprivation. *Applied Animal Ethology* 8, 487–495.

Vestergaard, K.S. (1994) Dustbathing and its relation to feather pecking in the fowl: motivational and developmental aspects. Doctoral thesis, The Royal Veterinary and Agricultural University, Copenhagen, Denmark.

Vestergaard, K. and Hansen, L.L. (1984) Tethered versus loose sows: ethological observations and measures of productivity. I. Ethological observations during pregnancy and farrowing production results. *Annales de Recherches Vétérinaires* 15, 245–256.

Vestergaard, K.S. and Lisborg, L. (1993) A model of feather pecking development which relates to dustbathing in domestic fowl. *Behaviour* 126, 89–105.

Vestergaard, K.S., Kruijt, J.P. and Hogan, J.A. (1993) Feather pecking and chronic fear in groups of red jungle fowl: their relations to dustbathing, rearing environment and social status. *Animal Behaviour* 45, 1127–1140.

Vilà, C., Leonard, J.A., Götherström, A., Marklund, S., Sandberg, K., Lidén, K., Wayne, R.K. and Ellegren, H. (2001) Widespread origins of domestic horse lineages. *Science* 291, 474–477.

Vitale, A.F., Tenucci, M., Papini, M. and Lovari, S. (1986) Social behaviour of the calves of semi-wild Maremma cattle, *Bos primigenius taurus*. *Applied Animal Behaviour Science* 16, 217–231.

Wade, J.R. (1996) Restraint and handling of the ostrich. In: Tully, T.N. Jr and Shane, S.M. (eds) *Ratite Management, Medicine and Surgery*. Krieger Publishing Co., Malabar, Florida, pp. 37–40.

Waiblinger, S., Menke, C. and Coleman, G. (2002) The relationship between attitudes, personal characteristics and behaviour of stockpeople and subsequent behaviour and production of dairy cows. *Applied Animal Behaviour Science* 79, 195–219.

Waiblinger, S., Boivin, X., Pedersen, V., Tosi, M.-V., Janczak, A.M., Visser, E.K. and Jones, R.B. (2006) Assessing the human–animal relationship in farmed species: a critical review. *Applied Animal Behaviour Science* 101, 185–242.

Walshaw, S.O. (2000) Behaviour problems. In: Flecknell, P.A. (ed.) *BSAVA Manual of Rabbit Medicine and Surgery*. British Small Animal Veterinary Association, Gloucester, UK, pp. 81–87.

Walshaw S.O. (2006) Behaviour problems. In: Meredith, A. and Flecknell, P. (eds) *BSAVA Manual of Rabbit Medicine and Surgery*, 2nd edn. British Small Animal Veterinary Association, Gloucester, UK, pp. 137–143.

Waran, N.K., Clarke, N. and Farnsworth, M. (2008) The effects of weaning on the domestic horse (*Equus caballus*). *Applied Animal Behaviour Science* 110, 42–57.

Waring, G.H., Wierbowski, S. and Hafez, E.S.E. (1975) The behaviour of horses. In: Hafez, E.S.E. (ed.) *The Behaviour of Domestic Animals*. Baillière Tindall, London, pp. 330–369.

Warriss, P.D., Kestin, S.C. and Robinson, J.M. (1983) A note on the influence of rearing environment on meat quality in pigs. *Meat Science* 9, 271–279.

Wascher, C.A.F., Scheiber, I.B.R., Weiss, B.M. and Kotrschal, K. (2009) Heart rate responses to agonistic encounters in greylag geese, *Anser anser*. *Animal Behaviour* 77, 955–961.

Watts, J.M. and Stookey, J.M. (2000) Vocal behaviour in cattle: the animal's commentary on its biological processes and welfare. *Applied Animal Behaviour Science* 67, 15–33.

Wauters, A.M. and Richard-Yris, M.A. (2002) Mutual influence of the maternal hen's food calling and feeding behaviour on the behaviour of her chicks. *Developmental Psychobiology* 41, 25–36.

Webb, N.J., Ibrahim, K.M., Bell, D.J. and Hewitt, G.M. (1995) Natal dispersal and genetic structure in a population of the European wild rabbit (*Oryctolagus cuniculus*). *Molecular Ecology* 4, 239–247.

Webster, A.J.F. (1970) Direct effects of cold weather on the energetic efficiency of beef production in different regions of Canada. *Canadian Journal of Animal Science* 50, 563–573.

Webster, A.J.F. (1974) Heat loss from cattle with particular emphasis on the effects of cold. In: Monteith, J.L. and Mount, L.E. (eds) *Heat Loss from Animals and Man; Assessment and Control*. Butterworths, London, pp. 205–231.

Webster, A.J.F. (1981) Weather and infectious disease in cattle. *Veterinary Record* 108, 183–187.

Webster, A.J.F. (1987) Cited by Clark, J.A. and McArthur, A.J. (1994) Thermal exchanges. In: Wathes, C.M. and Charles, D.R. (eds) *Livestock Housing*. CAB International, Wallingford, UK, pp. 97–122.

Webster, A.J.F., Chlumecky, J. and Young, B.A. (1970) Effects of cold environments on the energy exchanges of young beef cattle. *Canadian Journal of Animal Science* 50, 89–100.

Webster, J.R. and Matthews, L.R. (2006) Behaviour of red deer following antler removal with two methods of analgesia. *Livestock Science* 100, 150–158.

Weeks, C.A. and Nicol, C.J. (2006) Behavioural needs, priorities and preferences of laying hens. *World's Poultry Science Journal* 62, 296–307.

Weeks, C.A., Danbury, T.D., Davies, H.C., Hunt, P. and Kestin, S.C. (2000) The behaviour of broiler chickens and its modification by lameness. *Applied Animal Behaviour Science* 67, 111–125.

Wehrhahn, E., Klobasa, F. and Butler, J.E. (1981) Investigation of some factors which influence the absorption of IgG by the neonatal piglet. *Veterinary Immunology and Immunopathology* 2, 35–51.

Weisser, M. and Randler, C. (2005) Parental investment and brood rearing in feral Greylag Geese *Anser anser*. *Ornithologischer Anzeiger* 44, 1–8.

Wells, S.M. and Goldschmidt-Rothschild, B., von (1979) Social behaviour and relationships in a herd of Camargue horses. *Zeitschrift für Tierpsychologie* 49, 363–380.

West, B. and Zhou, B.-X. (1989) Did chickens go north? New evidence for domestication. *World's Poultry Science Journal* 45, 205–218.

Westin, R. and Algers, B. (2006) Effects of farrowing pen design on health and behaviour of the sow and piglets at farrowing and lactation – a review. *Svensk Veterinärtidning* 58, 21–27.

Westra, A.R. and Christopherson, R.J. (1976) Effects of cold on digestibility, retention time of digesta, reticulum motility and thyroid hormones in sheep. *Canadian Journal of Animal Science* 56, 699–708.

Whitehead, C.C. (2004) Skeletal disorders in laying hens: the problem of osteoporosis and bone fractures. *Welfare of the Laying Hen* 27, 259–278.

Whitten, W.K. (1985) Vomeronasal organ and the accessory olfactory system. *Equine Stud Farm Medicine* 14, 105–109.

Wiberg, S. and Gunnarsson, S. (2009) Health and welfare in Swedish game bird rearing. In: Aland, A. and Madec, F. (eds) *Sustainable Animal Production*. Wageningen Academic Publishers, Wageningen, The Netherlands, pp. 395–407.

Wichman, A. and Keeling, L.J. (2008) Hens are motivated to dustbathe in peat irrespective of being reared with or without a suitable dustbathing substrate. *Animal Behaviour* 75, 1525–1533.

Widowski, T.M., Wong, D.M.A.L.F. and Duncan, I.J.H. (1998) Rearing with males accelerates onset of sexual maturity in female domestic fowl. *Poultry Science* 77, 150–155.

Winskill, L.C., Waran, N.K. and Young, R.J. (1996) The effect of a foraging device (a modified 'Edinburgh Foodball') on the behaviour of the stabled horse. *Applied Animal Behaviour Science* 48, 25–35.

Wood-Gush, D.G.M. (1971) *The Behaviour of the Domestic Fowl*. Reprinted by Nimrod Press, London (1989), 147 pp.

Wood-Gush, D.G.M. and Rowland, C.G. (1973) Allopreening in the domestic fowl. *Revue du Comportement Animal* 7, 83–91.

Wood-Gush, D.G.M. and Vestergaard, K. (1993) Inquisitive exploration in pigs. *Animal Behaviour* 45, 185–187.

Yang, J., Morgan, J.F.N., Kirby, J.D., Long, D.W. and Bacon, W.L. (2000) Circadian rhythm of the preovulatory surge of luteinizing hormone and its relationships to rhythms of body temperature and locomotor activity in turkey hens. *Biology of Reproduction* 62, 1452–1458.

Young, B.A. (1981) Cold stress as it affects animal production. *Journal of Animal Science* 52, 154–163.

Young, R.J., Carruthers, J. and Lawrence, A.B. (1994) The effect of a foraging device (the 'Edinburgh Foodball') on the behaviour of pigs. *Applied Animal Behaviour Science* 39, 237–247.

Zeuner, F.E. (1963) *A History of Domesticated Animals*. Hutchinson, London, 560 pp.

Index

Note: page numbers in *italics* refer to figures